Construction Cost Estimating

Construction Cost Estimating equips a new generation of students and early-career professionals with the skills they need to bid successfully on projects. From developing bid strategies to submitting a completed bid, this innovative textbook introduces the fundamentals of construction estimating through a real-life case study that unfolds across its 24 chapters. Exercises at the end of each chapter offer hands-on practice with core concepts such as quantity take-offs, pricing, and estimating for subcontractor work. Online resources provide instant access to examples of authentic construction documents, including complete, detailed direct work estimates, subcontractor work estimates, general conditions estimates, markups, and summary schedules.

Through its unique mix of real-world examples and classroom-tested insights, *Construction Cost Estimating* ensures that readers are familiar with the entire estimating process even before setting foot on the jobsite.

Len Holm is Associate Teaching Professor at the University of Washington. He has over 40 years' construction industry experience at all levels and owns his own construction management firm. He is the author of numerous books on construction, including *Cost Accounting and Financial Management for Construction Project Managers* and *101 Case Studies in Construction Management*, and coauthor of *Introduction to Construction Project Engineering* with Giovanni Migliaccio and *Construction Superintendents* with John E. Schaufelberger.

John E. Schaufelberger is Dean Emeritus of the College of Built Environments and Professor of Construction Management at the University of Washington. He is the coauthor of *Management of Construction Projects* and *Construction Superintendents* with Len Holm, *Construction Equipment Management, 2nd edition* with Giovanni Migliaccio, and *Professional Ethics for the Construction Industry* with Rebecca Mirsky.

"Holm and Schaufelberger have always been visionaries in the construction industry."

The following documents are available on the eResource for *Construction Cost Estimating*, First Edition, Routledge, 2021. Reference www.routledge.com/9780367902681

This first list is made available to students and instructors:
- Case study 1 design drawings
- Case study 2 design drawings
- Case study 1 schedule
- Live Excel versions of all estimating forms
- Sample preconstruction contract
- Estimating Process Figure 1.1
- Steel truss project erection photograph
- Reference materials for *The Guide*
- Reference materials for Sage Construction and Real Estate software

The following are also made available to instructors:
- Instructor's Manual, complete with answers to all of the review questions and many of the exercises. A select group of case studies from *101 Case Studies in Construction Management* (Routledge 2019) are also included.
- Power Point lecture slides for all 24 chapters, 24 separate files, 650 slides in total
- Detailed case study 1 general conditions estimate, Figure 17.2L
- Detailed case study 2 estimate
- Case study 2 photographs of project completion
- Case study 3 design drawings
- Case study 3 estimate
- Case study 3 summary schedule
- Case study 3 site logistics plan

Construction Cost Estimating

Len Holm and John E. Schaufelberger

LONDON AND NEW YORK

First published 2021
by Routledge
2 Park Square, Milton Park, Abingdon, Oxon OX14 4RN

and by Routledge
52 Vanderbilt Avenue, New York, NY 10017

Routledge is an imprint of the Taylor & Francis Group, an informa business

British Library Cataloguing-in-Publication Data
A catalogue record for this book is available from the British Library

Library of Congress Cataloging-in-Publication Data
Names: Holm, Len, author. | Schaufelberger, John E., 1942– author.
Title: Construction cost estimating / Len Holm and John E. Schaufelberger.
Description: First edition. | Abingdon, Oxon ; New York : Routledge/Taylor & Francis
    Group, 2021. | Substantial re-write from previous estimating text: Construction Cost
    Estimating, Process and Practices by Holm, Schaufelberger, Griffin, and Cole; Pearson,
    2005. | Includes bibliographical references and index.
Identifiers: LCCN 2020043751 (print) | LCCN 2020043752 (ebook) |
    ISBN 9780367902681 (paperback) | ISBN 9780367902711 (hardback) |
    ISBN 9781003023494 (ebook)
Subjects: LCSH: Building—Estimates.
Classification: LCC TH435 .H848 2021 (print) | LCC TH435 (ebook) |
    DDC 692/.5—dc23
LC record available at https://lccn.loc.gov/2020043751
LC ebook record available at https://lccn.loc.gov/2020043752

ISBN: 978-0-367-90271-1 (hbk)
ISBN: 978-0-367-90268-1 (pbk)
ISBN: 978-1-003-02349-4 (ebk)

Typeset in Bembo
by Apex CoVantage, LLC

Access the Support Material: www.routledge.com/9780367902681

Substantial re-write from previous estimating text: *Construction Cost Estimating, Process and
Practices* by Holm, Schaufelberger, Griffin, and Cole; Pearson, 2005

Contents

4 Introduction to construction project management   37

5 Preconstruction   48

Figures

Tables

Preface

The ability to predict the cost of constructing a project is an essential construction management skill. This book examines several types of construction cost estimates and the processes used in developing each type. Cost estimating is not an exact science. It requires an understanding of construction materials, construction methods and techniques, construction equipment, and construction labor. To be a good estimator, one also needs practical construction experience.

The first version of this book was developed for use as a textbook for undergraduate courses in construction cost estimating and as a reference guide for construction professionals. That book was titled *Construction Cost Estimating, Process and Practices* and was first published by Pearson in 2005. A second soft-cover edition of the same book was published by LAD Publishing in 2018. The first edition was coauthored with professional construction estimators and estimating instructors Dennis Griffin and Tom Cole. We very much appreciate their contributions to that early and successful book.

The foundation of this new book is built on the success of our first book, which was coverage of the complete 'processes' of construction estimating. While the specific procedures for cost estimating may vary from one company to another and also for different types of construction, there are fundamental principles and processes that are used. It is these principles and processes that are also addressed in this book. There are many cost estimating books on the market that address the quantity take-off process, others that provide estimating databases, and still others that discuss development of a lump sum estimate. There are none, however, that we know of that provide the coverage of this book. It illustrates the process for developing several different types of estimates, and the case study illustrated throughout concludes with a bid for a lump sum contract. The book also discusses analysis of subcontractor quotations as well as estimating jobsite general conditions and company overhead costs.

The previous version of the book had one introductory chapter each on quantity measurement, pricing of direct work, and general contractor in-house pricing of subcontractor work. Because these topics are preferred subject matter for many instructors and good teaching materials for students, we have expanded that coverage from three to now 11 detailed chapters in this book. The new book also features a current commercial case study project that has many more different types of construction materials and systems than did the first book. Two additional sets of case study drawings and estimates are included on the book's eResource that feature residential and industrial construction materials and techniques.

This book is organized into six parts; I through VI. Part I provides an overview of cost estimating and construction management basics and introduces the construction project that is used as a case study throughout the book to illustrate concepts and processes and to provide a context for student exercises. Although the construction company used in this book is fictitious, the project is actually just under construction in Seattle, Washington. The authors assume that readers have a basic understanding of the construction process and types of construction contracts as well as the materials and methods used in the industry. In addition to an introduction to estimating and construction management topics, this first part includes a detailed chapter on preconstruction processes, including budget estimating.

Part II dives deep into quantity take-off for all aspects of work generally self-performed by a general contractor, including Construction Specification Institute divisions 03, 05, 06, 08, and 10. Part III then develops pricing for all of the work taken off in Part II, including analysis of labor productivity, material unit pricing, and construction equipment cost. Part IV shifts to in-house general contractor development of most standard areas of subcontract pricing, also known as 'plug' estimates.

Part V completes the estimate including detailed jobsite general conditions, markups and fees, and subcontractor quotations. Part V concludes with a detailed description of the process for finalizing a bid on bid day. The final part contains advanced estimating topics, including ethics, technology tools, and other types of estimates and project management issues.

Each chapter has a similar organization. First, concepts are discussed and then illustrated using the case study project. Review questions are provided to emphasize the main points covered in the chapter, and an instructor's manual containing the answers to the review questions is available for instructors. Exercises also are provided to allow students to apply the principles learned. A listing of all abbreviations used in the text and many other industry standard abbreviations is in the book's front material. A glossary of cost estimating terms is included at the back of the book. Appendix A provides many useful estimating resources, followed by a list of some of our reference materials that many students and professionals in the construction industry may find helpful. All of the estimating forms used in the book were created in Microsoft Excel and can be customized to meet the needs of the user. Live versions of these forms and several of the figures used in the book are included on the book's eResource.

Acknowledgments

This book could not have been written without the help of many people. We wish to acknowledge the following: Developers Rob Dunn and Bill Parks, for allowing us to use their projects as our case studies; Flad Architects and Johnston Architects, LLC, for allowing us to use their drawings; and the many University of Washington students who have used previous versions of this book and provided many valuable suggestions for improvement. During the book's development we reached out to several construction industry professionals, including chief estimators and preconstruction managers with some of the country's most successful firms, and ran drafts of tables of contents, chapters, and exercises by them for real-time input. Their contributions have helped create an estimating book like no other we know of. University of Washington Construction Management graduate students and industry professionals Matt Wiggins and Andrea Perea contributed immensely in preparation of instructional materials, including answers to review questions and exercises and PowerPoint presentations. In addition, we especially want to recognize our following industry partners:

- Winnie Bachwitz, editor, *The Guide Building Construction Material Prices*
- Larry Bjork, construction management professional and estimating instructor, UW
- Robert Guymer, retired, chief operating officer, Foushée and Associates, GC
- Kirk Hochstatter, industry professional and lecturer, University of Washington
- David Holm, project engineer, Pence Construction, general contractor
- Edward Krigsman, EK Real Estate Group, Windermere
- Lensit Studio Photography
- Kelly McCandless, field engineer, Hensel Phelps, general contractor
- Jud Youell, field sales engineer, Sage Construction and Real Estate, estimating software
- Mark Young, general manager, Brundage-Bone Concrete Pumping

If you have any comments or recommendations for inclusions in future editions, please feel free to contact the publisher Routledge or us direct at holmcon@aol.com or jesbcon@uw.edu. We hope you enjoy the book.

Len Holm
John E. Schaufelberger

Abbreviations

Most of the following abbreviations and acronyms are used in this book, including many from our figures, tables, and boxed-in examples and calculations. Other abbreviations listed here are standard in construction estimating and in the construction management industry, and/or may also be found on construction documents from which an estimator must prepare an estimate. Knowledge of standard abbreviations is important for the construction professional.

3D	three dimensional (drawings)
A	architectural (drawings)
AB	anchor bolt
ABC	activity-based costing
AC	acre
ACE	assumptions, clarifications, and exclusions
ACT	acoustical ceiling tile
ADA	Americans with Disabilities Act
AGC	Associated General Contractors of America
AIA	American Institute of Architects
Allow	allowance
Arch	architect
Asst.	assistant or assistance
B&O	business and occupation (tax), also excise tax
BCF	bank cubic feet
BCY	bank cubic yards
BF	board feet or backfill
BIM	building information models or modeling
BOF	bottom of footing (elevation or depth)
BSF	building square footage
BTR	better (lumber grade)
C	one hundred or center or Celsius or civil (drawings) or channel (structural steel shape)
CA	carpenter
CAD	computer-aided design
Cap.	capillary
CCC	City Construction Company, fictitious case study general contractor
CCE	*Construction Cost Estimating* (book)
CD	construction documents

CDF	controlled density fill (lean concrete backfill)
CDX	construction grade (plywood)
CE	chief estimator or civil engineer or construction engineer
CEO	chief executive officer
CF	cubic feet
CHR	crew hour
CIP	cast-in-place (concrete)
CJ	control joint or construction joint, also saw joint
CLT	cross-laminated timber
CM	construction manager or management
CMAR	construction manager at-risk (delivery method)
CM/GC	construction manager/general contractor (delivery method), also CM-at-risk
CMU	concrete masonry unit
CO	change order or close-out (C-O), also company
Comp	compensation, (workers' compensation insurance), also complete
COO	chief operational officer
COP	change order proposal
CPFF	cost plus fixed fee (price)
CPPF	cost plus percentage fee (price)
CS	closure-strip
CSF	hundred square feet, see also SQ
CSFA	contact square foot area (for formwork), see also SFCA
CSI	Construction Specifications Institute
CY	cubic yards
d	penny (nail gauge)
D or d	depth or dimension
DB or D/B	design-build (delivery method)
DD	design development
De-mob	demobilization
Det.	detail
DF	Douglas fir (lumber species), also door frame
DFH	doors, frames, and hardware, also DFHW or DF&H
Div.	division
DL	direct labor
DM	direct material
Docs	documents
Dwg.	drawing
E	electrical (drawing) or east
EA	each
EF	edge form
EL	electrical lighting (drawing)
E-mail	electronic mail
EMR	experience modification rate (safety)
Engr.	engineer
EP	electrical power (drawing)
EQ or Equip.	equipment
Est	estimate or estimator

EW	each way or east–west
Ex	example, excavation, or exercise
Excl.	excluded
Ext.	exterior
FE	fire extinguisher or field engineer
FEC	fire extinguisher cabinet
FF&E	fixtures, furnishings, and equipment (often owner-supplied)
FICA	Federal Insurance Compensation Act (social security)
FOB	free on board or freight on board
FT	foot or feet
FTF	floor to floor (dimension)
GC	general contractor
GCs	general conditions or general contractors
GLB	glue laminated beam (lumber)
GMP	guaranteed maximum price (estimate or contract), also GMax or MACC
GSF	gross square footage
GWB	gypsum wallboard, also sheetrock or drywall
H	height
HD	hold downs (wood connection)
HF	hem-fir (lumber grade)
HM	hollow metal (door or frame)
HO	home office
HR	hour or human resource
HSS	hollow structural section (structural steel element), formerly TS
HV or HVAC	heating, ventilation, and air conditioning (duct or mechanical system)
IBC	International Building Code
IFB	invitation for bid, also ITB
In	inch
Incl.	included or inclusion
IPD	integrated project delivery
ISO	isometric (drawing)
ITB	invitation to bid or instructions to bidders
IW	ironworker
JV	joint venture
K	one thousand or kip or kilogram
KW	kilowatt, also kilowatt hour (KWH)
L	length or landscape (drawings), also angle (structural steel shape), also labor
L&I	labor and industry
Lab.	labor
LB	pound or labor burden
LC or L cost	labor cost
LCY	loose cubic yards
LDs	liquidated damages
LEED	Leadership in Energy and Environmental Design
LF	lineal foot or feet
LLC	limited liability company or corporation
Loc.	location

LOI	letter of intent
LS or L/S	lump sum (estimate or contract)
LVL	laminated veneer lumber (engineered lumber)
M	million or mechanical (drawings), also material
M&E	mechanical and electrical (subcontractors or scopes)
M&M	methods and materials
MACC	maximum allowable construction cost, also GMP
Matl.	material
MBE	minority- or woman-owned business enterprise
MBF	thousand board feet (wood measure)
M cost	material cost
MDO	medium density overlay
MEP	mechanical, electrical, and plumbing (building systems or subcontractors)
MH	man-hour
Mil	million, also m
Misc.	miscellaneous
MO	month
Mob	mobilization
MOCP	*Management of Construction Projects* (PM book)
MUP	master use permit
MXD	mixed-use development (real estate classification)
N	north
NA	not applicable or not available
NIC	not included or not in contract
NFPA	National Fire Protection Association
No.	number
NTP	notice to proceed
NTS	not to scale
OAC	owner-architect-contractor (meeting)
OC	on center
OD	outside diameter (pipe dimension)
OE	operating engineer or owner's equity
OH&P	overhead and profit, also fee
OIC	officer-in-charge
OM	order-of-magnitude (GC plug estimate for subcontracted work)
OSB	oriented strand board (plywood)
OSHA	Occupational Safety and Health Administration
OST	on-screen take-off
OT	overtime
P	plumbing or piping (drawings)
P&ID	piping and instrumentation diagram (industrial drawing)
PC	pre-cast (concrete)
PE	project engineer or pay estimate or project executive
Perf.	perforated (drainpipe)
PEx	project executive, also PE
Plam	plastic laminate
PM	project manager or project management
Pnls	panels

P.O.	purchase order
PPE	personal protective equipment
PPP	public-private partnership (delivery method)
Precon	preconstruction phase or service
PSI	pounds per square inch
PSL	parallel strand lumber (engineered lumber)
PT	post-tension (concrete reinforcement) or pressure-treated (lumber)
PVC	poly vinyl chloride (pipe material)
Q&A	qualifications and assumptions or question and answer
QC	quality control
QTO	quantity take-off
QTY	quantity or quality
R	insulation value
RB	rubber base
RCP	reflected ceiling plan (drawing)
Rebar	concrete reinforcement steel
Recap	recapitulation (pricing sheet)
Rep	representative
RFI	request for information
RFP	request for proposal
RFQ	request for qualification or request for quotation
ROM	rough-order-of-magnitude (budget estimate)
ROT	rule of thumb
ROW	right of way
S	structural (drawing) or south or slope (excavation)
S/C	subcontract or subcontract cost
Schd.	schedule
SD	schematic design or storm drain
SDS	safety data sheets, formerly MSDS
Sec	section
SF	square feet or square foot
SFC	square feet of column or square feet of contact
SFCA	square foot of contact area
SFD	square foot of deck
SFF	square foot of floor
SFP	square foot of panels
SFR	square feet of roof
SFS	square foot of site or slab
SFW	square foot of wall
SHT	sheet
SJ	concrete saw joint or control joint
SOG	slab-on-grade
SOMD	slab on metal deck
SOV	schedule of values (bid form or pay request)
Spec	specification or speculation
SPM	senior project manager
SQ	square or hundred square feet, also CSF
SS	sanitary sewer

ST	subtotal
Sub	subcontractor or subcontract
Supt	superintendent, also super
SVP	senior vice president
SY	square yard
T	thickness or ton
T&M	time and materials
TCY	truck cubic yards, also loose yards
TI	tenant improvement
TJI	Trus Joist International (engineered wood shape, I beam)
T-O	take-off (quantity measurement)
TOW	top of wall (height or elevation or dimension)
TN	ton or tonne or tonnage (weight)
TNG	tongue and groove (wood to wood connection)
TQM	total quality management
TS	tube steel, also HSS
TVD	target value design
TVM	time value of money
Typ	typical
U	unit or units
UMH	unit man-hour or unit per man-hour
U Mos.	unit-months (combination of time and quantity)
U.N.O.	unless noted otherwise
UP	unit price
U.S.	United States of America
USGBC	United States Green Building Council
UW	University of Washington
VE	value engineering
VCT	vinyl composition tile
Vol.	volume
VP	vice president
W	width or wide flange steel beam shape or west
W/	with
WA	Washington State
WBS	work breakdown structure
WF	wide flange steel beam shape, also W or H or I
WK	week
W/o	without
WWF	welded wire fabric
YD	yard
YR	year

Part I

Introductory concepts

1 Introduction

The estimating process

Estimating can be defined as the process of determining the approximate cost of performing a work task based on the design information available, which is not always complete. For the general contractor (GC), estimating and their ability to do it accurately and reliably is a critical component to the success of each project and ultimately the success of the company. Many construction companies have suffered bankruptcy because of estimates that were not done properly or completely. Note we will utilize the terms general contractor and construction manager (CM) throughout this book, and although we define differences between the two terms in Chapter 4, we consider them synonymous with respect to the process of preparing construction estimates.

This book will provide students with the information, techniques, and forms required to determine accurate quantities, apply reasonable pricing for labor, materials, equipment, and subcontractor scopes, and determine appropriate markups, with the end result being accurate and complete estimates that will result in profitable construction projects. The book will provide detailed analysis for GC self-performed work such as cast-in-place concrete, concrete reinforcement, structural steel, rough and finish carpentry, doors and windows, and architectural specialty materials as well as methods for determining accurate costs for many areas of subcontracted scopes of work.

We have included detailed explanations of the different types of estimates that a contractor will perform in connection with the types of contracts that are used in the industry and how to accurately determine the jobsite general conditions necessary to successfully manage a construction project. In addition, risk management, estimate markups, and use of contingencies along with the important issue of ethics that all GCs must face are threaded throughout the book.

Estimating is not a simple task or a process that should be taken lightly but rather could be argued as the single most important objective for GCs to perform at the highest level. Without qualified estimators and successful estimates, contractors do not land construction contracts. Hopefully, this book will provide the reader with the skills necessary to begin to develop highly accurate and reliable construction cost estimates.

Construction estimates are prepared throughout the design process of a project, beginning with programming and conceptual phases followed by schematic design, design development, and construction documents. The estimates that are prepared at each of these design phases are different in their level of detail and completeness but similar in the process of developing any cost estimate. It is important for the contractor to be able to communicate accurate and reliable estimate information to the project owner and designers throughout the design process so that the design, as it progresses, can be constructed for the budget established by the owner. Subsequent chapters in this book will describe in detail the process for preparing estimates at each phase of the design and how those estimates would be used.

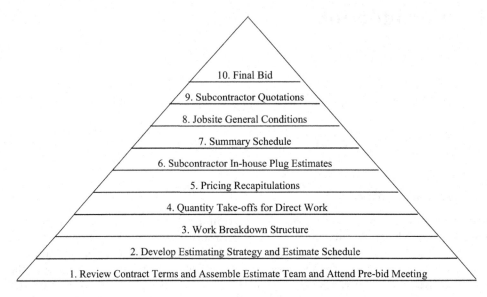

Figure 1.1 Estimating process

The estimating process as described in this book is a series of calculated steps leading to the preparation of a complete and accurate total price for bid day or a proposal on a negotiated project. These steps are reflected in Figure 1.1, which will be referred to through this book. It is recommended that the reader bookmark this figure for future reference. An additional copy has been included on the book's eResource, which we recommend readers print out and use as a bookmark. Two of the very important first steps undertaken by the estimator are the preparation of a schedule to complete the estimate and a work breakdown structure (WBS), both of which are discussed in the next two sections.

Estimate schedule

When possible, the project manager, experienced project engineer, and superintendent should be responsible for developing the estimate or, at a minimum, work as integral members of the estimating team. Their individual inputs regarding constructability and their personal commitments to the estimating product are essential to ensure not only the success of the estimate but also the ultimate success of the project. One of the first assignments for the estimating team is to develop a responsibility list, strategize their approach to the estimate, and to schedule the estimate. The estimating process should be scheduled for each project beginning with the dates for the pre-bid or pre-proposal conference and the end date the bid or proposal is due. With these milestones established, a short bar chart schedule should be developed that shows each step and assigns due dates to the estimating tasks depicted in the estimating process, Figure 1.1. The estimate schedule itself is included in step 2 of that figure. Familiarity with the steps or building blocks is essential in developing the schedule. Each team member is relying on the others to do their jobs efficiently and accurately and timely. Similar to a construction schedule, if one of the individuals falls behind on any one activity, the completion date may be in jeopardy or the quality of the finished estimate will be affected unless other resources are applied. An example of an estimate schedule for our case study project is included as Figure 1.2.

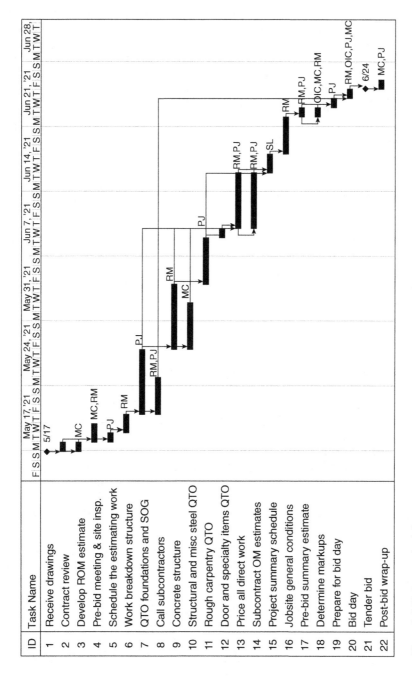

ID	Task Name
1	Receive drawings
2	Contract review
3	Develop ROM estimate
4	Pre-bid meeting & site insp.
5	Schedule the estimating work
6	Work breakdown structure
7	QTO foundations and SOG
8	Call subcontractors
9	Concrete structure
10	Structural and misc steel QTO
11	Rough carpentry QTO
12	Door and specialty items QTO
13	Price all direct work
14	Subcontract OM estimates
15	Project summary schedule
16	Jobsite general conditions
17	Pre-bid summary estimate
18	Determine markups
19	Prepare for bid day
20	Bid day
21	Tender bid
22	Post-bid wrap-up

Figure 1.2 Estimate schedule

Work breakdown structure

When the officer-in-charge decides to bid a project, the first job of the estimator is to plan and strategize the estimating process. One of the best organizational tools is the work breakdown structure. This is reflected as step 3 in Figure 1.1. This is simply an outline of the activities necessary to construct the project. The benefit of the WBS is that it takes a seemingly large complex project and turns it into many easy-to-manage segments. The WBS will be used, upon winning the project, for planning and scheduling, which in turn influences many of the project management activities. Making a comprehensive outline at this point is thus beneficial to later operations as well as to the estimating process. As a prelude to discussing the WBS, certain terms need to be defined. These are:

- Element or activity: A description of work such as concrete footings, steel erection, structural excavation, etc.
- Task: Specific tasks that need to be done to complete an activity. Form, fine grade, install reinforcing steel, place concrete, and finish and strip forms are six of the tasks that are required to complete the concrete footings.
- Basic outline: The basis of the WBS is the Construction Specification Institute (CSI) numbering system. The basic outline is therefore the CSI division and section numbers.
- First level expansion: The WBS expanded to show activities within a CSI division.
- Second level expansion: A second expansion of the WBS to task level.

Further expansion of the WBS beyond the second level by the project estimator results in such incremental detail that it is not always useful. Within estimating, and throughout our approach to construction management, we practice the 80-20 rule. This means that 20% of the systems, assemblies, tasks, or work items make up 80% of the cost or direct labor hours and therefore 80% of the risk for the contractor. It is those 20% that the estimator must pay the closest attention to. The first two levels of the WBS are used when planning and scheduling a project; thus their use in estimating helps to prepare the project team for the work after winning the bid. Expansion to the task level provides a comprehensive guide and checklist for the estimating process and should be referred to frequently to ensure that all items have been quantified and priced.

When creating a WBS, a good approach is to visualize how the project will be built. The CSI MasterFormat that is used for most estimating and technical specifications provides the general order of the work. For example, the first activities might be to clear and grade the site and install site utilities. This work use to be specified in CSI division 02 but is now in 31–33. Chapter 3 includes tables comparing the previous to new CSI division assignments. The substructure, which encompasses the structural excavation and concrete foundations, is found primarily in CSI division 03. The superstructure activities can be found in divisions 03 through 06. Divisions 07 through 14 are basically nonstructural items such as thermal and moisture protection, finishes, and various equipment and specialties, including elevators. Mechanical and electrical systems use to be specified in CSI 15 and 16 but are now in divisions 21–27. This basically is the order of construction of most projects. Figure 1.3 shows a partial WBS for the book's case study project.

City Construction Company, Inc.
Work Breakdown Structure

Dunn Lumber Project

CSI	Description
2	Demolition
3	Concrete:
	Cast-in-Place Concrete:
	Site Concrete
	Foundations:
	Spread Footings
	Continuous Footings
	Grade Beams
	Slab-on-Grade
	Columns
	Slab on Metal Deck
	Elevated Slabs:
	Shoring
	Form
	Rebar
	MEP and Stair Blockouts
	MEP Embeds
	Place Concrete:
	Pump
	Vibrators
	Screed
	Trowel Finish
	Curing Compound
	Remove Shoring and Formwork
	Re-Shores
4	Masonry
5	Structural Steel
6	Carpentry
7	Waterproofing
Continued . . .	

Figure 1.3 Work breakdown structure

When starting on a WBS for a new building an estimator may organize it in major divisions, such as: (1) foundations, (2) substructure, which could include foundations, (3) above-grade superstructure, (4) enclosure, (5) roofing and others. The substructure includes the cast-in-place concrete footings, foundation walls, pits, machine foundations, and any other concrete that is generally below the ground level of the building. For our case study project, we chose to further define foundations as spread footings, continuous footings, and grade beams.

Division 03 within the substructure includes the slab-on-grade (SOG). In many projects, the SOG is commonly placed prior to the steel erection, thus giving the ironworkers a stable working platform. Where very heavy equipment is needed for structural steel erection, the SOG may be placed after the framing and roof panels are in place. Buildings comprised of tilt-up concrete construction usually need the SOG as a place on which to cast the panels. A closure strip for the SOG is then placed after the tilt-up panels have been set and secured. It can be argued that the SOG may be either a part of the substructure or of the superstructure. It does not matter as long as there is consistency from project to project. For our case study project, we considered the SOG as part of the substructure because it is two stories below grade.

The superstructure is all of the structural work above the ground. This includes structural concrete and steel from the foundation to the roof, including elevated floor slabs and pre-cast concrete items. Pre-cast concrete elements can be either constructed by the GC or purchased from a supplier. Either way, the GC usually installs them. Generally, most pre-cast elements are purchased, but tilt-up walls are constructed on the project site and lifted into place.

A miscellaneous category may be included in CSI division 03 to list activities for exterior concrete work. This includes sidewalks, extruded curbs, the curbs and gutters along pavement, as well as concrete pavement. Some estimators may include site concrete with the site work portion of the WBS and estimate, as that is where the work is shown on the drawings and typically when it will be constructed. Depending on the work specified, these activities may be self-performed by the GC or subcontracted. Subcontractors frequently have specialty equipment for building these curbs and walks and thus may be more cost effective than the GC. If, however, the work is only flat work such as pavement and sidewalks, the GC may be able to do the work more efficiently. We have chosen to include the estimate for site concrete with other site work activities. So in a sense there are 'rules' with respect to where work activities might show up in an estimate, but each estimator and each construction company modify those rules to fit their specific business and custom.

The mechanical portion of the project consists primarily of the plumbing, heating, ventilating, and air conditioning (HVAC) and the fire sprinklers. Rough-in is done during the superstructure construction while the installation of fixtures, diffusers, setting the sprinkler heads, etc., all commonly known as trim work, are activities and tasks of the finish phases. The activities for bringing the utility services to the building are part of site work.

Some larger mechanical subcontractors perform all work in CSI divisions 21–23, while others specialize in only one set of activities. It is common for the mechanical contractor to do the plumbing and HVAC while a separate subcontractor does the fire sprinklers. All three activities should be listed on the WBS. Underground site utilities frequently are installed by a separate subcontractor.

Electrical work, CSI divisions 26 and 27 (old 16), includes two types of work: (1) power, distribution, and lighting and (2) controls, alarms, and data. It is common for a single subcontractor to perform all of these activities, but on larger projects, two or even three different subcontractors may be employed. The estimator needs to review the specifications to determine the extent of the work and list each major activity regardless of whether the work will be done by one or several subcontractors.

Figure 1.3 shows a task level expansion of the WBS for elevated concrete slabs showing the specific work items required to complete an activity. While the experienced estimator already knows what tasks are needed, this level provides a checklist to ensure that all work has been quantified. After winning the project, this level is used for the detailed planning and subsequent scheduling by the construction team.

City Construction Company
Project Item List

Project:	Dunn Lumber				Date: 5/19/2021.
Estimator:	Paul Jacobs				

			Provider			
			GC		Sub	
Line	CSI Div	Cost Item Description	Matl	Lab	Matl	Lab
1	w/3	Structural Excavation			X	X
2	3	Form, Place, and Finish Concrete	X	X		
3	3	Reinforcing Steel Supply	X			
4	3	Reinforcing Steel Installation				X
5	4	Masonry Walls			X	X
6	5	Structural and Misc. Steel Fab	X	X		
52	10	Specialties	X	X		
74	23	HVAC			X	X
80	26	Electrical			X	X
88	32	Site utilities			X	X
		Continued . . .				

Figure 1.4 Project item list

Another version of a WBS allows the estimator to designate which work will be performed by the GC and which will be performed by subcontractors. Figure 1.4 is a partial *project item list* as used by the estimator to list work activities and indicate whether an activity is to be done as direct work or is to be subcontracted. A live blank template of the project item list is included on the book's eResource. A separate column under each heading indicates whether the work is labor and/or material. This delineates that material may be furnished by a supplier to the jobsite for installation by the GC's own forces, furnished by a supplier for installation by a separate labor-only subcontractor, or furnished and installed by a subcontractor. For example, a rebar supplier may furnish the material, and a subcontractor may be contracted to install it. A single line item such as electrical with both the subcontractor labor and material spaces checked means that the same subcontractor will furnish and install the materials. The project item list is similar to the WBS, but it identifies who will supply the material and labor for each work activity. It is used after contract award to select subcontractors and suppliers. The remaining steps in the estimating process (Figure 1.1) are discussed throughout this book.

Introduction to the book

Part I of the book includes this chapter and other introductory CM chapters that prepare the estimator for the detailed estimating process. Chapter 2 describes several different types of estimates, from budgets, to guaranteed maximum price, to lump sum bids and how they relate to the different phases of design. Chapters 3 and 4 include many introductory contract and CM topics including procurement, contract documents, construction organizations, and individual contractor team member responsibilities. Preparation of budget estimates is one of the major functions of a contractor during the preconstruction phase, and in Chapter 5 we introduce preconstruction responsibilities, contracts, and fees.

In Part II we dive deep into the preparation of quantity take-offs (QTOs) for many different systems of potentially self-performed GC work, including foundations and slab-on-grade (Chapter 6), cast-in-place and pre-cast concrete superstructure (Chapter 7), structural steel (Chapter 8), and rough and finish carpentry, including doors and specialties in Chapter 9. The QTO effort is likely the most time consuming part of detailed estimating and is reflected as step 4 in Figure 1.1.

Once all of the GC direct work is quantified, then the estimator applies historical labor productivity rates, current wage rates, and market material pricing utilizing industry-standard pricing recapitulation sheets, as reflected in step 5 of Figure 1.1. Part III, Chapters 10, 11, and 12, covers all of this in detail for all of the systems we quantified in the previous chapters plus others. Chapter 13 also discusses the process to develop equipment pricing, some of which will be included with direct work packages and others with job-site general conditions, such as tower cranes and personnel and material hoists.

General contractors would be wise to develop reasonably approximate estimates for subcontracted work in-house before bid day. This allows them to better anticipate final markups, including fees, and to analyze the highs and lows of bid day subcontractor quotations. Chapters 14, 15, and 16 in Part IV discuss methods for GCs to prepare subcontractor pre-bid day plug estimates. Chapter 16 especially dives deep into a variety of civil scopes of work, from earthwork and shoring to site utilities and landscaping. Preparation of in-house subcontractor plug estimates is reflected as step 6 in Figure 1.1.

Once all of the direct and subcontracted scopes have been estimated, the GC will begin estimate completion processes, including development of a detailed jobsite general conditions estimate (Chapter 17), establishing competitive markup percentages (Chapter 18), and preparing the estimate summary page (Chapter 19), and Chapter 20 discusses in detail all of the very exciting – and sometimes stressful – activities associated with an actual bid day. All of these activities comprise our final steps 7 through 10 of Figure 1.1.

Some estimating students and instructors may have additional experience or interest or time available to delve into more advanced topics. We have assembled several advanced topics in Part VI including ethics (Chapter 21) and technology tools (Chapter 22). A variety of other types of estimates including residential, heavy civil, change orders, and others are in Chapter 23. Our final Chapter 24 bridges the next logical step from estimating to construction project management applications of estimates.

Case studies

We are utilizing several case studies in this book to provide examples of the material covered and advanced exercises for additional learning. We choose to use real projects as our case studies to connect all of the estimating elements together. The QTO from one chapter will be priced in another and general conditions, subcontractor pricing, and markups added to the same scopes of work in subsequent chapters. In this manner, the reader can 'connect the dots' and follow the project from when it first hit the estimator's desk until a bid is tendered or a proposal is submitted. Although these are real construction projects that your authors have been involved with professionally, including real estimates and schedules, the individual and company names have been changed and the estimates slightly modified. Any resemblance to actual built environment participants is a coincidence.

Case study 1

The first case study is the basis for most of the examples, including figures and tables, used in the book. For this project we are choosing to utilize a lump sum bid procurement process and a traditional general contractor delivery method where the owner has two separate contracts, one with the GC and one with the architect. All of the parties entered into standard American Institute of Architects contract agreements. The owner also chose to engage an agency CM as a consultant. The project owner is Dunn Lumber Enterprises, and the lead designer is Flad Architects. The bidding and eventually successful GC is City Construction Company.

This is a large mixed-use development (MXD) that includes a lumber yard, offices, a small restaurant, delicatessen, small grocery store, and two floors of underground parking and utilizes many sustainable enhancements. The project is being built by the developer on speculation of attracting tenants. It is considered a shell and core construction style with minimal tenant improvements (TI) planned at the outset. The interiors will be built out as TI once the tenants are identified.

The building structure includes a variety of cast-in-place concrete systems including elevated slabs and slabs on metal deck. The engineer transitioned from all concrete to all steel as the building rose out of the ground. It has extensive excavation and shoring elements, a complicated siding system, green roofs and terraces, and several elevators. The mechanical and electrical (M&E) design is minimal in a shell and core project. Design-build subcontractors will bid to the established criteria, and the M&E designers will continue throughout construction on a design-assist basis.

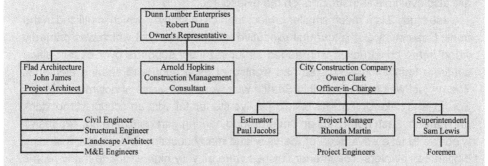

Figure 1.5 Case study organization chart

Figure 1.6 Rendering of Dunn Lumber case study project
Source: Rendering courtesy of Flad Architects

The total estimate at schematic design was approximately $50 million and is reflected in the next chapter in Figure 2.1, and the construction duration was planned for 20 months. Most of the figures in this book are connected with this project. An organization chart of the project team and rendering of the project are included as Figures 1.5 and 1.6. Several design drawings are included on the book's eResource that provide backup to the quantities reflected in the book's estimate examples.

Case study 2

Neither of our second or third case studies is featured in the book per se. The book's *eResource* includes drawings for both that are accessible for instructors to share with students. Many of the exercises in the book require the reader to access these drawings in order to develop solutions. Complete estimates for both projects, which include solutions to all of the review questions and many of the exercises, are also available to instructors on the eResource.

Case study 2 is a much smaller project than case study 1, which is utilized in the book. Case study 2 is a four-unit executive townhome project which was primarily wood frame construction with added structural steel supports over cast-in-place concrete foundations and also had significant excavation and shoring elements. The project was built on a hill in Seattle with wonderful views of downtown, Puget Sound, and Lake Union. The building is five stories tall with an occupied roof deck and garden. Each home has an underground parking garage. There are expansive windows to take advantage of the view and many sustainable features, including biofiltration planters. The interior finishes were all very high end, and many rooms have exposed structural elements which double as architectural features.

The name of the actual project is Lee Street Lofts. The development entity was a joint venture between the previous property owner and the GC. They established the name of Lee Street Lofts, LLC, as their one-project development company. The architect was Ray Johnston of Johnston Architecture and Parks Construction Company was the GC. Parks primarily employs subcontractors but employs a full-time superintendent and has a few craftsmen, including carpenters, who pick up the scopes that do not fit neatly into subcontract packages. Typical with most residential construction, this GC uses merit shop labor, which costs $15–$20 less per hour and is often more flexible with task assignments. Wages for case study 1 are included in Chapter 10 and wages for case study 2 are included with its detailed estimate on the web. The total estimated cost of construction, exclusive of land purchase, demolition, and soft costs, was $3.3 million. This was a very successful negotiated project, and the townhomes ended up selling for approximately $2 million each.

An actual photograph of the completed project was provided courtesy of the developer and is included as Figure 1.7. We appreciate this team for allowing us to use this successful project as an example in our book. Several additional photographs of the actual project are included on the book's eResource. Sufficient drawings have also been included for the student to be able to answer many of the book's advanced exercises and prepare a complete estimate. Instructors also have access to the complete detailed estimate.

Figure 1.7 Photograph of Lee Street Lofts case study project

Edward Krigsman | EK Real Estate Group
Photography courtesy of Lensit Studio

Case study 3

The third example case study is an industrial public works project that utilizes a competitively bid lump sum procurement and pricing method. The project was cancelled after it was fully designed but makes for several great exercises within an estimating book. The size of the facility is 14,800 SF and the site is just over 6 acres. The project's purpose is for vehicle maintenance for the highway department. The building is one-story, plus partial mezzanine, structural steel frame with metal siding. The large site requires an abundance of site utility work including storm water retention systems. The bid form submitted by Sound Construction Company (fictitious GC for use in the book) is included as Figure 1.8.

Vehicle Loop Maintenance Facility

Specification Section 00300

Bid Form

Scope of work: Construction of Vehicle Loop Maintenance Facility and associated site work to service tanker trucks which deliver much needed biosolid fertilizers from manufacturing facilities on the west side of the state to farms on the east side of the state.

The undersigned, as bidder, declares that it has examined the contract documents and will contract with the project owner on the agreement form provided in the special conditions of the specifications for the price(s) indicated below and according to all of the terms and conditions included in the contract documents.

Name of bidding contractor: *Sound Construction Company*

Name of signing authority: *Kelli McCloud*

Title: *President*

Signature: *Kelli McCloud*

Date of bid: *June 15, 2020*

Total lump sum bid amount: *$6,691,595.00*

Bid alternates: *None*

Addenda acknowledged: *1 through 3*

Five percent bid bond: *Attached*

Selected subcontractors:

Mechanical subcontractor:	*Balanced Air Systems, Inc.*
Electrical subcontractor:	*Grasley Electric Company*
Site utility subcontractor:	*Hanson Brothers, LLC*

Figure 1.8 Case study #3 bid form

The total construction estimate for this project is $6.7 million, and the estimated duration is 14 months. Example drawings for this case study are included on the book's eResource available for instructors along with the detailed estimate, summary schedule, and site logistics plan. Exercises are threaded throughout the book based on this case study, as with the previous two. Instructors could develop an unlimited variety of exercises on any of these three case study projects.

Alternate systems

In the first version of this book, we provided examples and exercises exclusively related to our single case study, which was a concrete tilt-up building. For this book we have chosen the Dunn project as our first case study because it includes many more building systems and materials that we can apply estimating principles to. The two additional case studies on the web also have additional systems and materials. But in addition to those three case studies, we have included other building systems and materials as boxed-in examples and alternate figures threaded throughout the book. These examples are not connected to a real project and do not relate to any of our other three case studies. For example, we have included detailed estimates of concrete tilt-up walls and wood framed structures from other projects we have worked on and complete tower crane and personnel-material hoist estimates to include with a jobsite general conditions estimate on larger projects.

Summary

Preparation of construction cost estimates is a series of logical steps in which each one relies on the successful and accurate and timely completion of the work ahead of it. One tool many estimators use is to schedule the estimate process itself, which is different than scheduling the estimate work. If any step in the estimate process falls behind schedule, the quality of the entire estimate may be jeopardized. An additional tool that most professional estimators and schedulers use is a work breakdown structure. This is essentially an outline of all construction work activities needed to build the project. A variation of the WBS is a project item list which further divides work activities into labor versus material and subcontractor performed versus self-performed with the contractor's own in-house direct crews.

Review questions

1   Which step in the estimate process (Figure 1.1) do you feel will take the longest?
2   Why not take every scope of work in a project down to the WBS task level?
3   Who should attend the site pre-bid meeting?
4   What is the difference between a WBS and a project item list?
5   What is the difference between a construction schedule and an estimate schedule?

6   We have portrayed the estimating process in a series of logical steps, 1 through 10. Why can't the construction schedule be prepared at the beginning of the process? Why can't subcontractor bids be received earlier?

7   Generally, which items included in either a WBS or a project item list will be in more detail?

8   Prepare a WBS for either of our other two case study projects. For the Lee Street Lofts project, expand the WBS to the second level for CSI division 06. For the Vehicle Maintenance Facility, expand the WBS to the second level for the site utility work.

2 Types of estimates

Introduction

As mentioned in Chapter 1, contractors are required to prepare cost estimates throughout the design process of a project. The completeness and accuracy of each of these estimates will largely depend on the completeness of the design documents and, as a result, will include varying levels of contingencies and risk assessment. This chapter will discuss the different types of estimates, how and when they are prepared, how they would be used by the team, and the level of detail in each. Estimate types often correlate with the phases of design and may be labeled as:

- Programming or conceptual design budget estimate,
- Schematic design budget estimate,
- Design development estimate, and
- Construction document estimate.

The word 'estimate' has different connotations to members within the built environment, including contractors, designers, and project owners. Even the court system has argued the differences between budgets, estimates, and bids. For this book budget estimates are typically developed early in the design process, whereas firm and detailed estimates, such as guaranteed maximum prices (GMPs) or bids, are developed later or at the completion of design.

Programming budget estimate

A program estimate is prepared at the very early stages of a project with little or no design information except for project location, area of the building, anticipated number of floors, and usage of the building. Estimates prepared at the program level are done by applying historical unit prices for similar buildings to the area of the project with reasonable allowances for site development and improvements based on the location of the project. Primary uses for this type of estimate would be:

- Project feasibility: Can the anticipated revenue justify the investment?
- Project financing: Owners and developers often go to lenders or investors with this estimate information to determine if financing will be available.

At the programming stage of a project the owner will need to use the general contractor's (GC's) program estimate to populate their project budget. A project owner that is developing a construction project as a business or for profit is also known as a 'developer' and will utilize the GC's construction estimate along with several other owner costs to prepare a financial *pro forma*, which is elaborated on in Chapter 23. Some of the items included in the pro forma include:

- Cost of property purchase,
- Construction budget,
- Design and engineering fees,
- Permits and entitlements,
- Applicable state and local sales taxes,
- Testing and inspection fees,
- Bonds and insurance,
- Fixtures, furniture, and equipment (FF&E),
- Off-site improvements,
- Moving expense,
- Contingencies,
- Interest on the construction loan,
- Legal fees,
- Development fees,
- Tenant improvement allowances, and others.

The programming budget may also be known as a conceptual budget. In this case, there are likely only a few early conceptual design drawings available for the contractor that represent the project owner's 'concept' of what a completed project might look like. An early conceptual budget for our case study project would have factored in the use of the building, square foot of floor (SFF) area, cost per SF, and cost modification indices for location and size. The initial rough order of magnitude (ROM) estimate developed by a GC when a potential project first comes into the door would likely have the same amount of detail as reflected in the following calculation. Note that in early budget estimates, contractors will round the values to reflect how 'rough' or approximate these figures are.

265,000 SFF @ $200/SFF (mixed-use office building) = $53 million
$53 million × 1.07 (location modifier) × .95 (size modifier) = $54 million

---

Schematic design budget estimate

Schematic design (SD) estimates are prepared when the design is approximately 30% complete. All of these percentages of design completion are 'relative' and have a range of +/−10% and are subject to a variety of variables and opinions. It is at this point that the contractor's estimate format is established and will be used throughout the estimating process. The most common way of formatting cost estimates is through the use of building systems, or cost assemblies. The Uniformat method of organizing cost budgets at the first level is very broad and would work only for very early ROM estimates. This format is reflected in the schematic design budget estimate shown in Figure 2.1. Different estimators will expand this format by splitting out broad categories – for example roofing from

City Construction Company
1449 Columbia Avenue
Seattle, WA 98202
206–447–4222

Schematic Design Construction Budget Estimate: Uniformat

Project: Dunn Lumber   Estimator: Paul Jacobs   Estimate Date: 3/18/2021

Div	System	Quantity	Units	Unit Price	Budget
2	Substructure (foundations and SOG)	54,000	sf footprint	$25.00	$1,350,000
3	Superstructure (elevated concrete and steel)	265,000	sf of floor	$29.00	$7,685,000
4	Enclosure	265,000	sf of floor	$11.35	$3,000,000
5	Roofing	49,000	sf of roof	$11.14	$546,000
6	Finishes, including doors and equipment	265,000	sf of floor	$26.79	$7,100,000
7	Elevator	23	stops	$53,000	$1,219,000
8	Mechanical, including plumbing and fire protection	265,000	sf of floor	$16.40	$4,346,000
9	Electrical	265,000	sf of floor	$12.00	$3,180,000
10	Sitework, including demolition	54,000	sf of site	$129	$6,966,000
	Subtotal				$35,392,000
11	Contractor markups:				
	Jobsite general conditions				$5,760,000
	Fees, taxes, and other markups				$3,494,343
	Subtotal				$44,646,343
	Construction contingency				$5,357,561
	Subtotal				$50,003,904
	Rounding adjustments				–$3,904
	Total Schematic Design Budget:				**$50,000,000**

Figure 2.1 Schematic budget estimate: Uniformat

enclosure – and others will have listed site work or general conditions as either the first or last line items. The estimator creating this budget has also slightly modified the systems for his presentation to the project owner.

Another early method of formatting budget estimates is by Construction Specification Institute (CSI) categories rather than Uniformat systems. In this method, the estimate is categorized by CSI divisions, which more closely aligns with technical specifications, also known as MasterFormat. The MasterFormat estimate is more detailed than is the Uniformat, as reflected in Figure 2.2, and will be used later in this book for our case study summary estimates. Note that both of these SD budget estimate formats are more detailed and therefore more accurate than the ROM provided earlier.

Depending on the level of the design and the information available, contractors will quantify as much as they can in order to develop an accurate cost estimate and will apply historical unit pricing to the measured material quantiles. For example, the documents should give enough information on the exterior closure that accurate quantities could be determined for each enclosure component and historical industry unit pricing applied. At this level, a GC will need to depend on guidance, often from notes on the

City Construction Company
1449 Columbia Avenue
Seattle, WA 98202
206–447–4222

Schematic Design Budget Estimate: MasterFormat
Dunn Lumber, Project # 9821

CSI Div	Description	Quantity	Units	Unit Price	Total
01	Jobsite General Conditions	12% of cost			$5,760,000
02	Demolition	54,000	SFS	$9	$486,000
03	Concrete	265,000	SFF	$22	$5,830,000
04	Masonry	1	Allow	$20,000	$20,000
05	Structural & Misc. Steel	265,000	SFF	$12	$3,180,000
06	Rough and Finish Carpentry	265,000	SFF	$0.50	$132,500
07	Thermal and Moisture and Roof	265,000	SFF	$13	$3,445,000
08	Doors and Windows	265,000	SFF	$19	$5,035,000
09	Finishes	265,000	SFF	$5.25	$1,391,250
10	Specialties	265,000	SFF	$1.50	$397,500
11	Equipment	1	Allow	$300,000	$300,000
12	Furnishings	by owner			Excluded
13	Special Construction	by tenants			Excluded
14	Elevators	23	stops	$53,000	$1,219,000
21	Fire Protection	265,000	SFF	$3.40	$901,000
22	Plumbing	265,000	SFF	$4.00	$1,060,000
23	HVAC and Controls	265,000	SFF	$9.00	$2,385,000
26	Electrical	265,000	SFF	$12.00	$3,180,000
27	Low-Voltage Electrical	w/CSI 26			Included
31	Sitework	54,000	SFS	$120	$6,480,000
	Subtotals:				$41,202,250
	Labor burden:	Included with above			Included
	State Excise, B&O Tax	1.00%		$412,023	$41,614,273
	Liability Insurance	1.00%		$485,061	$42,099,334
	Builder's Risk Insurance	by owner			Excluded
	Fee	5.00%		$2,104,967	$44,204,300
	GC Bond (Owner Alternate)	1.00%		$442,043	$44,646,343
	Contingency	12.00%		$5,357,561	$50,003,904
	Rounding Adjustments				−$3,904
	Total Schematic Design Budget:				**$50,000,000**

Figure 2.2 Schematic budget estimate: MasterFormat

drawings or in narrative form, from the architect and engineers to determine the work scope, such as:

- Quantity of reinforcing in pounds per cubic yard,
- Structural steel unit weight per square foot of building,

- Closure elements,
- Types of roofing,
- Level of finishes, and
- Mechanical and electrical system descriptions.

The indirect costs of the contractor such as general conditions, taxes and insurance, bonds, and fee are added generally as a percentage of the cost of the SD estimate summary. Another important element of the GC's estimate is the level of contingency that is included, which varies depending on the completeness of the documents. At the schematic level, contingencies which are included in the contractor's estimate are:

- Design contingency: Costs for unknown scopes that are not designed yet are allowed for at 7 to 10% of the estimated cost of the work.
- Construction contingency: Costs for unknown conditions that the contractor would be responsible for need to be allowed for at a rate of 5 to 7% of the estimated cost of the work.
- Estimating contingency: Because firm quantities and quotations cannot yet be determined for all building systems, the contractor may add +/−5% for accuracy of the proposed costs.
- Additional contingencies will need to be added on the project owner's side of the ledger for items outside of the contractor's control, such as permits and discovery of unknown conditions.

As can be seen in the following chapters, the levels of these contingencies will decrease as the level and completeness of the design progresses. The early SD budget estimate can be used for:

- Establishing an initial construction budget for the project owner;
- Supporting the application for a master use permit;
- Evaluating alternative design concepts, structural systems, exterior closure systems, and mechanical, electrical, and plumbing (MEP) systems;
- Identifying value engineering alternatives (see Chapter 5); and
- Communication tool between the contractor and designers to determine optimum building systems that will meet the project owner's construction budget.

Design development estimate

Design development (DD) estimates are prepared when the design is approximately 70% complete and the scope of the work is generally identified. These types of estimates are also referred to as semi-detailed estimates. It is at this point that the contractor will continue to use the systems format and will perform detailed quantity surveys for the scopes of work in each estimate system, with particular attention paid to the contractor's self-performed work such as concrete, steel, and carpentry. The structural elements of the project (concrete or steel or wood framing) will be clearly identified at this phase, such that quantities can be obtained and the contractor's historical unit pricing can be applied.

It is also at this point in the design that the contractor can solicit help from the subcontracting community in determining the cost of other building systems. Subcontractors can provide the GC with unit pricing to be applied to the GC's quantities (see

Chapters 14–16), or subcontractors may provide preliminary budget pricing based on the information in the documents. This is usually done at the discretion of the contractor. In either instance, it is important to get reliable information from subcontractors for complicated systems such as:

- Exterior closure,
- Vertical transportation,
- Mechanical, electrical, and fire protection, and
- Specialty items and finishes.

Another important element of the DD estimate is the proper use of allowances. Allowances are defined as costs for known scopes of work that are not designed yet. Good examples include landscaping, signage, floor covering, and FF&E. The contractor should be very specific with the project owner as to what scopes of work they have included as allowances and clearly communicate them in a separate document, as discussed later. Many times allowances become part of the contract documents until the design is completed.

It is at this estimate stage that the GC will do a detailed estimate of their indirect costs or jobsite general conditions. See Chapter 17 for a detailed explanation of jobsite general conditions. At this point in the estimate process, the GC understands the construction schedule and the anticipated staff, so the estimator will be able to prepare a detailed general conditions estimate rather than a percentage add-on, as was done with the SD estimate. This substantially reduces the risk for the GC for the general conditions portion of the estimate.

Contingencies will also be adjusted because the design is more complete. Design contingency will be included at approximately 3–5% and construction contingency at 2–3% during the DD phase. The DD estimate can be used for:

- Evaluating alternative design concepts, structural systems, exterior closure systems, and MEP systems;
- Establishing a GMP contract amount;
- Confirming the owner's construction budget;
- Identifying value engineering alternatives;
- Making commitments to subcontractors and suppliers for long-lead items such as structural steel, curtainwall systems, elevators, and mechanical equipment;
- Application for building permits; and
- Application for construction loans.

Construction document estimate

The construction document (CD) estimate is based on 100% design documents and will include detailed and complete quantity take-offs for all of the contractor's self-performed work and competitive bids for all of the subcontractors' work scopes and major material supply items. As with the DD estimate, the contractor's general conditions will be estimated in detail based on their schedule and anticipated jobsite staff requirements.

Contingencies will be much less or nonexistent given that the design is complete. The contractor may include reasonable levels of construction and estimating contingency based on negotiations with a private owner. This will vary with the project and will in most cases not be utilized in competitive lump sum bids. If contractor A includes significant contingencies and contractor B does not, contractor B will typically be the successful low bidder. The CD estimate can be used for:

- Establishing GMP contract amounts,
- Preparing lump sum bids,
- Application for construction loans,
- Incorporation into a contract agreement, and others.

In this book we will go through the process of preparing a fully detailed competitive estimate for our case study project utilizing 100% complete design drawings and specifications.

Qualifications and assumptions

One of the most important documents for each of the estimates already described is a list of qualifications and assumptions (Q&A) which describe what the contractor has used in preparing its estimate. Some contractors also refer to this document as assumptions, clarifications, and exclusions (ACE). At the earlier design stages, this document is more extensive, given the lack of design information, and becomes shorter as the design progresses and more information is available to the contractor, especially through the design team's development of technical specifications. The Q&A or ACE document will include:

- Assumptions made by the contractor in preparing its estimate due to lack of complete and detailed information;
- Information used that was provided by the designers and engineers but not shown on the documents;
- Allowances included in the estimate;
- Scopes of work that are excluded from the estimate either specific to the project scope or items that are generally excluded from contractor estimates and are customarily accounted for by the project owner. These may include:
    - Applicable state and local sales taxes,
    - Design and engineering fees,
    - Permits and entitlement fees,
    - Soils engineering,
    - Site survey,
    - Bonds and builder's risk insurance,
    - FF&E,
    - Signage, and
    - Other items specific to the project.

It is important for the contractor to include a detailed qualifications and assumptions document with their estimate so there is no confusion as to what is included or excluded from the estimate regardless of the level of design that it is based on; it is a good communication tool. This Q&A document, however, is not used on a competitive lump sum bid proposal. Inclusion of any qualifications on a public works bid would typically be grounds for rejection of that contractor's bid, but they are common for negotiated privately financed projects. An example of a Q&A or ACE document is included here as Figure 2.3.

City Construction Company

Project: Carpenters Training Center
Estimator: Paul Jacobs
Estimate Date: August 18, 2021

Schematic Design Budget Estimate Narrative

Qualifications and Assumptions:

- This estimate assumes a basically flat site without export of soil.
- Building will be founded on conventional spread footings without pile supports and will have a six-inch concrete slab-on-grade.
- The exterior walls will be tilt-up concrete and painted.
- Roof structure will be structural steel and deck supported by tube columns and ledger angles, all with one-hour rated fireproofing.
- Roofing will be either a mechanically adhered single ply or 4-ply build-up with cap sheet over R-30 semi rigid insulation.
- The interior clear height will be 15 feet.
- An estimate of $21.50 per square foot has been used since finishes are not shown.
- Mechanical estimate assumes natural gas air handling units with constant air volume.
- All areas will be fully fire protected with a standard wet-type system.
- Four restrooms that meet disability requirements are included.
- Classroom lighting will be drop-in indirect and laboratory lighting will be pendant indirect.
- Allowance for fire alarm, clocks, security, and closed-circuit television are included.
- Site work is an allowance, since it is undefined.

Exclusions:

- Window treatments
- Telephone and data systems
- Building signage
- Furniture and fixtures
- Demolition of existing structures on site
- Soil remediation or removal of underground structures

Figure 2.3 Contractor's estimate qualifications and assumptions

Summary

The four major design phases include programming, schematic design, design development, and construction documents. Different types of estimates are prepared by contractors at each of these phases, from ROMs to budgets to GMPs to lump sum bids. These estimate types have similarities and differences to contract pricing methods, which are project owner choices, as described in the next chapter. The accuracy of design improves as the design progresses, as does the accuracy of contractor-generated construction cost estimates. Estimating contingencies are also reduced as both design and estimating accuracy improve. Contractors often prepare qualification and assumption documents to accompany cost estimates, especially early in the design and budget phases. Qualifications are typically not allowed to accompany competitively bid public works lump sum or unit price projects. We appreciate the help of industry professional and estimating lecturer Larry Bjork, who contributed to early drafts of this chapter.

Review questions

1   At which design level (SD, DD, or CD) would a door, frame, and hardware schedule be included?
2   Match the three levels of design (SD, DD, and CD) with the three types of construction cost estimates (GMP, bid, and budget).
3   Which type of estimate is associated with programming or conceptual design?
4   Which of the schematic design budget estimates presented in this chapter would be considered the most accurate, and why that one?

Exercises

1   List five items which may be considered part of an owner's FF&E.
2   Why would inclusion of Q&A or ACE in a public bid be grounds for bid rejection?
3   What does the development phrase 'does the project pencil?' mean, and at which design phase (program, SD, DD, or CD) is that question addressed?
4   What would it take for a design team to produce a set of 100% complete and accurate drawings and specifications, without error, and change order proof?

3 Contract considerations

Introduction

This book is not a project management (PM) or a construction contracts book, but the approach to construction cost estimating is dependent on several issues discussed here, including the contract documents. We would be remiss to jump right into quantity take-offs (QTOs) and pricing and bidding without providing the reader with a brief overview of the construction management (CM) delivery and contracting process. Project owners choose the delivery methods (who contracts with whom), procurement method (bid or negotiated), and the contractor's pricing method (cost-plus, lump sum, and others). All of these have an impact on the estimator's approach and risks imposed on the contractor.

There is almost an infinite number of 'construction documents', of which the contract is one. Construction documents include requests for information, meeting notes, submittals, permits, shop drawings, quality control reports, safety inspections, photographs, transmittals, etc. The construction contract is the most important construction document. The contract can take on many different forms. In the next chapter we will provide a brief overview and introduction to current CM processes, including construction organizations and team member responsibilities. In this chapter we discuss:

- Project delivery methods,
- Procurement,
- Pricing methods,
- Contracts, and
- Contract documents.

Much of the material included in this chapter has been abbreviated from *Management of Construction Projects, a Constructor's Perspective*, by Schaufelberger and Holm. Our focus of CM here is on the role of the estimator and project manager as individuals, not companies. Many concepts and terms from one (CM or PM or estimator) apply to the other. We use a lot of abbreviations in construction; many of them are utilized in this chapter and are also listed in the front material.

Throughout this book we will examine estimating and project management from the perspective of the general construction contractor (GC). Other estimators and PMs typically are involved in a project representing the owner and the designer, but our focus is on the knowledge, skills, tools, and techniques needed to be successful as a PM or estimator for a construction company. Our context in this book will be that of an estimator for a mid-sized commercial GC. The principles and techniques discussed, however, are equally applicable on residential, industrial, and infrastructure or heavy civil construction projects, as well as for specialty subcontractors. Residential and heavy civil construction estimating will be introduced in Chapter 23.

Project delivery methods

The principal participants in any construction project are the project owner or client, the designer (architect or engineer), and the GC or CM. The relationships among these participants are defined by the delivery method used for the project. The choice of delivery method is the owner's, but it has an impact on the estimator's approach and the scope of responsibility of the contractor's PM. Owners typically select project delivery methods based on the amount of risk they are willing to assume and the size and experience of their own in-house contract management staffs. In this section, we will examine the four most common delivery methods that are used in the United States.

The *traditional delivery method* is illustrated in Figure 3.1A. The owner has two separate contracts with both the designer and the general contractor. There is no contractual relationship between the designer and the general contractor. Typically, the design is completed before the contractor is hired in this delivery method. The contractor's PM is responsible for obtaining the project plans and specifications, developing a cost estimate and project schedule for construction, establishing a project management system to manage the construction activities, and managing construction. This is the model we will use throughout this book.

In the *agency construction management delivery method*, the owner has three separate contracts (one with the designer, one with the general contractor, and one with the construction manager), as illustrated in Figure 3.1B. The CM acts as the owner's agent or owner's representative and coordinates design and construction issues with the designer and the GC. The CM usually is the first contract awarded and is involved in hiring both the designer and the GC. In this delivery method, the general contractor may not be hired until the design is completed. The general contractor's PM has similar responsibilities to those for the traditional delivery method. Responsibilities of the key team members are discussed in the next chapter. The primary difference is that the GC's project manager interfaces with the agency CM instead of the owner, as is the case in the traditional method. The agency CM is sometimes referred to as the owner's representative.

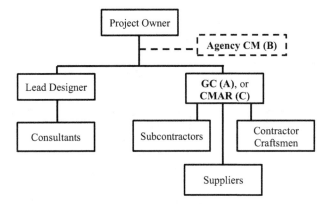

Figures 3.1 Traditional, agency CM, and CMAR delivery methods

In the *construction manager-at-risk (CMAR) delivery method*, the owner has two contracts (one with the designer and one with the construction manager/general contractor). The organization chart for the CMAR delivery method is very similar to that of the traditional delivery method except the position of 'GC' is replaced with 'CM', as reflected in Figure 3.1C. This delivery method is also known as the construction manager/general contractor (CM/GC) delivery method. One difference between a traditional GC delivery method and CMAR is a GC often employs some skilled craftsmen such as carpenters and laborers to construct concrete foundations, whereas the pure CM primarily employs subcontractors.

In the CMAR delivery method, the designer usually is hired first. The CM typically is hired early in design development to perform a variety of preconstruction services, such as cost estimating, constructability analysis, and value engineering studies. The preconstruction process is described in upcoming Chapter 5. Once the design is completed, the CM-at-risk constructs the project. In some cases, construction may be initiated before the entire design is completed. This is known as *fast-track* or phased construction. The contractor's PM and/or lead estimator interfaces with the designer and manages the execution of preconstruction tasks. Once construction starts, the PM's responsibilities are similar to those in the traditional method.

In the *design-build (D/B) delivery method*, the owner has a single contract with the design–build contractor for both the design and construction of the project, as illustrated in Figure 3.2. The design–build contractor may have design capabilities within its own organization, may choose to enter into a joint venture (JV) with a design firm, or may contract with a design firm to develop the design. Construction may be initiated early in the design process using fast-track procedures or may wait until the design is completed. In this delivery method, the contractor's PM and/or lead estimator is often responsible for interfacing with the owner and managing both the design and the construction of the project. This method is very similar to the ancient master-builder concept, where one individual was in charge of both the design and construction. The integrated project delivery method (IPD), as depicted in Figure 3.3, is a relatively new collaborative delivery concept where all three prime parties sign the same agreement and share in the same risks.

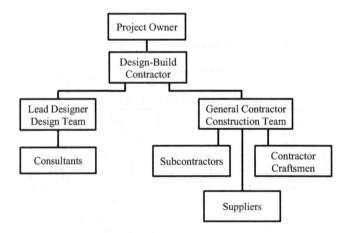

Figure 3.2 Design–build delivery method

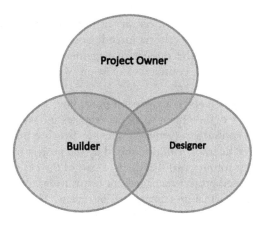

Collaborative Delivery

Figure 3.3 IPD delivery method

Procurement

Project owners procure contracts using either a bid or a negotiated procedure. Public owners, such as government agencies, use public solicitation or procurement methods. These owners may require potential contractors to submit documentation of their qualifications for review before being allowed to submit a bid or proposal, or the owners may open the solicitations to all qualified contractors. The first method is known as prequalification of contractors, and only the most qualified contractors are invited to submit a bid or proposal. Private owners can use any method they like to select a contractor. Private owners often use contractors they have had good experience with in the past and may ask a select few or even only one contractor to submit a proposal.

Bid method

Bid contracts generally are awarded solely on price. The owner defines the scope of the project, and contractors submit lump sum bids, unit price bids, or a combination of both. The project owner requests pricing with a vehicle such as a request for quotation (RFQ) or invitation to bid (ITB). The owner awards the contract to the contractor submitting the lowest total price for the project. The pre-bid conference usually is held in the designer's office or at the project site to resolve any contractors' questions relating to the project or the contract. A lump sum contract may require a single price for the entire scope of work or require separate prices for individual portions of the scope of work. Some contracts may have separate additive or deductive items that must be priced during the bidding process. These are known as bid alternates. The owner selects which combination of additive and/or deductive alternates to award once the bids have been opened. Bid alternates, along with allowances, are discussed in Chapter 23. We will assume the lump sum bid method throughout most of this book.

Negotiated method

Negotiated contracts are awarded based on any criteria the project owner selects, especially if it is a private client. Typical criteria include cost (or fee in the case of a cost-plus contract), schedule, expertise of the project management team, plan for managing the project, contractor's safety record, contractor's existing work load, and contractor's experience with similar projects.

Some negotiated contracts involve a two-step procedure. First, prospective contractors are prequalified after review of their prior work experiences and safety records. Then a short list of the most qualified contractors (generally three to five) are invited to submit proposals containing project specific information required by the project owner. This invitation is often issued with a request for proposal (RFP). As a part of the evaluation procedure, owners may require the proposed project management teams to brief their plans for managing the project, often during a formal presentation or interview. This may include preparation of a project schedule and budget. The owner then selects the contractor submitting the best proposal and negotiates a contract price. The pre-proposal conference is similar to the pre-bid conference used in a bid procedure. The major difference in a negotiated procedure is the opportunity for the owner to discuss the contractors' proposals, modify contract requirements, and clarify any issues before executing a contract. The owner selects the contractor submitting the best-value proposal, which may not be the least cost.

Some owners use a more informal negotiating procedure, particularly if they have long-term relationships with their contractors. A private owner may simply ask one or a few contractors to submit proposals. After reviewing the proposals, the owner would negotiate contract terms with the selected contractor. The negotiated proposal is also discussed in Chapter 23.

Pricing

There are several methods for pricing contracts used in the construction industry. The choice of which to use on a particular project is also made by the owner after analyzing the risk associated with the project and deciding how much of the risk to assume and how much to pass on to the contractor. Contractors want compensation for risk they assume; the higher the risk, the higher the price. The most common pricing methods include:

- Lump sum,
- Unit price,
- Cost-plus fixed fee or cost-plus percentage fee, and
- Cost-plus with a guaranteed maximum price (GMP).

These four primary pricing methods have similarities and differences with the estimate types presented in the last chapter. Lump sum contracts are awarded on the basis of a single lump sum estimate for a specified scope of work. Unit price contracts are used on heavy civil or industrial projects when the exact quantities of work cannot be defined. The designer estimates the quantities of work, and the contractor submits unit prices for each work item. The actual installed quantities required are multiplied by the bid unit prices to determine the final contract price. Cost-plus contracts are used when the scope of work cannot be fully defined. All of the contractor's project-related costs are reimbursed by the owner, and a fee is paid to cover profit and home-office overhead. A GMP contract is a cost-plus contract in which the contractor agrees not to exceed a specified cost. We will hold off until Chapter 23 for a more detailed discussion of these other contract pricing methods.

Contract agreement

The prime contract is the agreement between the GC or CM and the project owner. It is a legal document that describes the rights and responsibilities of the parties. Our focus here is on the prime contract agreement, but many of the documents and processes described apply to subcontract agreements and supplier purchase orders as well. Five things must be aligned for a contract to exist:

- An offer to perform a service,
- An acceptance of the offer,
- Some conveyance, that is transfer of a completed building for money,
- The agreement has to be legal, and
- Only authorized parties, for example the chief executive officer, can sign the agreement.

The terms and conditions of the relationship of the contracting parties are defined solely within the contract documents. These documents should be read and completely understood by the contractor before deciding to pursue a project and prepare a bid. They also are the basis for determining a project budget and schedule. To manage a project successfully, the GC's PM must understand the organization of the contract documents and understand the contractual requirements for his or her project. This should all be resolved in steps 1 and 2 of the estimating process (Figure 1.1).

The contract documents describe the completed project and the terms and conditions of the contractual relationship between the owner and the contractor. Usually there is no description of the sequence of work or the means and methods to be used by the contractor in completing the project. The contractor is expected to have the professional expertise required to understand the contract documents and select appropriate subcontractors or qualified tradespeople, materials, and equipment to complete the project safely and achieve the quality requirements specified. For example, the contract documents will specify the dimensions and workmanship requirements for elevated concrete slabs but will not provide the design for required formwork or methods for temporary shoring or re-shoring. The contract usually includes at least the following five essential elements. All of these documents must work together.

- The agreement,
- Special or supplemental conditions,
- General conditions,
- Drawings, and
- Specifications.

Contracts are either standard or specially prepared agreements. Most government agencies use internally developed standard formats as construction contract documents. Federal and state agencies typically have standardized general conditions and agreement language. Many local government agencies and private owners use contract formats developed by the American Institute of Architects (AIA) or the new family of construction contracts from ConsensusDocs headed up by the Associated General Contractors. Contracts should not be signed until they have been subjected to a thorough legal review. This is to ensure that the documents are legally enforceable in the event of a disagreement and that there is a clear legal description of each party's responsibilities. The advantage of using standard contract forms is that they have been developed by individuals skilled at contract law and have been tested in and out of courts. The contract documents have a significant impact on the responsibilities of the PM and superintendent. Many requirements are contained in the general conditions, but project-specific requirements are defined in the special conditions of the contract.

Contract documents

For this discussion, we are assuming the reader already has had an introduction to plan (or drawing) reading. This section is just a brief refresher and shows the connection of these documents to the contract agreement and the estimating processes.

Drawing types

The word 'plan' has many different connotations in construction. To 'plan' is a broad term which means to look ahead or to anticipate what is coming, which is a trait of construction leaders. Construction contractors develop several 'plans' or preconstruction 'plans' in how they anticipate managing the project. Preconstruction will be discussed in Chapter 5. The construction schedule, or bar chart, is thought of as a 'plan' where the GC graphically shows all of the construction activities, including their anticipated start and completion dates. In the built environment we often use the term 'planning and scheduling' to describe that process. Design drawings are generically referred to as plans; a rolled up set of drawings will be called a set of plans. But within that set of drawings there are actually 'plan' views, as well as several other views. So for this book we will attempt to stay with the term 'drawings' to make that distinction.

Design disciplines: As indicated earlier, the master-builder used to be both the designer and the builder. The design element then split off from the building element, and contractors arose. The designer further split between the architect and the engineer. Today there are several design disciplines that have different specializations, but the architect is still the prime 'designer', especially in commercial work. In civil work the civil engineer will be the prime designer. Within a typical set of drawings, each design entity contributes their respective sheets, and their documents all have different letter prefixes to indicate their authors – for example, A for architect, S for structural engineer, C for civil engineer, and many others.

Views or types of drawings

Also within that set of drawings there are a multitude of 'views'. Some of those are 'plan' views, but it would be too generic to call them all 'plans'. Each of the different design disciplines will include a variety of these views as well.

- Plan: Looking down,
- Reflected ceiling plan: Looking up at the ceiling,
- Exterior Elevation: Looking horizontally at the outside of the building,
- Interior elevations: Looking at the inside walls,
- Building sections: A vertical cut through the building,
- Wall sections: A vertical section through just a wall,

- Details,
- Riser diagrams: Single-line mechanical and electrical drawings,
- Rendering: Graphic representation,
- Isometric or three-dimensional drawings, and
- Piping and instrumentation diagrams.

A variety of *schedules* are included within the drawing set as well. These are not the same as the contractor-generated time line schedule. These are often Microsoft Excel spreadsheets. Some examples include:

- Footing schedule, hold-down schedule, and shear wall schedule,
- Room finish schedule and door and hardware schedules,
- Window schedule and wall type schedule,
- Mechanical equipment, plumbing fixture, and light fixture schedules,
- Landscape plant schedule, and others.

Drawings and specifications serve a variety of purposes for the client or developer, designer, and contractor teams, including:

- Pro forma (see Chapter 23), permit applications, bank loan approvals,
- Basis for contract, communication and understanding with all parties, and
- Estimating and bidding.

Technical specifications

Similar to distinguishing between construction and contract documents, it is worthwhile to clarify the difference between a *project manual* and a *specification book*. The project manual is a larger document that includes the technical specifications, but it can also include many other documents such as:

- RFP or RFQ or ITB;
- Supplemental or special conditions;
- General conditions (usually AIA A201 or public entity generated);
- The contract agreement, such as A101, A102, ConsensusDocs, or home-grown;
- Addenda and bid form;
- Subcontractor identification form and Davis-Bacon or prevailing wage rates;
- Geotechnical (soils) report and jobsite visit verification form; and
- Room finish, light fixture, or plumbing fixture schedules.

The technical specifications define the products to be used along with installation standards and quality control expectations. They do not define the 'means and methods' of assembly, as this is up to the contractor. Up until 2004, the MasterFormat specifications had 16 major divisions, plus 00, which were the special conditions, as represented in Figure 3.4.

A typical specification section number was a total of five digits, where the first two digits were the division number; for example, section 06410 was the specification section for wood cabinets and 06 was the CSI division number. In 2004, an expanded MasterFormat was issued with up to 50 major divisions. Some of these divisions are placeholders for when new products and processes are popularized. Some of the significant differences between the old 16 divisions and the new 50 divisions are as follows in Figure 3.5.

Previous 16 CSI Divisions

00:    Supplemental conditions
01:    General requirements
02:    Site construction
03:    Concrete
04:    Masonry
05:    Metals, including structural steel and metal studs
06:    Wood and plastic, including rough and finish carpentry
07:    Thermal and moisture protection
08:    Doors, door frames, windows, and door hardware
09:    Finishes
10:    Specialties, including toilet accessories and partitions, fire extinguishers
11:    Equipment, including dock levelers
12:    Furnishings, including window coverings and manufactured casework
13:    Special construction
14:    Conveying systems, including elevators
15:    Mechanical, including HVAC, plumbing, and fire protection
16:    Electrical

Figure 3.4 Previous 16 CSI divisions

Current CSI Divisions

00–14:   Pretty much the same, except site work was moved from 02 to 31, 32, and 33.
02:    Existing conditions, including demolition
15–20:   Reserved for future use
21:    Fire suppression (formerly included within 15)
22:    Plumbing (formerly included within 15)
23:    HVAC (formerly included within 15)
24, 29, 30:   Reserved for future use
25:    Integrated automation (mechanical controls)
26:    Electrical (formerly 16)
27:    Communications (formerly included in 16)
28:    Electronic safety (formerly included in 16)
31:    Earthwork (formerly 02)
32:    Exterior improvements (formerly included in 02), including landscaping
33:    Utilities (formerly included in 02)
34:    Transportation
35:    Waterway and marine construction
36–39:   Reserved for future use
40–45:   Industrial construction
46, 47, 49:   Reserved for future use
48:    Electrical power generation

Figure 3.5 Current 50 CSI divisions

The new MasterFormat sections are six digits, such as 064023 is for interior architectural millwork. Note that the division 06 is the same for both versions in this instance. A few design firms are still using the prior 00–16 divisions; that is how their internal systems are set up, and that is how many of their clients and contractors are accustomed to working. Eventually, all will likely transfer over to the new 50 divisions. The estimating student should be familiar with both and be prepared to see both on construction projects in the foreseeable future.

Summary

There are four major project delivery methods used in the United States. The primary differences among them are the relationships between the project participants. Owners select contractors by one of two methods, bidding or negotiating. The owner also selects the pricing method the contractors must use to develop the bid or proposal, including lump sum, unit price, cost-plus fee, or guaranteed maximum price. The construction contract describes the responsibilities of the project owner and the contractor and the terms and conditions of their relationship.

A thorough understanding of all delivery, procurement, pricing, and contractual requirements is essential if an estimator and/or project manager expects to prepare a competitive estimate and complete the project successfully. Contracts are either standard or specifically prepared documents. Standard contracts generally are preferred because they have been legally tested in and out of courts. A particular project could utilize many potential combinations of delivery, procurement methods, pricing, contract formats, and construction organizations, which are introduced in the next chapter.

Review questions

1   What is the difference between the traditional and the agency construction management delivery method?
2   What is the difference between the construction manager-at-risk and the design-build delivery method?
3   Why is it essential that an estimator fully understand the requirements and procedures specified in the contract documents?
4   What five types of preconstruction 'plans' might a project manager or estimator or superintendent prepare during the preconstruction phase, especially as it relates to (A) contract development and (B) estimating?
5   List three items/documents included in a project manual beyond the technical specifications.
6   What is the difference between 'general' and 'special' or 'supplemental' conditions?
7   Which design entities/firms prepare 'plan' views?
8   Who should sign a contract, and why them?
9   What are the procurement, pricing, and contracting methods chosen by the owner for the book's case study project?

Exercises

1   Utilizing an outside case study project, draft one or two sentences each describing the project's delivery, procurement, pricing, and construction organization.

2   Who assumes risks associated with design errors, unforeseen site conditions, cost control, schedule control, quality, and safety with respect to (A) traditional delivery method and (B) IPD delivery method?

3   Obtain a blank copyrighted contract format such as an AIA A101. Complete the contract for the book's case study project or another outside project. Make whatever assumptions are necessary.

4   Research a material item or two, such as hollow metal door frames or carpeting or plumbing fixtures or storm system catch basins, and look up the prior CSI specification division and section number (five digits) and the new numbers (six digits).

5   If you were a GC PM and could assemble any combination of delivery, procurement, and pricing methods and contract format from this chapter, along with organization styles from upcoming Chapter 4, what would you choose, and why?

6   Prepare a spreadsheet comparing types of estimates presented in the last chapter and the pricing methods discussed in this chapter. Which ones go with which, and what are the similarities and differences?

7   Add procurement methods and contract types to the spreadsheet prepared in Exercise 6. Which ones are the most common matches?

8   Which firm pays for post-bid document discrepancies in a design-build delivery method?

4 Introduction to construction project management

Introduction

This chapter and the last have been included in *Construction Cost Estimating* to provide a brief overview and introduction to current construction management delivery and contracting processes. In the last chapter we described how project owners choose delivery methods, procurement and pricing choices, and contract forms. In addition to other aspects of construction project management, in this chapter we discuss construction organizations, team member responsibilities, subcontracting strategies, subcontractor prequalification, and risk assessment.

Introduction to project management

Project management is the application of knowledge, skills, tools, and techniques to the many activities required to complete a project successfully. In construction, project success generally is defined in terms of document control, safety, quality, cost control, and schedule control. These project attributes can be visualized as depicted in Figure 4.1. The project manager's (PM's) and superintendent's challenge is to balance quality, cost, and schedule within the context of a safe project environment while maintaining control of the many construction documents. While cost and schedule may be compromised to produce a quality project, there can be no compromising regarding safety, and proper documentation is required to ensure compliance with contract requirements.

In this book, we focus on the estimating activities of a general contractor's (GC's) lead estimator, which may also be the project manager assigned to the project if the bid or proposal is successful. Other estimators and PMs typically are involved in a project representing the owner and the designer, but our focus is on the knowledge, skills, tools, and techniques needed to be successful as a PM for a prime construction contractor. Our context will be that of an estimator and/or PM for a commercial GC. The principles and techniques discussed, however, are equally applicable on residential, industrial, infrastructure or heavy civil construction projects, and for specialty subcontractors.

The project manager is the leader of the contractor's project team and is responsible for identifying project requirements and leading the team in ensuring that all are accomplished safely and within the desired budget and time frame. To accomplish this challenging task, the PM must organize his or her project team, establish a project management system that monitors project execution, and resolve issues that arise during project execution. In the previous chapter, we introduced contracting and other delivery and procurement methods the PM and estimator must operate within. In successive chapters, we will discuss the many tools that a PM and lead estimator will utilize in development of a

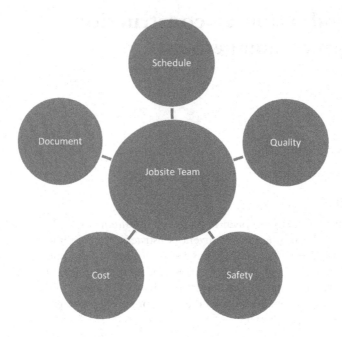

Figure 4.1 Project management attributes

complete project estimate. In our last chapter (Chapter 24) we introduce several project management applications which rely on a properly prepared construction cost estimate. They may not apply to every project, but the project team must select those that are applicable for each project.

Construction organizations

The size and structure of a contractor's jobsite construction organization depends on the size of the project, its complexity, and its location with respect to other projects or the contractor's home office. The cost of the project management organization is considered jobsite overhead and must be kept economical to ensure the contractor's cost is competitive with other contractors. The jobsite overhead costs are also referred to as indirect or general conditions costs. The goal in developing a project management organization is to build the minimum organization needed to manage the project effectively. If the project is unusually complex, it may require more technical experts, such as a jobsite cost engineer or specialty superintendents, than would be required for a simpler project. If the project is located near other projects or the contractor's home office, technical personnel can be shared among projects, or backup support can be provided from the home office. If the project is located far from other contractor activities, the jobsite office must be self-sufficient.

General contractors typically organize their project management teams in one of two models. In one organizational concept, estimating and scheduling are performed in the contractor's home office, as illustrated in Figure 4.2. In an alternative organizational structure, estimating and scheduling are the PM's responsibilities, as illustrated in Figure 4.3. Notice how the function or position of estimating or estimator is different between these

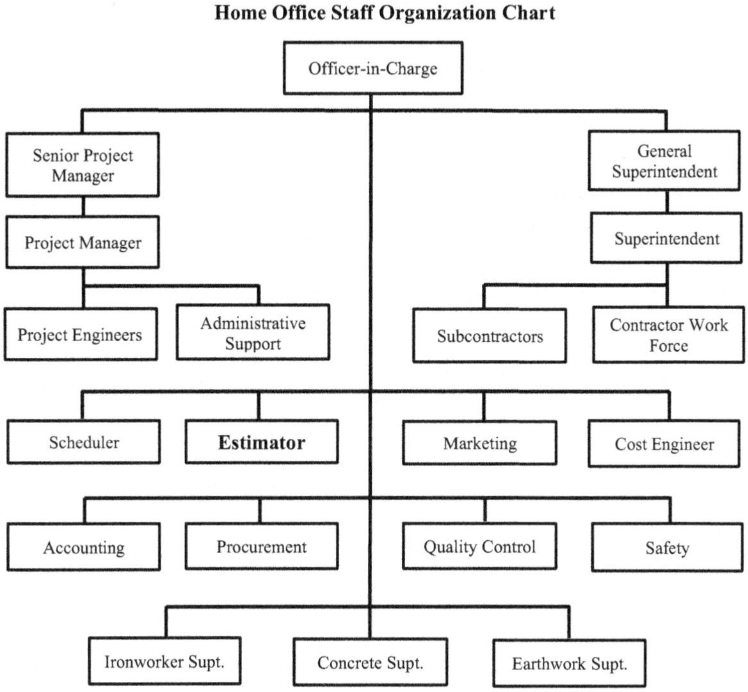

Home Office Staff Organization Chart

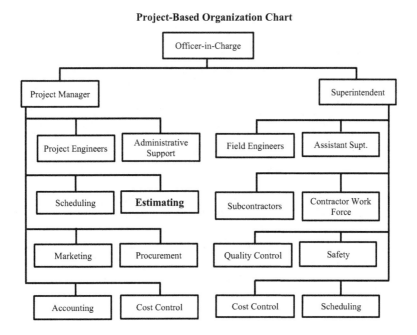

Figure 4.2 Home office staff organization chart

Project-Based Organization Chart

Figure 4.3 Project-based organization chart

two organizational styles. The choice of organizational structures depends on the contractor's approach to managing projects. The officer-in-charge is the PM's and estimator's supervisor. He or she may have various titles, as described in what follows.

Project team member responsibilities

Individual team member responsibilities may vary from contractor to contractor and from project to project, but in general they are as described here. The *officer-in-charge* (OIC) is the principal official within the construction company who is responsible for construction operations. He or she generally signs the construction contract and is the individual to whom the owner turns in the event of any problems that need escalated attention. This individual may carry the title of vice president for operations, chief operations officer (COO), project executive, senior project manager, or chief executive officer or may be the construction company owner.

The *project manager* has overall responsibility for completing the project in conformance with all contract requirements within budget and on time. He or she organizes and manages the contractor's project team. Specific responsibilities of the PM include:

- Coordinating and participating in the development of the project budget and schedule;
- Developing a strategy for executing the project in terms of what work to subcontract;
- Leading buyout activities with best-value subcontractors and drafting subcontract agreements;
- Negotiating and finalizing contract change orders with the owner and subcontractors;
- Scheduling and chairing meetings;
- Submitting monthly progress payment requests to the owner; and
- Managing project close-out activities.

If the PM is also assigned as the project estimator for the bid or proposal, his or her duties will be similar to those of the estimator, as described later. But if they are in a support role, the PM will assist the chief estimator by attending the pre-bid meeting, quantity take-off (QTO) creation, pricing recapitulations, subcontractor identification, general conditions estimate, and many of the bid day activities. The PM will report to the OIC or COO.

The *superintendent* is responsible for the direct daily supervision of construction field activities on the project, whether the work is performed by the contractor's direct employees or craftsmen employed by subcontractors. Specific responsibilities of the GC's project superintendent include:

- Developing and updating the detailed construction schedule;
- Determining the construction building means and methods and work strategies for work performed by the contractor's own work force;
- Managing falsework choices (formwork and scaffold) and for selecting and directing hoisting operations;
- Planning, scheduling, and coordinating the daily activities of all craftspeople working on the site;
- Ensuring all work conforms to quality expectations included in the contract requirements; and
- Ensuring all construction activities are conducted safely.

During estimate creation, the superintendent may also assist by visiting the jobsite during the pre-bid meeting, developing QTOs for direct work, generating subcontractor interest, planning manpower, scheduling development, and reviewing the jobsite general conditions estimate. The superintendent may also report to the COO or a general superintendent.

The *project engineer (PE)* or *field engineer* is responsible for resolving or coordinating any technical issues relating to the project. On small projects, the project engineer's responsibilities may be performed by the project manager. On large projects there may be multiple PEs. There may also be a jobsite cost engineer or accountant. Specific PE responsibilities include:

- Processing submittals and requests for information and maintaining associated tracking logs;
- Preparing subcontract documents and correspondence and maintaining the contract file;
- Supporting the superintendent and foremen with material purchase and expediting, cost control work packages, documentation support, and foremen and safety meeting notes; and
- Reviewing and processing subcontractor invoices and requests for payment.

During estimate development, the PE can also assist with QTOs, calling subcontractors, and taking bids on bid day or potentially running the bid in.

The *foremen* are responsible for the direct supervision of the craftsmen on the project. The construction firm will assign foremen for work that is performed by the company's own construction workers. Foremen for all subcontracted work will be assigned by each subcontractor. Specific responsibilities of GC direct craft foremen include:

- Coordinating the layout and execution of individual trade work;
- Verifying that all required tools, equipment, and materials are available; and
- Preparing daily time sheets for all direct craft personnel.

Home office staff specialists: Depending upon the size of the construction firm, there may be many people in the home office who are experts in one facet of the business or another who will support all of the jobsite teams. This includes functions such as marketing, accounting and human resources. As it relates to estimating, the following specialists may support the jobsite team either in initial estimate preparation and/or support throughout the estimating cycle:

- Chief estimator and/or staff estimator,
- Staff scheduler,
- Quality control officer,
- Safety officer, and
- General or specialty superintendents such as concrete or structural steel.

The *chief estimator* or *project estimator* assigned to the bid or proposal, if not the PM, will be responsible for leading the GC's estimating team. Similar to the PM, the estimator reports to the COO or OIC. Responsibilities are discussed throughout this book and would include:

- Schedule the estimate and assign responsibilities to the estimate team,
- Work breakdown structure,
- Quantity take-offs for risky areas of work, such as structural steel,
- Direct work pricing recapitulations,

- Jobsite general conditions estimate,
- Recommendations for markups, and
- Setting up and management of the bid room on bid day.

Subcontracting plan

In Chapter 1, we introduced an alternate work breakdown schedule called a project item list. This document allows the estimating team to forecast what work will be self-performed by the GC's direct craftsmen and what will be subcontracted. Most of the choices of which work to self-perform versus subcontract are straightforward. For example, the GC will typically not perform plumbing or electrical work. But choices for other scopes such as rebar placement, concrete slab finishing, structural steel installation, and rough carpentry may not be made until bid day.

In the past, GCs may have performed 50% of the project scope with their craftsmen employees, but today self-performance is down to only 10% to 20% of the project scope for most commercial contractors. Industrial contractors employ more types of craftsmen than commercial contractors do and perform a larger percentage of construction projects with their own crews. Only a few residential custom builders employ any craftsmen; many subcontract almost 100% to trade specialists. Because a typical commercial GC subcontracts 80% to 90% of the construction project scope, subcontractors, also referred to as specialty contractors, are important members of the GC's estimating and project delivery team and have a significant impact on the GC's success or failure. A GC cannot be successful without its subcontractors also achieving success. Since subcontractors have such a great impact on the overall quality, cost, schedule, and safety success for a project, they must be selected carefully and managed efficiently. There must be mutual trust and respect between the GC and the subcontractors, because each can achieve success only by working cooperatively with the other. Consequently, estimators, PMs, and superintendents find it advantageous to develop and nurture positive, enduring relationships with reliable subcontractors. General contractors need to treat subcontractors fairly to ensure they remain financially solvent, not only to finish this project but to be available to provide competitive bids on future projects.

The use of subcontractors by the prime contractor is a risk management process. Subcontractors provide the GC with access to specialized skilled craftsmen and equipment that they may not have in-house. One of the major risks in construction is accurately forecasting necessary manpower and associated cost of labor required to build a project. By subcontracting significant segments of work, the GC can transfer much of that cost risk to its subcontractors. When the estimator asks a subcontractor for a price to perform that scope of work, the subcontractor bears the risk of properly estimating the labor, material, and equipment costs. Craftsmen experienced in the many specialized trades required for major construction projects are expensive to hire and generally are used on a project site only for limited periods of time. It would be cost prohibitive for a GC to employ all types of skilled trades as a part of their own full-time work force.

The PM and project superintendent should play an important role with critical decisions whether to subcontract work out or to perform it with their own direct labor forces. Some of the advantages and disadvantages in performing work direct versus subcontracting are listed here:

- The GC's jobsite team may be able to control the schedule and achieve safety goals better with their direct craft labor.

- The project superintendent may argue that his or her craftsmen can build it better with enhanced attention to quality control, but if the subcontractor performs the work and rework is required, that is the subcontractor's responsibility.
- If subcontracted, the subcontractor is liable for cost overruns.
- Conversely, if performed direct, and the estimate is underrun, the GC can improve its fee potential.
- Subcontractors have access to specialized craftsmen and specialized equipment.
- It may be an owner contract requirement to subcontract work.
- Employing subcontractors improves a GC's cash flow position because the subcontractors are not paid until the GC is paid and retention is held on subcontracted work until close-out is complete. The GC does not withhold retention from material suppliers or their own craftsmen's wages.

On negotiated projects, the GC has an opportunity to prequalify select subcontractors and provide them with detailed requests for proposal or quotation, which tells subcontractors exactly what to price. But on public lump sum projects, oftentimes, any subcontractor can provide a bid, and it may be difficult for the GC to analyze which companies have provided complete prices and are qualified to perform the work. Regardless if the project is negotiated or bid, it is recommended that the GC develop pre-bid day subcontractor order of magnitude plug estimates so that they are better prepared to analyze quotations once received. Subcontractor plug estimates are discussed in Part IV of this book. In the next chapter we dive deep into preconstruction planning. There are many elements the PM, superintendent, and estimator contribute to the preconstruction plan, including their approach to subcontracting.

Subcontractor selection

Once the GC is notified of the owner's intent to award a contract, a separate buyout meeting is held with each potential subcontractor and supplier before also awarding them the work. The project team should share prior project experiences with the estimator during subcontractor buyout so that only the best-value subcontractors are awarded work on this project or solicited for pricing on the next project. Some of the issues to be covered in the buyout meeting include:

- Compare the subcontractor's list of inclusions and exclusions ahead of time with the other bidders. Go through their list at the meeting to make the scope as comprehensive as possible.
- Make sure there are no scope gaps, no scope overlaps, and no conflicts.
- Did the subcontractor receive and understand all addenda?
- Make sure that this subcontractor has picked up all work that will be signatory to their crafts. For example, all plumbing work on the project should be awarded to the plumbing subcontractor.
- Slowly leaf through the drawings and review each relevant detail: "Is this work included?" An "I bid it per plans and specifications" response from the subcontractor is not sufficient; the GC's project team needs to dig deeper.

It is imperative that the PM and superintendent select quality best-value subcontractors if they are to produce a quality project, on time, safely, and within budget. The jobsite management teams must remember that poor subcontractor performance will reflect negatively on their professional reputations and their ability to secure future projects from project owners. In addition, a subcontractor with a bad attitude can be contagious.

General contractors who take pride in treating subcontractors with a heavy hand will not retain loyalty and success in the long run. A GC does not necessarily want a subcontractor to make a fortune on any one project, especially at the GC's expense, but they should want the subcontractor to make a fair profit so that they will be around for the next project. This is a collaborative and not a top-down approach to subcontracting. Subcontractors make up 80% or more of the workforce, and it is difficult to have a successful team if you are not including that 80%.

In addition to prequalification and selection of best-value subcontractors, the GC's project team continues with management of subcontractors during physical construction, project close-out, and warrantee management. These are all PM topics, many of which are introduced in Chapter 24.

Risk assessment

Construction is a risky business, as evidenced by the high number of construction firm failures each year. To minimize the potential for financial difficulty, a contractor should analyze each potential project to determine the risks involved and whether or not the potential rewards justify acceptance of the risk exposure. The sources of estimating risk on a project are extensive, as shown in Figure 4.4. The contractor needs to forecast the likelihood of such risks, the range of possibilities, and the impact of each on the contractor's ability to complete the project profitably. Many of these risks are owner-risks, but depending on contract language, they may also be imposed on the construction team. Before the estimator begins any of the hard work of taking off material quantities, the risks must be assessed by the GC's project team.

Selection of risk management strategies involves selection of the appropriate response to each of the identified risks. Internal risks must be identified also, and appropriate management strategies selected. The three most common internal risks are unrealistic cost estimates, unrealistic construction schedules, and ineffective project management, including cost and schedule control, material management, and subcontractor coordination. Contractors must adopt strategies to minimize the potential of these problems occurring. Often the basic issue to be addressed is the selection of qualified people to manage the project, particularly the PM and superintendent.

The greatest risk in developing a cost estimate is estimating the productivity of direct craft workers. This is where an experienced estimator has a great advantage. Other risks involve failure to include some element of work or double counting another element of work. To minimize the potential for making errors when developing an estimate, the estimator should:

- Rely on good estimating practices and procedures;
- Choose good in-house project management and supervision teams not only to manage the project but also to assist with the estimating;
- Choose qualified or best-value subcontractors and suppliers;
- Plan to build the project in less time than specified in the contract to save jobsite overhead expenses; and
- Be selective on which projects are chosen to bid. Familiarity with the client, designer, and building location and project type are important.

Potential Estimating Risks

- Unusually adverse weather
- Material cost inflation
- Owner's inability to finance the project
- Limited availability of skilled direct work craftspeople
- Overly optimistic estimate of labor productivity
- Poor labor productivity performance
- Limited availability of qualified subcontractors
- Lack of adequate subcontractor financing and potential for bankruptcy
- Inexperienced design team
- Incomplete design documents
- Lack of adequate quality control and cost of rework
- Excessive project size
- Vague geotechnical report and/or poor soils conditions
- Inadequate or improper tool and/or equipment choices
- Inexperienced in-house superintendent and project manager team
- Inadequate jobsite general conditions estimate
- Slow payment terms and/or excessive withholding of retention
- Lack of cash flow for owner, GC, and/or subcontractors
- Delay in obtaining the building permit
- Permit and inspection interpretation by the authority having jurisdiction
- Unfamiliar project location
- Unknown or unforeseen conditions such as building rot
- Union jurisdiction disputes and/or strikes
- Insufficient estimate allowance for markups such as fee and contingency
- Inadequate cost estimate
- Unreasonable contract terms
- Inexperienced project owner
- No prior relationship with owner or design team
- Lack of construction cost controls
- Unreasonable schedule duration or inadequate schedule plan
- Lack of schedule control
- Safety control and potential accidents
- Slow or delayed construction close-out phase
- Project constructability and complexity, and many potential others.

Figure 4.4 Potential estimating risks

The output of a risk assessment is a decision whether or not to pursue a project. If the construction team decides to proceed with an estimate, the amount of contingency to include in the bid or cost proposal, whether or not to joint venture (JV) with another firm, the portions of work to subcontract, and the type and amount of insurance to purchase must be factored before a final price is submitted.

Construction is a risky business, and estimators and PMs must carefully assess the risks associated with each prospective project. Once the risks have been identified, risk management strategies must be developed. In some cases, the risks can be mitigated by obtaining a JV partner or hiring subcontractors. In other cases, the risks are too great, and the project should not be pursued.

Summary

Contractors establish separate project management organizations to manage construction activities on each individual project. The project team typically consists of a project manager, superintendent, project engineer, foremen for self-performed work, and home office specialists such as estimators and schedulers, depending upon project size and complexity. The contractor's project manager is the leader of the jobsite management team. He or she is responsible for managing all the activities required to complete the job on time, within budget, and in conformance with quality requirements specified in the contract. The PM may take the point on development of the initial estimate, or this may be accomplished by the chief estimator.

Subcontractors make up 80% to 90% of the construction workforce on most commercial construction projects. In order for the GC to be successful in any aspect, be it cost, schedule, quality, and/or schedule, the subcontractors must also be successful. The GC's project team should develop a subcontracting plan, because they are estimating the project, and build upon that plan during subcontractor prequalification and selection of best-value subcontractors. Subcontractor management continues into buyout and throughout all construction phases, including contract close-out.

Any contractor, GC, CM, or subcontractor contemplating to bid or propose on a construction project must go through a thorough risk-assessment process. If a project is deemed too risky, the contractor should pass on the opportunity. If a decision is made to pursue a proposal, the contractor will make necessary adjustments in its fee and contingency, as will be discussed in Chapters 18 and 19.

Review questions

1   How do the responsibilities of the project manager differ from those of the project superintendent?
2   List these positions in order of who reports to whom: General superintendent, apprentice, project superintendent, COO, general foreman/assistant superintendent, journeyman, and helper.
3   List these positions in order of who reports to whom: Senior project manager, OIC, PM, PE, and intern.

Exercises

1   As a PM or PE or estimator, would you prefer to work on a project-based or a home office staff organization, and why that choice? Assume for this exercise that even with a jobsite organization, the construction still has a staff estimator for support.
2   Draw an organization chart for an outside case study including all of the companies and team members discussed here and whatever/whomever else you feel would be necessary to adequately staff the job.
3   Draw an organization chart of a project you have worked on.
4   Which of the project attributes in Figure 4.1 is (A) the most important to the project stakeholders (owner/designer/GC/subcontractors/city), and, conversely, (B) which is the least important to the project stakeholders (owner/designer/GC/subcontractors/city)?
5   What are the advantages of (A) a staff estimator taking the point on development of a project estimate over a PM or, conversely, (B) a PM taking the point on the estimate in lieu of a staff estimator?

6 List at least three estimating activities a PE can be involved with in development of a project estimate.

7 List at least three estimating activities a project superintendent can be involved with in development of a project estimate.

8 Why should the PM and the superintendent be directly involved with preparing the project estimate and schedule?

9 Which of the project attributes shown in Figure 4.1 can be eliminated and still consider the project successful?

5 Preconstruction

Introduction

Construction projects do not just simply start and end, but rather they have many clear and defined steps or phases. The major phases of a construction project include:

- Planning or preconstruction,
- Start-up,
- Controls: Cost, schedule, quality, safety, document control, and others,
- Close-out, and
- Post-project analysis.

During the planning phase, the general contractor's (GC's) project manager (PM) and estimator evaluate the risks that are associated with the project, particularly those related to safety, cost, quality, and schedule. Risk analysis and risk management are critical skills essential to successful project management and were introduced in Chapter 4 and will also be elaborated on in Chapter 18 regarding estimating markups. The PM develops the organizational structure needed to manage the project and the communications strategy to be used within the project management organization and with other project stakeholders. Material procurement and subcontracting strategies also are developed during the planning phase.

Our specific focus in this chapter is on preconstruction (precon). Preconstruction for the project owner includes everything from property purchase and entitlements to early pro forma estimates and assembling the design and construction teams. In addition to the design team's preparation of design documents during precon, they also are involved in zoning analysis and permit applications. There are essentially unlimited aspects of preconstruction, but we feel we have captured those most important for the construction project manager, estimator, and preconstruction team. Our discussion in this chapter includes:

- The preconstruction phase;
- Preconstruction services including estimating, scheduling, constructability review, value engineering (VE), early bid packages, and early quality and safety control planning;
- Preconstruction contracts; and
- Preconstruction fees or costs.

Estimating is not the primary focus of the construction team during the other four phases of a construction project, as listed here. But each of these other phases utilize the estimate and impact other future estimating aspects, as will be discussed in the last two chapters of this book.

Preconstruction phase

So when does the preconstruction phase occur? Obviously, it occurs 'pre' the construction phase as shown here.

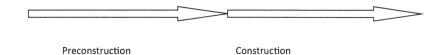

Preconstruction   Construction

How long does preconstruction last? Well, like the answer to most questions, that depends. We have seen preconstruction last as short as a month or two and as long as two years. Now the more important question: What is done during the preconstruction phase? That answer is quite long and is owner, architect, general contractor, and project specific. We have again included here what we feel is most relevant for the PM and estimator.

The most obvious activity that occurs during precon is to have the project designed. The role of the design team is to prepare construction documents for the contractors to the build the project. The major phases of design include:

- Conceptual design or programming,
- Schematic design (SD),
- Design development (DD), and
- Construction documents (CDs).

Not every project experiences a conceptual design, because many owners and architects 'know' what they want to build and proceed more quickly into the SD stage. Although the American Institute of Architects (AIA) has a general definition of what occurs during these distinct design phases, they are again project and project-team dependent and often overlap.

Preconstruction for contractors can occur any time during the preconstruction phase as diagrammed here. When (or if) a GC becomes involved in preconstruction depends upon the project owner and the lead designer and the type of project. A simple project that may be bid lump sum under a traditional delivery method likely may not have any role for a contractor during preconstruction, especially if the owner is a government agency. A complicated project such as a hospital may involve the contractor very early during the design process, possibly at the end of SDs. When would contractors like to be involved in preconstruction? As early as possible!

Project owners may engage either a general contractor or construction manager (CM) or a preconstruction services contractor or consultant to work during this phase. On some public projects, jurisdiction dependent, the owner may be required to engage a preconstruction agent who will not end up being the construction contractor. Private negotiated project owners have complete freedom whether to engage a preconstruction agent or not, often with guidance from its architect. Preconstruction services provided by a GC are more prevalent on privately funded projects than they are on public projects. This is because private owners are not restricted to the public bidding procedures that are required of many public owners. For the balance of this chapter's discussion, let's assume that a private owner has chosen to engage a GC to perform preconstruction services with full support of the project architect. Let us also assume that by participating in precon, the GC has not eliminated itself from the possibility of becoming the construction contractor – this being the case with many public projects.

Preconstruction services

The project owner may choose to select the construction contractor during the development of design and ask the contractor to perform preconstruction services. For contractors, this involves attending design coordination and review meetings and providing advice regarding the use of materials, systems, and equipment and cost and schedule implications of design proposals. It is customary to have weekly or twice-monthly preconstruction meetings, often chaired by the architect, which include agendas and action items and meeting notes, very similar to the weekly owner-architect-contractor meeting that occurs throughout construction of most projects, regardless of type. The GC's PM or estimator may also offer to chair preconstruction meetings. Owners often hire GCs or CMs to perform preconstruction services to provide construction expertise during the DD phase to optimize cost, schedule, and constructability input prior to bidding and during construction. These services may include:

- Preliminary budgeting or estimating;
- Precon activity scheduling and preliminary construction scheduling;
- Constructability analysis;
- Value engineering;
- Early release of subcontractor and supplier bid packages to facilitate long-lead material procurement; and
- Planning, including development of project-specific quality control (QC) and safety control plans, jobsite layout plan, traffic plan, environmental protection plan, and others.

Each of these six areas is expanded here. As indicated in our introduction to this chapter, there are numerous functions a precon agent or contractor may assist with. Experienced contractors may also participate in a variety of more-advanced preconstruction services such as:

- Sustainability planning;
- Design document quality control reviews;
- Management of the design team;
- Material recommendations and selections, including mechanical and electrical systems;
- Permit coordination and expediting;
- Lean processes such as target value design (TVD) development and management; and
- Other cost-related considerations such as cash flow predictions and life cycle cost analysis.

Budget estimating

One of the main preconstruction contributions the GC makes is with early budgeting. The project owner needs to know approximately (within 10–20% accuracy) what the project will cost before they commit financial resources to progress the design to the next phase. Preliminary cost estimates are developed using conceptual cost estimating techniques and refined as the design is completed to ensure the estimated cost of the project is within the owner's budget. Some architects used to provide early budgets to the owner, but now most rely on estimating consultants or contractors. One of your authors is a former contractor, and our bias is that contractors provide the most accurate estimates of anticipated construction costs. The contractor knows what each item in the last project cost (see as-built estimates in Chapter 23) and has access to current subcontractor and supplier market pricing, even during early design and budget development. Subcontractors want to assist the GC in hopes of having an opportunity to either negotiate or bid the project once the design is completed. By providing early budgets, subcontractors are at least guaranteeing themselves the opportunity to be on a short list of bidders.

The quantity of budgets (one, two, three, or more) produced by the construction/estimating team should be defined in the preconstruction contract, as discussed later. It is customary for the GC to produce a budget after completion of each of the design phases. Once the DD drawings are finished, the GC will take three to four weeks and develop as detailed an estimate as possible. This estimate will be presented to the owner and design team with a list of changes from the prior budget and a list of recommendations to enhance the design either for constructability or cost savings. The owner will then approve the budget and the design team progresses to the next level of design, in this case, CDs.

Often the contractor will maintain a *budget options log* throughout the design process, which is very similar to the value engineering (VE) log discussed later. The owner and architect may ask the GC to price options or propose changes in the design at each of the weekly design meetings. In this manner, everyone is kept as up to date as possible with design progression and will not be surprised with large budget swings at the next formal design submission.

The contractor also has an incentive to develop detailed and accurate budgets, even if the design is not yet complete, and make them as true as possible. A construction contractor is not solely in the business of performing preconstruction services; rather, their end goal is to attain a construction contract on this project. The best way for them to do so is to be cooperative team players and produce accurate budgets with minimal swings in value, that is, no surprises when the final budget, which is often a guaranteed maximum price (GMP), is presented.

Target value design is an early cost control method of lean construction. The concept of TVD is to establish a project budget first, such as $50 million, and then design to the budget. Each of the major elements of the project, such as foundations, roofing, windows, electrical, and others, are assigned a portion of the budget; for example, elevators may be budgeted at $1,260,000. Essentially, each of the major work areas is akin to a piece of the pie as represented in Figure 5.1. Then during the course of design if an element exceeds its budget, say elevators are now $1,400,000, then another element must be reduced such that the total budget, or the pie, stays the same size.

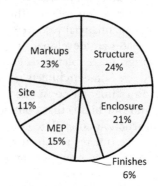

Figure 5.1 Target value design

Scheduling

Production of construction schedules follows the same course as production of budget estimates. The more detailed the design is, the more detailed and accurate the schedule will be. A preliminary construction schedule may be developed to assess the time impacts of design alternatives. In addition to scheduling the procurement and construction activities, the contractor/scheduler will also offer to incorporate design and permit activities into their schedule. Some of the types of construction schedules contractors prepare include:

- Preconstruction activities schedule, including design and permits;
- Detailed construction schedule or contract schedule;
- Summary schedule; also may be the contract schedule;
- Three-week look-ahead schedules;
- Subcontractor, area (floor or wing), pull planning, and phased mini-schedules.

The final construction or contract schedule will not be prepared until the final set of drawings is issued. Ideally the complete and final cost estimate will have been prepared, which includes direct labor hours for the GC's crew and firm subcontractor prices or quotes. Once the subcontractors have been selected, they also will input to the GC's schedule. The contract schedule is one of many CM products produced from the construction estimate, as discussed in Chapter 24.

Constructability review

One important contribution contractors make during the preconstruction phase is to review progress drawings from the design team and comment and make suggestions regarding their 'constructability'. This is not to say that the design team's documents are not constructible, but rather, can they become more constructible or easier to build? Constructability analysis involves reviewing the proposed design for its impact on the cost and ease of construction. These proposals are often as simple as changing a welded structural steel connection to bolted, which can be assembled in the field faster and safer, or having steel gusset plates welded on to the columns in the fabrication shop rather than in the field, which is also safer and ensures better quality control. An example with wood framing is to change dimensional lumber to engineered lumber, which will be straighter and not shrink. A popular change with wood-framed mixed-use development apartment buildings today is to have the wall and floor systems 'panelized' and built in a fabrication shop, which improves quality and enhances the schedule. These prefab panels are then flown in with a tower crane and connected with fewer field connections. Many of these types of changes save cost but not all. Some constructability changes may actually increase cost, but improve the schedule, quality, safety, energy efficiency, and/or long-term building maintainability.

Along these same lines, the construction team acts as another set of eyes and helps edit or QC the drawings. The GC can help mitigate potential subcontractor change order opportunities while reviewing progress drawings. Most designers are not keen on contractors finding errors in the drawings during the construction process and drafting requests for information (RFIs) followed by change order proposals (COPs), but they are often appreciative of the GC when errors are corrected before the documents have been let to subcontractor bids.

Value engineering

The value engineering process involves a systematic evaluation of a project design to obtain the most value for the cost of construction. It includes analyzing selected building components to seek creative ways of performing the same function as the original components at a lower life cycle cost without sacrificing reliability, performance, or maintainability. Value engineering studies may be performed by consultants during design development, as a contractor-performed preconstruction service, or by the GC during construction. The most effective time to conduct such studies is during early design development. Some lump sum construction contracts may contain a VE incentive provision that allows the contractor to share in the savings that result from approved value engineering change proposals.

Value engineering studies are conducted to select the highest value design components or systems. The essential functions of each component or system are studied to estimate the potential for value improvement. The VE study team needs to understand the rationale used by the architect and their team in developing the design and the assumptions made in establishing design criteria and selecting materials and equipment. The intent of VE is first to develop a long list of alternative materials or components that might be evaluated, often through a brainstorming process. Preliminary cost data is generated, and functional comparisons are made between the alternatives and the design components being studied. The intent is to determine which alternatives will meet the owner's functional requirements and provide more value to the completed project. Estimated life cycle cost data is developed for each alternative. The advantages and disadvantages of each alternative are identified, and the ones representing the best value are selected for refinement in preparation of presentation to the architect and the owner.

The final step is the preparation of formal value engineering proposals. The VE proposal looks very similar to a post-contract change order proposal, including all detailed costs and markups and substantiation. Detailed technical and cost data that is developed to support the recommendations should be included. The advantages and disadvantages of each recommendation are described. Each VE proposal is tracked in a log, similar to other document control or tracking logs managed by the construction team, including RFIs, submittals, COPs, and close-out documentation. The VE proposals are submitted to the designer and the owner for approval. If approved, the proposals are incorporated into the design. Value engineering proposals approved after the construction contract is awarded must be incorporated into the contract by formal contract change order. The preparation of change order estimates is also introduced in Chapter 23.

To some, VE may appear the same as TVD, but there are differences. Target value design establishes the budget up front, and each of the design disciplines are tasked to design within their share of the budget. Value engineering happens after some or all design is completed, which requires redesign. Lean construction is focused on minimizing waste, and re-design to some extent is a wasted effort.

Early release of subcontractor and supplier bid packages

An additional service a general contractor brings to the preconstruction team is the ability to solicit pricing from long-lead material subcontractors and suppliers. General contractors rely on positive long-term relations with these companies to receive this valuable early input. Some of the subcontractors or suppliers the GC estimator may engage early include:

- Elevators,
- Curtainwall,
- Mechanical equipment,
- Electrical equipment, including switchgear and emergency generators,
- Laboratory equipment and casework, if applicable,
- Concrete reinforcement steel,
- Structural steel, and others.

Ideally the contractor will not issue subcontracts or purchase orders until it has also received a contract from the owner. But sometimes these long-lead items must be ordered before that can happen. Some vendors will accept letters of intent (LOI), which may offer financial reimbursement for early submittals and shop drawings and a promise of a forthcoming construction contract if the project proceeds as planned. If a financial commitment has to be made to a supplier, then a similar commitment will need to be made from the owner to the contractor and incorporated into the precon fee and contract, as discussed later.

Planning

The term 'planning' has many different connotations in construction. There are many drawings produced by just about every different design discipline that are 'plan' views, or looking down on the building or site; those were discussed in Chapter 3. We go through a planning phase to produce the final construction schedule, which includes evaluation of logic and deliveries and manpower. This chapter focuses on the preconstruction plan. Presented here are just a few other plans that contractors may prepare as part of their preconstruction services.

Very similar to the constructability review, early *quality control planning* can have significant impacts on the contractor's ability to meet schedule and cost goals. Contractors can input to early design development documents regarding their ability to meet the client's and architect's intentions. It is important that the QC plan is project specific and not generic. It should include items such as prequalification of subcontractors and suppliers, preconstruction meetings with subcontractors and suppliers, submittals and mockups, inspectors and inspections, and early in-process punch lists.

Development of an active – not passive – *safety plan* is also a preconstruction activity. Passive safety (and quality) efforts involve fixing problems after accidents occur. A proactive plan prevents issues from happening. Some considerations the safety plan will address involve full-time safety inspectors, prequalifying subcontractors, Monday morning safety meetings, personal protective equipment requirements in subcontract agreements, and changing building design to make it safer to build. An example is raising a roof parapet just a few inches such that those craftsmen working on the roof both during construction and after would not need to be tied-off or require spotters. Another example is including shop-welded gussets on columns with eye-holes in them allowing cables to be strung serving as tie-offs and guard rail while the structural steel is being erected.

The superintendent does not wait until day one of construction to plan his or her site logistics. A *jobsite layout plan* will be developed during preconstruction that will consider material laydown or staging, site access and traffic flow, crane locations, dumpster and trailer locations, storm water control, and others. The superintendent is the proper person to develop this plan, as it is his or her site to manage. The site logistics plan for the industrial case study project is included on the book's eResources. There are many contributions the superintendent can make throughout the preconstruction process. An expanded discussion of the role of the project superintendent, both during preconstruction and throughout construction, is included in *Construction Superintendents, Essential Skills for the Next Generation*. The interested reader may wish to look to a resource such as that for additional superintendent roles and responsibilities. These and many other preconstruction 'plans' developed by the project team have some effect on the anticipated construction cost and should be incorporated into early budgets by the estimator.

Preconstruction contracts

The adage that good fences make good neighbors applies to construction as well. Good contracts make good contractors and good construction projects, and likewise with preconstruction. During a slow economy, contractors will offer to perform preconstruction services for free and will not request a preconstruction contract from the project owner. This may be a mistake. The contractor's goal obviously is to get its foot in the door so that it at least has the first shot at a construction contract. A preconstruction services contract is a professional services agreement similar to a design services contract and is not a construction contract. A short preconstruction agreement should be drafted and signed by both parties that clearly defines the expectations from the client and the architect of the contractor and deliverables, as discussed earlier. The cost of these services should be clearly defined, along with what promises are made, if any, involving a potential construction contract. Some of these precon contract considerations include:

- How many budgets are required and what is the timing, and in what detail?
- How many schedules are required, and are they preconstruction services schedules or construction schedules?

- How many meetings is the team to attend, for how long, and who will prepare meeting notes?
- Is travel expected to inspect potential material fabrication facilities?
- Are early material submittals expected?
- Are outside workshops such as partnering and lean planned, and who pays for these?
- What are the VE expectations?
- What is the anticipated duration of the preconstruction phase?
- Will the GC receive a construction contract at completion of a successful preconstruction phase?

There are copyrighted documents from AIA and ConsensusDocs for preconstruction services, but many contractors will offer to draft up a short proposal defining services and costs and timing, with space for the client to sign in agreement. In other cases, the owner will issue an LOI which states the contractor will be reimbursed, for, say, $50,000, for four months of preconstruction assistance, and it is the project owner's 'intent' to give them a construction contract at the completion of the preconstruction phase. These agreements often state that the contractor will prepare a GMP at completion of preconstruction, and if that GMP is acceptable to the owner, they will roll their precon costs into a construction contract. But if the parties cannot agree on a GMP, there is no obligation of a construction contract, and the contractor will be paid their precon fee and the two firms will part ways. If the GMP is approved, the LOI or preconstruction contract then should be attached to the prime contract as an exhibit once it is finalized. A sample preconstruction agreement is included on the book's eResource.

Preconstruction fees

Preconstruction has a cost. The more the contractor participates in the process, the more deliverables they provide, and the longer the preconstruction phase, the more it costs. The amount the contractor charges and the amount the client wants to pay also follows economic cycles. During a slow market, contractors will offer to perform preconstruction for free just to get a shot at negotiating a construction contract. During a busy market they may charge $100,000 for six months of effort, again depending upon the detail expected. Some refer to the preconstruction cost as a 'fee', similar to designers charging a fee for design services. Contractors prepare an estimate of the preconstruction fee similar to estimating any other work that factors in scope, hours, wages, and required materials or resources. The variations of the preconstruction fee options are very similar to estimate types, including:

- Lump sum fee, say $80,000 for eight months of work;
- Time and material wages with a loaded (including labor burden and home office overhead and materials and profit) hourly fee or rate of $120 for the PM, $130 for the superintendent, $95 for the estimator, etc.;
- Hourly fees with estimated hours to come up with a budget for preconstruction services;
- GMP of $75,000 based on hourly rates and hours and quantity of meetings and definition of deliverables expected;
- Preconstruction fees for design–build or design–assist mechanical and electrical subcontractors may be added to the GC's precon fee;

- Description of reimbursable costs that may be included with or in addition to, any of these fee options; and/or
- If long-lead materials are ordered, their costs may also be added to the preconstruction fee.

Regardless of the structure of the preconstruction fee, contractors rarely completely cover their cost; they generally do this work at a loss. Again, construction contractors are not in the business of performing preconstruction services; rather, they are looking for an opportunity to negotiate a construction contract. But owners who expect to receive these services for free and/or without a preconstruction contract often receive exactly what they pay for and are unhappy with the contractor's contribution toward design completion.

Summary

All projects realize a preconstruction phase in that design is accomplished pre-, or before construction, but not all projects involve preconstruction services from estimating and scheduling consultants or contractors. Contractors are not solely in the business of performing preconstruction services but often do so in hopes of negotiating a construction contract at the completion of the preconstruction phase. There are a variety of services a GC may perform during this process, including:

- Preparation of budgets and GMPs,
- Schedule creation,
- Constructability review,
- Value engineering proposals,
- Long-lead material supplier procurement,
- Quality, safety, and jobsite layout planning, and others.

Preconstruction contracts or letters of intent are good instruments to clearly define the amount the contractor will be paid for its work, the duration of the precon phase, and the amount of deliverables the client and architect expect from the construction team. Contractors should be paid for their services, the value of which is rarely sufficient to cover their cost. During a slow economy, the precon fee is minimal because the GC's goal is primarily to get a shot at a construction project. During a busy economy, GCs customarily perform preconstruction less often, and when they do, the amount they charge for their services increases.

Additional, more advanced construction management topics are typically covered in a construction project management book such as *Management of Construction Projects, A Constructor's Perspective*. These topics often also occur during the preconstruction phase and have an impact on project management and might include:

- Partnering,
- Permit expediting,
- Developer pro formas,
- Lean construction techniques,
- Building information modeling,
- Storm water control planning, and
- Sustainability and Leadership in Energy and Environmental Design.

This concludes the introductory concepts section of the book. In the next section we discuss the detailed quantity take-off process for many GC self-performed scopes.

Review questions

1   Why would a construction contractor perform precon for free? Why would they do so below cost? Would an estimating or scheduling consultant do this work at a loss?
2   Why would the GC offer to chair the weekly design coordination meetings and prepare meeting notes?
3   Why does estimating typically precede scheduling?
4   A VE log may also be known as a _____ log.
5   What is the difference between TVD and VE?
6   What is the difference between an LOI and a precon agreement?
7   As a GC, which would you prefer, a LOI, precon agreement, or construction contract?
8   What is the difference between a post-contract VE proposal and a COP?
9   Match these estimate types with the design phases. You can use each answer more than once or not at all, and some may have more than one answer.

Design phases	Estimate types
_____ Programming	(A) GMP
_____ Conceptual design	(B) Unit price
_____ Schematic design	(C) ROM budget
_____ Design development	(D) Lump sum
_____ Construction documents	(E) Cost plus

Exercises

1   Draft a preconstruction proposal with a stated scope and fee and duration and include a space for your client to sign it.
2   Prepare three value engineering ideas from one of our case study projects and include them in a VE log.
3   Provide a three-point argument why the client should hire the GC earlier rather than later during preconstruction.
4   As the architect or client, why would you NOT want to hire the GC early during design?
5   Other than the preconstruction services described earlier, what service might a consultant or contractor offer the client during design?
6   Assume a major subcontractor, such as mechanical or electrical or elevator or laboratory casework, is added to the precon team. Describe the process when and how they should be chosen, contracted, paid, and what their deliverables would include.
7   Building on Exercise 6, how does the precon process with a subcontractor conclude? How are these situations handled and what are the implications if (A) the subcontractor is dismissed or, alternatively, (B) they are employed as a member of the construction team?
8   How would your answers to Exercises 6 and 7 change if the mechanical, electrical, and plumbing subcontractors were (A) design–bid–build, with a separate designer

working for the architect, or (B) design–build, where the subcontractors performed their own design?

9   In your opinion, should GCs receive an incentive bonus for VE proposals, or, conversely, should they credit back the fee on VE cost deductions?

10   What might cause a project not to proceed 'as planned' such that the LOI does not roll into a construction contract? There are several possibilities.

11   Looking back to the organization charts presented in the previous chapters, and those that you may have prepared for an outside case study project, list three members from the GC's team who should participate on the preconstruction team and three members from outside of the GC's organization who should be on the team. There are many possibilities for this question.

Part II

Quantity surveying for general contractor direct work

6 Concrete substructure quantity take-off including foundations and slab-on-grade

Introduction

Chapter 6 is our first of four chapters in Part II, which focuses on the quantity take-off (QTO) process general contractors (GCs) utilize on normally self-performed work, including cast-in-place (CIP) and pre-cast concrete, structural steel, carpentry, doors/frames/hardware, and architectural specialties. Concrete is the single most important work item that most general contractors self-perform on a construction project, and as a result, estimators must be able to accurately quantify and apply pricing to a variety of concrete work items.

The QTO is one of the most important elements of the estimating process as it is here that quantities are determined and become the basis for the estimate. If quantities are not accurate, then the estimate accuracy is at risk regardless of the quality of the pricing. This and the next chapter focus on CIP and pre-cast concrete work normally performed by the general contractor, specifically foundations, columns, walls, and slabs. Superstructure concrete systems are quantified in the next chapter. Pricing for self-performed GC work will be discussed in Part III.

Quantity take-off process

This section will focus on the basic principles of quantity take-off that will apply to all work of the estimator, regardless of the scope being quantified. Although the description here is for concrete, these principles can be applied to the other chapters in the book pertaining to quantified self-performed work as well as many areas of subcontractor specialization.

Document review

Before starting the QTO process for concrete, estimators must familiarize themselves with the documents to get a general sense of the work. These documents will include:

- Structural (S) drawings: Most of the concrete will be shown on the structural drawings. A quick look at sections, details, and schedules is also helpful at this point to get more familiar with the project;
- Architectural (A) and civil (C) drawings: Site concrete and architectural concrete will be shown here;

- Mechanical (M) and electrical (E) drawings: Equipment pads, curbs, and similar concrete scopes are often shown on subcontractor drawings but will be excluded from their bids; and
- Technical specifications: Concrete strengths, forming, finishing, and curing requirements must be researched.

Marking the drawings

As the estimator is working his or her way through the documents and quantifying the work scope, it is necessary to mark the drawings to know what work has been quantified and what work remains. Marking can be done in a variety of ways, such as slashes or coloring detail numbers and drawing sections, depending on the preference of the estimator. The important thing to remember here is "when you count it or measure it, mark it on the drawings so you know what you've done" – and more importantly, what you haven't done. It is also important to be neat when marking the drawings so they can be read and understood after the estimate is completed.

There may also be several quantities taken from the same drawing; for instance, structural steel columns are typically shown on a foundation plan, so care must be taken to mark only that which is being quantified leaving room to add other marks as the QTO progresses. The steel columns for our case study begin above the upper elevated concrete deck and CIP columns sit below directly on the footings. The structural steel process for QTO is described in Chapter 8.

At the completion of the QTO process, the estimator should be able to go back and review the drawings and see if everything has been marked and quantified correctly. Any items that have not been marked will raise a red flag to check to see if anything has been missed and make corrections accordingly.

Automated quantity take-off

Today's estimator is likely using an automated QTO process such as Bluebeam, On-screen Take-off, and others. Even though the QTO is performed electronically on the computer rather than with paper drawings, the same markup procedures also apply. These software systems allow the estimator to utilize various colors to show what work has been quantified. Screen shots of the take-off can be printed if they need to be reviewed by a senior estimator or retained for record-keeping purposes. Use of technology in preparation of construction cost estimates is discussed throughout this book and is the focus of Chapter 22.

The quantity take-off form

The quantity take-off sheet is the form that will be used to quantify all work scopes and is organized to provide a consistent format for quantity survey activities. Estimators should use consistent and professional QTO sheets and not record quantities on the drawings or on yellow notepads. The QTO sheet, as shown in Figure 6.1, is organized as follows:

- The title information at the top showing the project name and location, classification of the work being quantified, estimator's name, date, and sheet number.

- The left hand portion of the sheet indicates the location or description of the work being quantified, for instance, 'type' of footings or grid lines or floors or levels. It is here that the estimator can refer to sections and/or details from which the work was quantified.
- Adjacent to the description are the columns where quantities and dimensions of the work being quantified are noted. It is important to note that dimensions are recorded with significant digits in mind, such as: 157′, 10.5′, and 1.25′.
- The remaining columns calculate materials and scopes that have been extended from counts and dimensions taken from the drawings. Quantities in these columns should be noted as whole numbers using accepted rounding protocols, as well.

City Construction Company
Quantity take-off sheet

Project: Dunn Lumber
Location: 3800 Latona, Seattle
Classification: **CIP Concrete Foundations**

Sheet Number: 1
Estimator: PJ
Date: 5/21/2021

| Description | No | Dimensions | | | SF | SFCA | CF | Rebar | Rebar |
		L	W	H	Fine Grd	Form	Concr	#/EA	#
Spread footings:									
Type A	33	9	9	2.42	2,673	2,875	6,469	363	11,979
Type B	53	11	11	2.83	6,413	6,600	18,149	617	32,701
15/S401					9,086	9,475	24,617		44,680
					SF	SFCA	CF		#
							1.05 waste		2000#/T
							27CF/CY		**22.3**
							957		**Tons**
							CY		
								Rebar	
								3 ea #5	
Continuous footings:		1050	2.50	1.00	2,625	2,100	2,625	3,150	1.10 lap
17/S401					SF	LF	CF	LF	1.043 #/LF
							1.05 waste		3,465
							27CF/CY		#
							102		2000#/T
							CY		**1.7**
									Tons

Notes:

Many other ancillary cost items will use these quantities and units and have not all been repeated here, such as floating tops of footings (rough concrete finish) is the same SF as fine grade. These will be carried on recaps.

Figure 6.1 Footing quantity take-off sheet

Concrete take-off basics

In quantifying concrete, there are several basic things that an estimator needs to know before beginning their work. First of all, concrete is a volume take-off, regardless of what it is you are quantifying. The volumes of concrete are initially quantified in cubic feet and then converted to cubic yards, which are the standard unit of measurement for purchasing and placing concrete. There are 27 cubic feet of concrete in a cubic yard. Most concrete elements are standard in shape: Length (L) × width (W) × height (H) or depth (d) or thickness (t); however, there may be complicated shapes that would require special formulas to determine accurate quantities. Depending on the specific concrete scope that the estimator is quantifying, the QTO sheet should be set up to allow easy and quick quantities to be entered. In this chapter and each of the following chapters in this section of the book, partial or complete QTOs for our case study will be included as examples.

Concrete, like development of the construction schedule, should be quantified and developed in the same sequence as the building is anticipated to be constructed, from the bottom to the top. In this manner foundations should be first, followed by slab-on-grade, walls and columns, and other elements, moving up through the building. Roof curbs and mechanical pads would be the last elements to be quantified. It is also important to quantify all like items at once rather than going back and forth. Consistency is important in preparing QTO sheets so that anyone could look at the QTO and understand the estimator's work. The same format of estimating worksheets should be utilized by all estimators throughout the same construction company.

Building systems

General contractors will separate their estimates into building systems or assemblies to allow for accurate and consistent summaries of cost components and to provide similar comparisons from project to project. The systems that most contractors will use are:

- Foundations,
- Substructure (often including foundations and slab-on-grade),
- Superstructure (above grade concrete, steel, and/or wood framing),
- Exterior closure or envelope,
- Roofing and sheet metal,
- Interior construction,
- Mechanical (including plumbing and fire protection),
- Electrical, and
- Site work.

This method of estimate preparation and summary is known as the Uniformat method and is preferred by many general contractors. The work items that will be described in this book will begin using these building systems and continue to break them down to additional levels of detail. Foundation walls and some below-grade but elevated concrete slabs may also be technically categorized as substructure, but for this discussion we have moved them all to the next superstructure chapter, along with CIP columns and beams and pre-cast concrete elements. Structural steel, which is also part of the superstructure, will be quantified in its own Chapter 8, and all of the self-performed carpentry work will be discussed in Chapter 9. Uniformat is often associated with very preliminary rough order of magnitude estimates, as introduced in Chapter 2, but most contractors will provide additional detail, such as separating the three types of mechanical systems and breaking site work into several cost entries, especially with more advanced levels of estimates.

Spread and continuous footings

There are several types of concrete foundation systems, including:

- Spread footings (sometimes called pier footings or spot footings or pad footings);
- Pile caps (similar to spread footings but often deeper, with more rebar, and associated with piling);
- Continuous footings (sometimes called strip footings or running footings);
- Grade beams (similar to continuous footings but deeper and with more rebar, and often associated with shear walls or connection to pile caps);
- Large mat footing; and
- Other specialty foundations such as drilled piers and caissons, which are typically performed by subcontractors.

Depending on the project, there could be other foundation elements that the estimator will need to individually identify. Figure 6.1 is a completed QTO sheet for spread and continuous footings for our case study project. Spread and continuous footings, along with grade beams, could all be quantified on separate QTO sheets, but in this book we have sometimes combined multiple assembles to allow us to present additional material examples. Note that the quantity columns include the major work items that will need to be quantified, as listed here. The vertical description column on the left indicates the type of footing being quantified. This can be a mark number from a foundation schedule on the 'S' drawings or a specific location on a drawing to allow the estimator to know where that element is on the drawings, for example 'detail 15/S401'. The specific work items to be quantified for spread and continuous footings that will be discussed in this section include:

- Concrete,
- Formwork,
- Finish top,
- Hand excavation and backfill or 'fine grade',
- Concrete reinforcement, or 'rebar',
- Embedded steel,
- Pump, and
- Structural excavation and backfill.

Spread footings and pile caps

The volume of *concrete* in each type of footing is first quantified in cubic feet (CF) but converted to cubic yards (CYs), as that is the unit of measure in which it is purchased. The conversion to CYs is not done until all of the concrete has been quantified for each category in order to minimize the number of calculations and also the chance for an error in the conversion process. Waste is then added. Waste and lap factors for several materials are included in Appendix A, but vary depending on the type of work and the contractor's means and methods of installation.

The area of the *formwork* necessary to form the footings is an important calculation and often ends up with the largest quantity of direct man-hours; therefore, it is one of the riskiest for the estimator and the contractor. Formwork is also known as falsework, or temporary structures, and can be thought of as a temporary mold to hold the wet concrete until it has properly cured. The quantity or area of formwork is sometimes referred to as square foot of contact area (SFCA). This calculation needs to be performed based on the shape of each concrete element, in this case the spread footings. If an element is 12″ or less in height or depth, then the unit of measure is lineal foot (LF), if greater than 12″, then the unit of measure is in square feet (SF). The reason for this is that if the depth is less than or equal to 12″, the footing is formed with dimensional material, such as a 2 × 8, whereas if deeper than 12″ the forms are metal or built from plywood and other dimensional lumber materials.

The operation of finishing the footing tops includes 'rodding off' or scraping the excess concrete off the top of the footing to provide a reasonably level surface to place future structural elements such as columns or walls on top of the footing. Although typically not a significant source of man-hours, there is a specific labor activity included in this operation, so the quantity should be identified and will be equal to the area of the top of the footing.

A separate measurement or calculation is typically not performed for hand excavation, which may also be known as fine grading. The backhoe digging the footings cannot perform the structural excavation shown in Figure 6.2 exactly, so a laborer will need to get into the hole with a hand shovel and rake to finish the work. The quantity is often just the square foot of footing, which will be the same SF as noted for rodding off. Counts and dimensions for each type of footing are entered, and quantities are calculated from the dimensions and rounded to whole numbers. Decimals should not be used in the quantity columns.

Continuous footings and grade beams

Continuous footings or running footings and grade beams are continuous in nature and are quantified in a manner similar to spread footings. In the event that a continuous footing is less than 12″ high, the formwork quantity would be shown in LF; if more than 12″, then a SF calculation is determined and entered in the appropriate column. Most general contractors will have unit pricing for both LF and SF quantities. Continuous footings are quantified by determining the running length of the footing rather than counting 'each', as is the case with spread footings. Continuous footings will generally be located on the building perimeter and often along grid lines or between spread footings, especially if they support bearing or shear walls. The estimator should mark the length of the footing on the drawing after they have determined the dimensions to know that they have quantified that particular footing. Most of the other areas to be quantified, the process, and the units are the same as discussed earlier with the spread footings.

An experienced estimator knows that quantities, once measured and extended, if done properly, can be applied to other building elements without recalculation. For example, continuous footings are often associated with and have similar lengths as foundation walls. Perimeter continuous footings will often have a *footing drain system* that includes a perforated pipe, drain gravel, and likely drain fabric. These quantities do not need to be separately quantified, as they are often simply the outside LF of the footing. Ten percent may be added to the pipe quantity for waste. Drain gravel does not expand or contract as with dirt, as discussed later, so the quantity measured is the same as is needed to purchase. As with spread footings, the estimator should show the drawing number and details as to where the footings are located. After the foundation elements have been quantified, the estimator should go back and check the drawings to make sure that all elements have been marked to avoid missed items.

Reinforcing steel

A separate QTO sheet is often used to quantify concrete reinforcing steel, or rebar. This is customary because the work is measured differently and is also done by a different crew. Figure 6.1 includes the reinforcing steel for the case study footings. An estimator should read and understand the notes in the specifications and on the drawings prior to starting this quantification. Perhaps most important are two items: (1) clearance to the edge of the concrete and (2) the length the bars will lap each other. This information is needed to determine the length of the reinforcing steel.

When reinforcing steel is specified on a spacing basis, instead of the specific number of bars, one more bar must be added to the number of spaces for the end bar. For our example here a type F10 spot footing is 10-foot square and may require two mats of #5 rebar at 6″ on-center (OC) running each way (EW), with a minimum 2 inches of clearance required on all sides. Because the builder must start with and end with a bar on each side, this will require 21 bars for each mat, and each bar will be 9′-8″ long. The total length of #5 rebar required for each footing is shown in the following calculations. This length will then be multiplied times the total count of type F10 footings for the project.

---

10 feet/.5′ = 20 bars + 1 more for the end = 21 bars for each mat
Bar length: 10′ – 2″ – 2″ = 9′8″ = 9.67′.
21 bars/mat × 2 mats × 9.67′/bar = 406 LF #5 rebar for each spot footing

---

When setting up a separate rebar QTO sheet, the estimator will first differentiate the types of concrete elements (spread footings versus foundation walls) and sizes of rebar (#5 versus #8). Lengths are then extended for all similar bar sizes and multiplied by their weight in #/LF. The weights are then summed, and the total is converted to tons. Waste or lap is added for rebar in slabs, walls, and running footings but not spot footings, as that is purchased fabricated to length. One shortcut for calculating rebar lap is to simply add 10% to the measured quantified length. Published estimating references usually have a table of unit weights for reinforcing steel. We have included the most common sizes and weights of rebar in Table 6.1.

At the time of letting our case study project to bid, the structural engineer had not yet sized all of the concrete reinforcement but instead directed the bidding contractors to include rebar weight allowances for some concrete elements, such as 363# and 617# of rebar for each of the two different sizes of spot footings. Other members such as continuous footings and slab-on-grade included specific rebar sizes and spacing.

The quantity take-off for concrete reinforcing steel can be very time consuming to take off for an entire project. If the bidding time is short, a reasonable estimate may be made by determining the unit weight of rebar per cubic yard of concrete for a given concrete system. Representative unit weights can be calculated from portions of the footings, SOG, and most CIP walls within a reasonable accuracy to use for labor costing. Prorated

Table 6.1 Concrete reinforcement steel weight conversions

Concrete reinforcement steel (Rebar) weight conversions	
Bar size*	Weight (pounds) per LF
2	0.167
3	0.376
4	0.668
5	1.043
6	1.502
7	2.044
8	2.670
9	3.400
10	4.303
11	5.313
14	7.650
18	13.60

★ Note: Rebar sizes reflect 1/8″ in diameter, such that a #4 bar is ½″ thick and #8 = 1″

rebar weights for all concrete elements on one project are okay for budget estimates but should not be used on any other project bid because of differing engineering requirements. Shortcuts such as this are common within the construction estimating industry but are also risky. If the estimator on our case study wanted to use a standard quantity of rebar in pounds per cubic yard of concrete for the entire project, he or she would draw on detailed historical estimating data such as the following. It is easy to see that drawing any conclusions from these would be risky.

- Spot footings at 47 #/CY,
- SOG at 53 #/CY,
- Columns at 275 #/CY, and
- Tilt-up walls at 114 #/CY.

Embedded steel

Embedded steel is placed such that it becomes an integral part of a concrete member. It usually is supplied by the structural steel fabricator, which provides all of the structural steel in CSI division 05. We could account for it when we perform the QTO for the structural steel, discussed in Chapter 8, but since the embedded steel is often shown on the foundation drawings and will be installed by carpenters when forming the foundations, we account for it with the cast-in-place concrete. The most common embedded items are anchor bolts (ABs) used to secure structural steel columns, pressure-treated wood plate, and some pieces of permanent building equipment. Anchor bolts are either hooked on the end or have washers to secure them into the concrete and thus securely anchor the member to which it is bolted. Anchor bolts that connect a structural steel column to a spot footing are often in a pattern, such as four or six ABs per column and spot footing. Anchor bolts embedded in running footings or on top of a CIP foundation wall are found at intervals of 4 feet on center or so and tie down pressure-treated wood plates for shear walls. Similar to counting reinforcing bars, an additional AB needs to be added for the end of each wall. Other embedded items include trench or pit edge angles, weld plates in walls, and tilt-up connection angles and lifting anchors.

Carpenters install embedded steel such as ABs as part of, and while, building concrete formwork. An alternative to ABs is expansion bolts. They are installed after the concrete has cured rather than being cast into the footing. They are drilled in exactly where they are needed and secured with epoxy. The trade that uses them for securing their material or equipment installs these types of bolts. Bolts used for the structural steel will be installed by ironworkers. Similarly, a plumber will install ones that are used to hang piping and electricians those for securing conduit. If there is a significant quantity of embedded steel, it may be shown on a separate quantity sheet that becomes part of the concrete QTO, but if only a few ABs are required, they will often be included with the spot footing QTO.

Most commercial projects pump all of the concrete. A separate calculation and QTO is not necessary, but a simple repeat of the amount of concrete to be purchased will be noted and priced separately on the pricing recap sheet discussed in Chapter 11.

Structural excavation

A separate quantity sheet is also often used to calculate the required structural excavation and backfill and off-haul of surplus spoils. Although this construction work is performed before footing formwork and rebar, the QTO is typically later, as footing dimensions are necessary. Structural excavation and backfill could also be quantified and estimated with site work, as these steps in the concrete process are often performed by subcontractors. When an estimator is calculating the structural excavation, the first thing to determine is the depth of the footing. Looking at the case study drawings, there are different depths depending on the specific footing and the adjacent ground elevations. Detail 15/S401 shows the top of the spot footings to be 8 inches below the floor surface. The SOG is 4 inches thick, and the soils report requires a 4-inch granular base on the subgrade. When the footings are constructed, the ground will have been leveled to the subgrade. The technical specifications require the footings to be placed on undisturbed earth. A backhoe and a little fine grading by hand will provide an undisturbed surface.

On some projects, the GC project team may choose to construct the footings 'neat'. This means that instead of building forms, the contractor can dig a neat hole and place the concrete directly into it without further forming. It is customary to increase the footing excavation size to compensate for dirt contamination or sluffing at the sides. This process requires fairly firm soil conditions; clean sand will not hold firm enough. To determine which is more economical, an estimator must compare the cost of building forms and over-excavating against the cost of purchasing, placing, and finishing the extra concrete. This is a means and methods choice of the contractor. It is customary to estimate conservatively and include formwork.

During structural excavation, construction equipment such as a backhoe or track hoe removes undisturbed earth, and the quantity is designated as either bank cubic feet (BCF) or bank cubic yards (BCY), which reflects the volume of the undisturbed bank or ground. Soil swells or expands upon excavation. Some estimating references discuss swell and suggest swell factors to use for various types of soil. The estimator needs to make an informed judgment as to what to use. There are many terms used for the volume of soil after it has been excavated, such as truck cubic yards (TCYs) or loose cubic yards (LCYs). Assuming that the soil at the location of the case study project swells about 30%, the estimator will then multiply bank cubic yards by 1.3 to get the volume of earth that has to be handled. When working with any below-grade concrete, 2 feet must also be left clear beyond each side so that carpenters have room to set and brace the forms. The slope (S) of the excavation must also be accounted for, as discussed later. The excavation volume (Vol.) is:

$$\text{Vol.} = (\text{Length} + 2' + 2' + \text{S}) \times (\text{Width} + 2' + 2' + \text{S}) \times (\text{Depth}) = \text{BCF}$$

A simple sketch such as Figure 6.2 is a helpful estimating tool for all types of quantity measurements and recording of assumptions. These sketches would be most helpful if included on the actual QTO sheet, but this sketch is included here as a separate larger-scale figure for clarity.

The slope of the excavation will be accounted for in the previous formula by adding the depth of the excavation to each of the length and width dimensions and accommodating the angle of the slope. Most excavations are on a 1:1 slope; therefore, the slope is a triangular shape, and ½ of the depth of excavation is figured, but on two sides, so multiplied times 2, as shown in the next formula. A vertical excavated cut may be possible with firm soil and shallow excavations, but again the conservative method is to estimate with a 1:1 sloped setback.

$$\text{S} = \text{Depth} \times \tfrac{1}{2} \times 2 \text{ sides} = \text{LF for both sides of a footing}$$

The results of the excavation at the spot footings are shown in the excavation column on the quantity sheet. The units are kept as bank cubic feet, and only their total is converted to BCY and then TCY to keep conversion calculations to a minimum.

$$\text{BCF}/27 \text{ CF/CY} = \text{BCY}$$
$$\text{BCY} \times 1.3 \text{ swell} = \text{TCY}$$

Backfill is calculated by subtracting the volume of concrete from the excavation volume. Units are still in bank cubic feet, as shown in the backfill column heading. After summing the columns, the totals are then multiplied by the swell factor. An accounting crosscheck that compares the total backfill to the result of the excavation minus the concrete is made to ensure all calculations are correct. Earthwork volumes are rounded to the nearest whole cubic yard. The estimator must check the soils report, or geotechnical report, to verify that on-site excavated materials, if protected from the weather, can be utilized for backfill. If not, select import must be purchased and trucked in and placed, and all of the excavated materials must be off-hauled and properly disposed. This can be very expensive.

$$\text{Total excavation (BCY)} - \text{Total concrete volume (CY)} = \text{Total backfill (BCY)}$$
$$\text{Backfill (BCY)} \times 1.3 \text{ swell} = \text{Backfill required (TCY)}$$

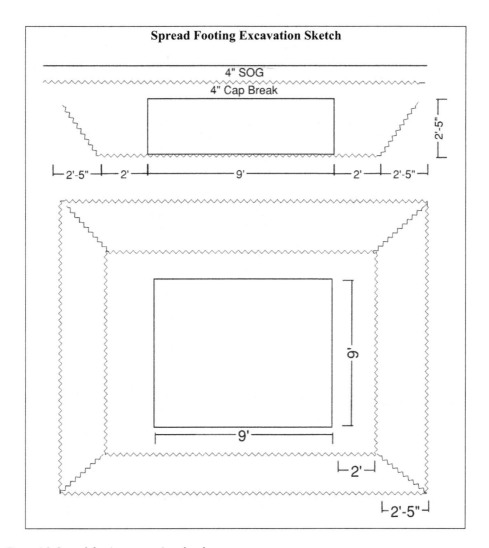

Figure 6.2 Spread footing excavation sketch

What remains to be calculated is the amount of excavated materials to be disposed of. For this, an estimator needs to do the following calculation:

(Excavated material × Swell) − (Backfill × Swell) = Disposal
Excavated (TCY) − Backfill (TCY) = Export (TCY)
Crosscheck: Export (TCY)/swell percentage = Concrete CY

Surplus excavated materials must be hauled off the site and disposed of at an acceptable location. This can cost the contractor significantly and must be accounted for by the estimator. Checking the geotechnical report again will determine if the exported material has value and can be sold, or if unsuitable, the GC has to pay to off-haul and dispose, which can be expensive. The common error regarding excavation and backfill is the state of the material during the various calculation processes. Special care needs to be taken to ensure that they are all made in terms of bank cubic yards or in the swelled volume. A mix-up will result in a major error in earth volumes, which can have a significant effect on the bid price. The structural excavation and backfill quantities for our three case study foundation systems is shown in Figure 6.3.

City Construction Company
Quantity take-off sheet

Project: Dunn Lumber
Location: 3800 Latona, Seattle
Classification: **Structural Excavation and Backfill**

Description	No	L	W	H	Structural excavation and backfill (BCF)
		\multicolumn{3}{Dimensions}	
Spread footings:					
Type A	33	9	9	2.42	$33 \times [9' + 2' + 2' + (2.4' \times 2 \times 1/2)] \times$ $[9' + 2' + 2' + (2.4' \times 2 \times 1/2)] \times 2.4'$d $= 33 \times 15.4' \times 15.4' \times 2.4' = \underline{18,783\ BCF}$
Type B	53	11	11	2.83	$53 \times [11' + 2' + 2' + (2.8' \times 2 \times 1/2)] \times 53 \times$ $[11' + 2' + 2' + (2.8' \times 2 \times 1/2)] \times 2.8'$
957 CY of concrete, A & B					$= 53 \times 17.8' \times 17.8' \times 2.8' = \underline{47,019\ BCF}$
Continuous footings:		1,050	2.50	1.00	$1,050' \times [2.5' + 2' + 2' + (1' \times 2 \times 1/2)] \times$ $1'$d $= \underline{6,825\ BCF}$
102 CY of concrete					
Grade beams:		750	4	2.00	$750' \times [4' + 2' + 2' + (2' \times 2 \times 1/2)] \times$ $2'$d $= \underline{15,000\ BCF}$
233 CY of concrete					
Total foundation concrete:					$957 + 102 + 133 = 1,192$ CY of concrete
Structural excavation recap:					$18,783 + 47,019 + 6,825 + 15,000 = 87,627$ BCF 87,627 BCF/27 CF/CY = 3,245 BCY 3,245 BCY $\times$ 1.3 swell = $\underline{4,219\ TCY}$ excavation
Structural backfill:					3,245 BCY − 1,192 CY of concrete = 2,053 BCY backfill
* Excluding accounting for 95% compaction					2,053 BCY $\times$ 1.3 swell = $\underline{2,669\ TCY}$ needed for backfill *
Export excess:					4,219 TCY total excavation − 2,669 TCY backfill = $\underline{1,550\ TCY}$ export

Figure 6.3 Structural excavation quantity take-off sheet

A more advanced calculation that an estimator will consider is accounting for the ability of earth to be re-compacted to its original condition. Commonly, specifications state that a 95% compaction rate must be attained. This means that the backfilled earth will be compacted until it reaches 95% of the density of the original undisturbed material. For the estimator, this means that 95% of the excavated material will be used in backfilling. It must be remembered, however, that the earth has expanded and that the swelled amount is being handled by the equipment.

When footings are fairly deep, for example five feet below the subgrade, an estimator has to consider safety issues. A vertical trench wall is very dangerous and generally illegal due to the possibility of collapsing. The geotechnical data will frequently provide information for safe slopes for trenches depending on the soil type. If nothing is shown, the estimator should use slopes no steeper than 1 to 1. This greatly increases the amount of excavation and backfill.

It is a good practice for all estimators to include their assumptions as notes on the QTO sheets and accompany those with sketches indicating how they assumed the work item would be constructed. This will be a good communication tool with the jobsite team if the contractor is successful and the estimate needs to be transformed into a planning and cost control tool.

Other foundations

The process for quantifying and pricing other foundation systems, such as auger cast piling, drilled piers, and caissons, will be discussed in Chapter 16. These systems are installed by specialty subcontractors; however, the estimator must be able to quantify the appropriate quantity and develop pre-bid day OM estimates or to apply unit pricing provided by subcontractors on bid day. This is typically accomplished by:

- Noting the type of piling,
- Noting the pile size or diameter and length or depths,
- Counting like-sized piles,
- Mathematically extending the lengths of the like-type and size piles, and
- Applying unit pricing by the each for certain types and sizes or by the total LF.

Slab-on-grade

We have chosen to include the process for quantifying the slab-on-grade (SOG) with this substructure chapter but have grouped foundation walls with other CIP walls in the next superstructure chapter. The SOG QTO sheet is set up to record the size of the slab (L × W) and thickness (t) of the SOG, and columns are established to calculate the quantity of edge form, concrete volume, reinforcement, flatwork finish, and other significant cost categories. Our case study SOG is specified to be 3,000 psi strength, 4 inches thick, and reinforced with #4 bars at 18″ OC (1.5′) EW. Note that the abbreviation EW introduced previously is defined as 'each-way' and not east-west, although the direction east-west versus north-south is also common in construction.

Drawing S-201 for our project shows the layout of construction joints or control joints (CJ) and saw joints (SJ) in the SOG. Construction joints are formed edges that, in this design, are running along the longitudinal column lines. The control joints are put into the slab to control where the SOG cracks as it shrinks during curing. The designed control joints are shown as 1-inch-deep saw cuts, which will be made after the initial curing of the concrete. Control joints will be quantified for pricing but otherwise do not affect the placement of the SOG concrete. Large slabs, such as our case study, cannot be placed all in one day. The estimator should consult with the superintendent to determine the size of slab placement lanes or quadrangles (often checkerboard) for planning purposes.

As stated, the generally accepted placement method for commercial concrete is to use a concrete pump. Some superintendents will place all concrete with a pump regardless of the size or location. Part of a winning attitude is to decide if there is a more economical way to place concrete in various applications. The SOG is one place that should be studied, and discussions with a superintendent may be helpful for determining whether it is more cost effective to use a pump or to place the concrete directly from the truck chute. Generally pumping is the most cost effective way of placing large quantities of concrete for commercial projects when compared to all other methods. A concrete placing 'bucket' may be used for isolated smaller pours such as a CIP column. Smaller pours and smaller projects may be able to chute their footings and slabs, and this is often the case with residential construction.

Preparations for the SOG include laying down a base layer of aggregate and/or sand and installing a vapor barrier. Edge forms are then set, rebar or wire mesh is installed, and the slab is ready to place. It is the superintendent's job to select placement methods that can be done within the cost parameters of the estimate. Edge forms for slabs and other shallow concrete sections are constructed from dimensioned lumber such as 2 × 6, 2 × 8, etc., and quantified as lineal feet. When edge forms are deeper than 12 inches, they are made using a sheet form material such as plywood or medium density overlay (MDO) board. These are quantified on a square foot basis, as has been done for the footings. Some estimators prefer to quantify all form material by the square foot, but most published and historical estimating databases provide both lineal feet and square feet pricing. An estimator should review the units prior to quantifying these forms to minimize dimensional conversions.

There are several configurations of construction joints between slab sections, and most are designed to minimize vertical movement between them. Some joints are keyed either with a preformed metal form or with wood. Keyway is quantified by measuring LF. Dowels of smooth reinforcing bars are inserted through the forms to keep adjacent slab sections aligned. These are quantified by counting each one and furnished as part of the reinforcing steel package. Both the construction joints and the dowels are cost factors that an estimator must consider. The SOG quantities for our case study are reflected in Figure 6.4.

An estimator needs to look for additional work scopes in the SOG that will affect the cost. The case study project has 4 inches of free draining capillary break below the vapor barrier. Closure strips and column blockout fillers will be placed after the CIP columns have been cast and stripped. The work includes backfilling between the placed SOG, forming, keying or doweling into adjacent slab edges, pumping, placing, and trowel finishing the slab. This work is still part of the substructure, and the estimator should perform the QTO work in that section so it is not forgotten. Because the work is done later, it may be accounted for on a separate QTO sheet.

Control joint sawing may be done by the general contractor's own forces but is more likely to be done by a concrete sawing subcontractor. The estimator should quantify the length of the control joints, obtain a unit price from a subcontractor, and develop an independent estimate. This will then be used as an OM estimate in Chapter 16. Saw cut

City Construction Company
Quantity take-off sheet

Project: Dunn Lumber   Sheet Number: 3
Location: 3800 Latona, Seattle   Estimator: PJ
Classification: **Concrete slab-on-grade**   Date: 5/21/2021

		Dimensions			SF	LF	CF	SF	Rebar
Description	No	L	W	H/t	Fine Grd	Edge form	Concr	Finish	
SOG		270	192	0.33	51,840	732	17,107	51,840	# 4 @ 1.5'
Trapezoid shape ~	0.5	192	45	0.33	4,320	237	1,426	4,320	OC EW =
S401					**56,160**	**969**	18,533	**56,160**	270 * .67
					SF	**LF**	CF	**SF**	* 192 =
							1.05 waste		34,733
Visqueen vapor barrier					56,160		27 CF/CY		192 * .67
					1.1 lap		**721**		* 270 =
					61,776		**CY**		34,733
					SF				69,466
									LF #4
Diamond col. blockouts	78								1.10 lap
4/S402	EA								.668 #/LF
									2,000 #/T
Expansion joints		2,400	LF						**25.5**
Control joints/sawcut		4,000	LF						**Ton**s
Construction joints		600	LF						

Notes:

Many other ancillary cost items will use these quantities and units and have not all been repeated here, such as pumping concrete and protect and cure are the same as finish. These will be carried on recaps.

Figure 6.4 Slab-on-grade quantity take-off sheet

joints are also often filled with caulking by a CSI division 07 subcontractor. This caulking can be quantified here with the SOG or later with subcontractors as long as the work is not missed and not quantified twice.

Concrete finishes are an item that the estimator should pay particular attention to. Different types of finishes include leveling to the top of the forms (rodding off); float finish; steel trowel finish (hand or machine); and broom finish, and some exterior slabs will have an exposed aggregate finish. In addition, there are several surface treatments such as a curing compound that retards water evaporation, spray-on hardeners, integral colors, and the seeding of aggregates for architectural or wearing purposes. An estimator needs to note on the QTO sheet which type of slab finish is specified. These different finishes all have different pricing factors. Concrete finishing applies not only to the SOG but also elevated slabs and tilt-up walls, both discussed in the next chapter.

Summary

Although some of the processes in preparation of a construction cost estimate involve creativity and managerial decisions, the quantity take-off process is very analytical.

The term 'stripping the drawings of quantities' is used in construction, which basically means take all of the material types and sizes and quantities and counts off of the drawings and place them on quantity take-off sheets such that they can be extended and later transferred to pricing sheets. The best process to use is to proceed in the same fashion as the building would be built, from foundations through the roof and completing with interior finishes and landscaping. As quantities are 'stripped' from the drawings, the drawings should be marked-up, whether by hand or electronically, to indicate what work has been 'taken-off'.

The GC estimator will put much more QTO effort in the work that will be self-performed, such as CIP concrete, than with subcontracted scopes. This chapter discussed the process to prepare detailed QTOs for spread and continuous footings, concrete reinforcement steel, structural excavation and backfill, and slabs-on grade. Additional cast-in place and pre-cast concrete elements are coming up in the next chapter, and structural steel and wood framing in subsequent chapters. The methods of counting and measuring and extending quantities for one building system can be applied to most others, as we will continue to demonstrate. Early outlines and drafts and reviews of this chapter were contributed by estimating lecturer and industry professional Larry Bjork. We appreciate his contributions.

Review questions

1   What happens if an estimator assumes neat-cut footings, but the superintendent chooses to form them?
2   Why would a contractor prefer to use expansion bolts for steel columns versus anchor bolts?
3   What would happen to the contractor if the estimator assumed all of the foundation excavation material could be reused as backfill, but then it was rained on and the inspector rejected it?
4   Match the labor craft (laborer, concrete finisher, ironworker, carpenter) with the concrete activity:

   a   Hard trowel finish
   b   Install rebar or welded wire mesh
   c   Install embedded weld plates
   d   Build wood forms
   e   Build metal forms
   f   Place concrete in form
   g   Fine grade
   h   Place anchor bolts

5   Place the work items from the bullet list in question 4 in the logical order that the work will be performed.

Exercises

1   Anchor bolt templates are used to locate ABs before placing concrete and often hold them in place until the concrete is set. This is not an error-proof process. What would happen if an AB was out of place? What is the remedy? When should this be discovered and resolved?

2 What would happen to the superintendent who, instead of disposing of the surplus excavation materials properly, dumped them in nearby wetlands?

3 What size of lumber is typically utilized to form a 10″ deep continuous footing? What is the lumber's actual (or planed) size? How is the difference accommodated by the formwork crew?

4 Perform a QTO for a substructure system introduced in this chapter for one of our case study projects included on the book's eResource. Electronic copies of all of the estimating forms used in this book are also included on that source.

7 Concrete superstructure

Introduction

In the last chapter we introduced basic quantity take-off (QTO) procedures and applied them to those areas of direct work that many general contractor (GC) estimators begin their QTO with, the concrete substructure, including the foundation and slab-on-grade (SOG) systems. In this chapter we continue with GC self-performed direct work and pick up all of the remaining cast-in-place (CIP) concrete systems and pre-cast concrete systems, including site-cast tilt-up concrete walls. Our specific coverage includes:

- Cast-in-place concrete elements and systems, including:
  - Columns,
  - Beams,
  - Walls,
  - Elevated CIP slabs,
  - Elevated slab on metal deck, and
  - Post-tension (PT) concrete;
- Pre-cast concrete elements and systems, including:
  - Columns,
  - Beams,
  - Slabs,
  - Walls, and
  - Tilt-up walls.

For our analysis, the concrete foundations and SOG are considered substructure and were included in the previous chapter. Additional concrete systems such as foundation walls and columns and beams and elevated decks may also occur below grade and would then technically be classified as substructure as well. But these systems also exist above grade and would therefore be considered superstructure. We have included these scopes of work in this chapter for convenience and to avoid duplication. Additional superstructure materials such as structural steel and wood framing are covered in the next two QTO chapters.

Cast-in-place concrete elements and systems

With all concrete elements, we have many of the same items to measure and count as we did with the foundations and SOG, including:

- Formwork,
- Concrete reinforcement steel (rebar) or welded wire mesh,
- Embedded steel,
- Purchase and pump and place concrete,
- Finish concrete,
- Protect and cure, and
- Patch and sack if necessary.

Columns

The estimating process for CIP columns includes the previously mentioned types of work scopes and is similar to many other concrete elements but also has some differences. The column forms are typically steel because the pressure of the wet concrete toward the bottom of the column is significant. These forms are often 'engineered' by a consultant employed by the GC. The column forms will be reused on multiple occasions to make rental or purchase cost effective. Because of the expense of the forms, a GC will encourage the structural engineer to use the same size columns, such as 24″ or 30″ square, as often as possible. This will occur as part of 'constructability' and was discussed in Chapter 5, "Preconstruction". The forms require significant bracing to hold them in place. Chamfer strips are used vertically within the forms on the column corners because it is difficult to hold a perfect 90 degrees outside corner in concrete and the chamfer minimizes chipping after stripped.

Rebar is assembled in 'cages' and is often prefabricated in a specialized assembly area somewhere on the site, and the cages are flown in with a tower crane. Rebar in a column is of larger diameter than in other concrete elements, and the quantity of rebar, both vertically and horizontally, is quite dense. The concrete is typically very high strength, maybe 6,000 or 8,000 PSI, and will therefore not be mixed with other pours such as foundations or slabs. The concrete must be pumped but is often placed with buckets suspended from the tower crane. The tops of the columns will need to accommodate other concrete elements either with anchor bolts (ABs) or embedded steel plates to connect with steel or pre-cast concrete elements (discussed later) or rebar extensions to connect with other CIP concrete elements. An example QTO for our case study CIP columns is included as Figure 7.1.

Beams

Estimating elevated structural slabs and beams presents some different estimating techniques. Both require bottom forms that are comprised of heavier temporary construction materials and thus costlier. The difficult item is the support of the forms. Concrete weighs approximately 150 pounds per cubic foot or over 4,000 pounds per cubic yard. An elevated beam that is 12 inches wide by 2 feet deep then weighs

City Construction Company							
Quantity take-off sheet							

Project: Dunn Lumber   Sheet: 23
Location: 3800 Latona, Seattle   Estimator: PJ
Classification: **CIP Concrete Columns**   Date: 5/28/2021

Description	No	Dimensions			Forms SFCA	Concr CF	Rebar 275#/CY	Patch & Sack
		L *	W	H				
Type A columns	66	12	1.50	1.50	4,752	1,782		
Type B columns	106	12	2.00	2.00	10,176	5,088		
S501					14,928	6,870		14,928
					SFCA	CF		SF
						+ 1.05 waste		
						@ 27 CF/CY		
						267	73,471	
						CY	#	
							@ 2,000 #/Ton	
							36.7	
							Tons	

Notes: * Column height varies, used 12' average
Other pricing activities carried on recap sheets with common quantities not shown here

Figure 7.1 CIP concrete columns quantity take-off sheet

300 pounds per foot of beam length, and if it were 20 feet long, the total weight is 6,000 pounds. The forms alone will not hold this weight, and supplemental supports are required. These temporary supports are termed 'shoring'. The size and type of shoring depends upon the location of the concrete element. A beam that is 6 feet above a ground surface may only need pole jacks placed at selected intervals, while an elevated floor slab 20 feet up will require a system of heavy duty scaffolding. The ability of an estimator to determine the support system for elevated concrete depends on experience. There is an extreme risk associated with these support systems, and the estimator should seek outside structural engineering help to determine the correct system. The project owner's design team will not assist with this means and methods evaluation.

Concrete walls

Cost factors for CIP walls are different than those for other concrete sections; thus, an estimator will quantify them on a separate QTO sheet. The quantity sheet is set up essentially the same as for other concrete members but will have a column for de-fin and patch and sack. Forming CIP concrete walls is very labor intensive and therefore one of the riskiest elements for the estimator to quantify. Typically, wall forms are calculated in square foot of contact area (SFCA). The height and length of the walls is noted, and multiplying these two dimensions by two for both sides of the wall results in total formwork. A small amount of forms for end conditions is also added. The GC's estimating team must determine the type of forming system and whether forms are to be stick-built on-site or rented. Snap tie patterns and count are quantified. Many formwork material unit prices will allow for snap ties, whalers, form oil, and wedges, in which case additional counts are not necessary.

In general, when constructing walls, two concrete finish items must also be considered: Wall tops and the vertical surface. Finish other than leveling the tops and de-fin and patching of rock pockets is not required. Foundation walls exposed on the outside to earthwork will typically be waterproofed, and therefore patching of snap tie holes and rock pockets is required. Other estimating elements such as volume of concrete, embeds, and rebar are taken off similar to CIP concrete columns.

Elevated slabs

As indicated earlier, estimating elevated structural slabs and beams present some different estimating techniques. Both require bottom forms that are built from heavier construction methods and thus are costly. These forms must be 'shored' up and supported by the slab below – assuming it has cured sufficiently to reach design strength. The method of shoring is either with steel pole shores or wood posts such as 4 × 4s. Both construction of the formwork and shoring is often subcontracted to specialty firms that own these forms. They typically will utilize small hard-tired 'warehouse' forklifts to lift and lower forms.

After shoring is accomplished, much of the remaining categories to include with quantifying elevated slabs are similar to the process of taking off the SOG. This includes volume of concrete, rebar, and finishing. A CIP elevated slab will have a significant amount of rebar or will include post-tensioning cables, as discussed later. These slabs are also often integral with CIP concrete beams. Because beams may drop down significantly below the slab and impact clearance for the occupied floor below, 'drop caps' are used, which in plan view look similar to spot footings except they sit atop concrete columns and not the earth. Our case study project uses the drop cap system.

It is expensive and time-consuming to core drill holes or saw cut rectangular penetrations in cast slabs, therefore, the mechanical and electrical subcontractors will place as many embeds and blockouts on the formwork before casting the slab as they can. Once the slab reaches design strength the formwork can be dropped and the shoring removed except 're-shores' are added to extend the curing period, especially if this new slab is to receive the weight of another slab above it. Re-shores are similar in design to the initial shoring system but are much more spread out. Again, this requires structural engineering input employed by the GC.

Slab on metal deck

Mezzanine and elevated floor slabs for offices are relatively simple structures. For a structural steel framed building such as the case study, structural steel girders support steel joists on which metal decking is welded. Steel erection includes installation of angles that are to be used as slab edge forms that become permanently embedded in the concrete. The combination of metal deck and concrete slab working together is known as slab on metal deck (SOMD) and is considered a composite structural slab. There may be some places where wood edge forms are required to have a clean concrete edge showing. These additional edge forms occur at blockouts for mechanical ductwork and stair and elevator shafts. An estimator needs to pay particular attention to the type of edge forms and make sure that they are all covered within the estimate.

The volume of concrete placed on a corrugated metal deck is less than the equivalent thickness placed on a flat surface. There are many configurations for decking that affect the concrete volume, and an estimator needs to use those for the specified decking. Some estimating references have quantities or formulas for representative shapes of decking, but manufacturer's data for the specified product will provide more accurate information. As mentioned previously, a good practice for an estimator is to include sketches and assumptions on the QTO sheets, as is shown in several of our examples.

Other than shoring and metal edge form that comes with the metal deck, and accounting for the differences in deck depth, most of the QTO elements are similar to that of the other elevated CIP decks and SOG. *Nelson studs*, also known as 'lugs', are welded to steel beams and create a secure connection with the elevated concrete slab, metal deck, and the structural steel. These are typically quantified with the structural steel take-off in the next chapter. An example QTO for SOMD for our case study is included as Figure 7.2.

Pre-cast concrete elements and systems

Pre-casting versus cast-in-place

Sometimes the designer will specify and draw whether certain concrete elements are site-cast or pre-cast, and other times they will be open to the contractor making a means and methods decision. Like many of our other construction management topics, a contractor needs to consider cost, schedule, quality, and safety factors when deciding to cast concrete in-place, pre-cast on site, or purchase elements from a prefabrication (prefab) supplier and ship to the jobsite for erection. Pre-cast concrete building elements will include significant steel embeds such that the different members can be connected to each other and with structural steel utilizing field welding. Pre-cast concrete requires preparation of shop drawings or fabrication drawings that should be processed through the structural engineer of record for approval before casting.

In some instances, pre-casting may be performed on the jobsite if a suitable casting bed and space can be made available. But most often, typical pre-cast concrete elements are fabricated off-site in a warehouse or yard where quality can be controlled better and the casting process is protected from harsh weather. These concrete elements are then

City Construction Company
Quantity take-off sheet

Project: Dunn Lumber   Sheet: 25
Location: 3800 Latona, Seattle   Estimator: PJ
Classification: **Composite slab on metal deck**   Date: 5/28/2021

| Description | No | Dimensions | | | Metal deck | Concr | 6 × 6 | Finish |
		L	W	H/t ★	SF ★★	CF	WWF	Concrete
Concrete slab over metal deck S601 1/S602		Various		0.354	95,000	33,630 CF + 1.03 waste @ 27 CF/CY 1,283		95,000 SF
							95,000 SF	
Edge form at blockouts		629						
		LF					+ 10% lap 104,500 SF	

Notes:

★ 3.5″ concrete over metal deck, total 5″ t, averages 4.25″ concrete

★★ Metal deck will likely have been taken-off with structural steel and transferred here.

Other pricing activities carried on recap sheets with common quantities are not shown here, such as shoring, re-shoring, and protect and cure.

Figure 7.2 Composite slab on metal deck quantity take-off sheet

shipped to the jobsite as needed, also known as just-in-time delivery, which is an aspect of lean construction techniques. Another schedule advantage for pre-casting is the concrete elements are fabricated in parallel with other preparatory work occurring on the jobsite.

Additional decisions need to be made by the construction team when deciding CIP versus pre-cast, and if pre-cast, whether site-cast or off-site fabricated. Some of the questions to consider will include: Which is most cost effective? Which will enhance our ability to meet or beat schedule the best? How can we best control quality, and are there any exceptional quality control standards specified in the contract documents? And will one choice or another make it safer for our craft workers?

Pre-cast systems

There are many different shapes that pre-cast concrete can take, but economy and efficiency of scale usually dictate that structural elements are of common sizes. Pre-cast concrete elements can be cast on site with a casting bed and erected or purchased from a prefabricator and hauled to the site and erected. Erection of pre-cast can be by the general contractor's own forces or with a subcontractor. If self-erected, it is common that the GC uses a combined crew of carpenters, ironworkers, laborers, operating engineers, cement masons, and surveyors. Some of the more typical concrete shapes or elements that are pre-cast include:

- Columns,
- Beams:

  - I beams
  - T's
  - Double T's,

- Slabs, including hollow-core planks, and
- Walls.

The installation of pre-cast concrete is very similar to structural steel. The members are unloaded from a truck with a crane and hoisted to their final location. Double handling is minimized, if not completely eliminated, to avoid damage and control costs. The concrete is furnished with embedded steel plates and fastened to other structural elements by field welding, performed by ironworkers. The craftsmen will often work off of scissor lifts or personnel hoists to make the connections. Some pre-cast elements will also require temporary shoring until installation of all adjoining members is accomplished. The system of pre-cast may only function safely when the whole assembly is installed. This may include elevated slabs above the columns and beams. After assembly, the joints may also require grouting, which will be accomplished by cement masons. The estimator must consider all of these differences between different types of concrete systems, including formwork, hoisting, and craft workers.

Tilt-up concrete

An economical structural and envelope method for construction of a building is the use of tilt-up concrete walls. These walls are pre-cast, as discussed earlier, but they are site-cast, usually very close to the location where they will be erected. Panels are cast on a flat slab, such as the SOG, and after curing are lifted and set (tilted up) onto the foundations. The economics compared to vertical CIP walls is in the forming. Since only edge forms are required, less form material is needed, and the craftsmen are always at ground level. These savings offset the cost of the equipment and manpower used to lift and set the panels, and the construction is much faster.

When starting the tilt-up QTO, an estimator should discuss the project with an experienced superintendent. Together, they can determine if the SOG is suitable for the work and what has to be done for preparation of it. If the SOG is not suitable, they will need to determine where a special casting slab can be constructed. It is common that some panels can be placed one on top of another, known as 'stacking', if relatively the same size.

Information about the tilt-up panels is primarily detailed on the structural draw-ings, but there is information needed to quantify from the architectural drawings as well. An estimator must look at all drawings and note where different information is located. For example, reveals are typically found on the architectural elevations or wall sections.

Of particular concern to an estimator are the embedded items. There are angles and plates that connect concrete to concrete and structural steel to the concrete. One signifi-cant item not shown is the lifting hardware. There are companies that supply all sorts of concrete accessories from form-ties to tilt-up anchors, and a product catalog should be in every estimator's library. The responsibility of engineering the sizing and placement of the lifting anchors lies with the contractor doing the work, in this case the general contrac-tor. The GC's estimator is not a licensed structural engineer. Therefore, he or she should seek professional help for sizing and locating the lifting anchors. Contractors who build tilt-up walls usually have an outside structural engineer available for this. The engineer, ironworker foreman, and anchor supplier can usually determine the quantity, size, and location of the anchors. The estimator must also add cost to the estimate for the panel engineering and detailing work.

When quantifying tilt-up work, an estimator will want to look for panels that are basi-cally the same. For example, if panels #1 and #2 are dimensionally the same, with the only difference being the embedded connection steel, they can be quantified on the same line as long as the different types of embeds are quantified separately. Most embedded steel is relatively similar in size, and it takes about the same effort to install each one. Unless there is an unusually large piece of embedded steel that takes more effort to install, they may be grouped together on a single line. Our case study is not a concrete tilt-up build-ing, but we have provided Figure 7.3, which exhibits a portion of a typical tilt-up panel QTO sheet.

Figure 7.3 demonstrates a feature that is helpful in the QTO. Simple panel sketches have been drawn so that someone else knows exactly what is being quantified. This is a good practice for the estimator to visualize reinforcing steel embeds and concrete shapes. Sketches may add pages to the estimate but provide valuable clarity to those who will use the estimate after the project is won.

As an estimator becomes more experienced, he or she looks for ways to save time and effort. The quantification of the tilt-up panels is just such a place where this can be done. A QTO for the embedded steel will need to be made for developing an order of magnitude estimate later on for supply of the fabricated steel. While the panels are being quantified, the estimator can start a separate sheet for the embedded steel and list the pertinent information such as the quantity, dimensions, and mate-rials. This saves time by not having to review the drawings again to search for the information.

There are many similarities in quantifying tilt-ups compared to CIP walls and SOG, but there are also significant differences, such as the quantity of embeds and additional engineering. Once the tilt-up panels have been cast and cured, they are ready to be tilted up and set in place. Labor cost factors include attaching the braces, rigging the panel for lifting, breaking the panel loose as lifting tension is applied, guiding them into place, securing the bracing, patching, and grouting the base. Material and equipment costs are included with all of these activities and the cost of the crane, rental of braces, shim packs, patching material, and grout must be added as well.

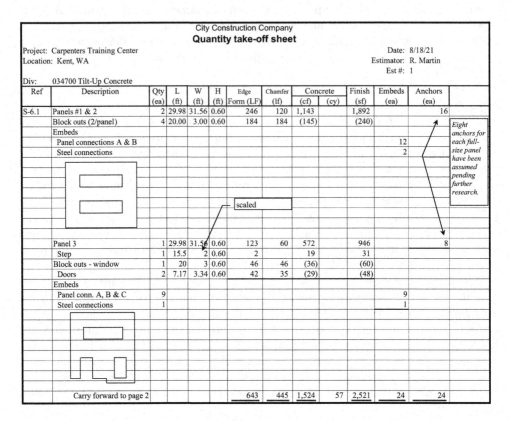

Figure 7.3 Tilt-up concrete quantity take-off sheet

Cranes are rented by the hour, and their rate usually includes the rigging for hoisting and setting the panels. There are, however, two additional cost items associated with them: (1) the travel time to and from the job, sometimes referred to as 'in and out' time, and (2) the cost of mobilizing the crane on site. Mobilizing a crane includes assembling and rigging the boom for lattice-boomed cranes and installing rolling outriggers. One tank of fuel is furnished, and if more is needed, the general contractor will furnish it. Cranes used for relatively heavy lifting will come to the jobsite with a trained crew of at least an operator, and if it is a very large crane, it will also have an oiler. Some large cranes also require an 'assist crane' to assemble and disassemble the tilt-up crane on-site. An estimator must call the crane rental company to get the hourly rate, in and out cost, and mobilization costs. The general contractor furnishes the labor to supplement the crane crew during mobilization, and the estimator needs to find out from the vendor the approximate time that this operation takes. We have provided a detailed example of a tilt-up crane estimate with Chapter 13.

The erected panels need to be supported until the structural steel is installed. Braces usually are rented from the same accessories vendor who furnishes the lifting embeds discussed earlier. The estimator must determine how many are needed and how long they will be required. Each panel requires at least two braces, and larger panels require more, say four braces. A superintendent and brace supplier can help with this. The braces for all of the panels of the building need to be rented for the entire time until the steel erection is completed.

Only after the panels have been completely tied into the other building structural systems may the braces be removed. The panels will be connected at the top to the structural steel roof and to floor systems if more than one story. Each of the panels will be welded on its side to adjoining panels. Some panels are connected to the footing by welding embedded plates together. In many cases, a small strip of the SOG was left out, called a closure strip. The void space can now be backfilled, and the closure strip will structurally connect rebar from the panels to the SOG, tying in the bottom.

Grout

Grout is a cementitious material applied under the base or between different elements to provide full bearing surfaces. It is applied under tilt-up panels and under the structural steel columns. The design drawings will indicate the approximate gap under the members of each application in which the grout is to be installed. Cement finishers install grout, and although amounts are small, the cost can be significant. Grout typically has high cement content and is specified as very high strength, such as 10,000 PSI.

An estimator should set up a separate QTO sheet for all of the grout on a project. There are only two cost items in the grouting operation, the grout material and installation labor. It is quantified in cubic feet but not converted to CY, and most estimating references have productivity rates that are reasonably accurate.

Post-tensioned concrete

Because project owners and their architects want to achieve as much building height as zoning code will allow and free up building interiors for horizontal runs of mechanical and electrical systems and still maintain high ceilings, they are always looking for vertically thinner building systems. Post-tension concrete replaces most of the standard rebar with cables, which are pulled or jacked and tensioned after the concrete has been cast. By placing stress into the concrete slab, thinner slabs can be used. Once other building loads or occupancy loads are placed on the cured slab, it relaxes into a more neutral condition. Cast-in-place beams and drop cops discussed earlier can be reduced in depth and spread further with the use of PT slabs. Sometimes the beams are also cast integral with the slabs and may have post-tension cabling in them as well. Many other concrete shapes can also be constructed using the PT method.

Many of the elements, such as shoring, formwork, and finishing and curing associated with elevated CIP slabs discussed earlier apply to PT slabs as well. And as with all concrete systems, there are differences. The installation, termination, and stressing of the cables is often another subcontract specialty. Shop drawings will be prepared and processed through the building structural engineer for approval before fabricating the cable. This work is all performed under the supervision of a third party test and inspection company. Conventional rebar is measured in lengths and converted to pounds and purchased by the ton. PT cables are counted by the each and extended for the lengths, but the final purchasable units are pounds, not tons. Mechanical and electrical embeds and other blockouts MUST be accurately located and installed before casting the slab. It is very difficult to safely core drill or saw cut a PT slab after placement. Care must be taken not to cut a cable. After the slab has cured to design strength hydraulic jacks are employed to stress the cables. Occasionally a cable breaks or will not tension to its design strength. Because of this, the engineer will have included a few additional cables as a factor of safety in the original design. Because of the difficulty of working with a concrete slab with embedded and tensioned cables, it is a good practice to take photographs of the cable location before placing the concrete and marking the underside of the slab with cable locations. Preparation of as-built drawings is critical as well.

Summary

Many of the estimating processes used for one type of concrete system, such as spread footings, apply to other concrete systems, just as the QTO processes are similar for other building materials such as steel and wood. But there are differences between different concrete systems as discussed in this chapter. Some of the estimating processes for concrete elements presented in this chapter will be undertaken by senior estimators, whereas the beginning estimator or project engineer will assist with straight-forward concrete foundations.

Concrete can be cast-in-place, prefabricated on a jobsite casting bed, or purchased from an off-site prefabricator and trucked into the jobsite and hoisted into place, often with the use of a tower crane. In many cases, the building's structural engineer will have made these determinations, sometimes with the input from the GC during the preconstruction phase. But in other instances the GC can make these choices as part of its means and methods, in which case the GC will need to contract with an outside structural engineer to assist with calculations and shop drawing development.

Some of the CIP systems we discussed in this chapter included columns, beams, walls, elevated slabs, and composite steel and concrete slabs. Just about any concrete shape that can be cast in place on the jobsite can be prefabricated. A GC estimator should work with the project superintendent and evaluate cost, schedule, quality, and safety considerations when choosing between CIP, on-site pre-cast, or off-site pre-cast. Concrete tilt-up construction is typically a project owner's choice and can be a very economical building system that combines structure with building envelope. There are many more estimating elements associated with tilt-up walls than CIP walls. Post-tension cables replace rebar and allow slabs to be thinner and lighter and improve interior building clearances. In our next two chapters, we will evaluate QTO techniques associated with structural steel and wood – those materials that often sit atop and connect with the concrete discussed in these last two chapters.

Review questions

1   If one pre-cast concrete element was to be connected in the field to another pre-cast element, how is that connection accomplished?
2   When would a contractor choose to cast concrete in place or pre-cast concrete?
3   When would a contractor choose to site-cast pre-cast concrete rather than purchase from a supplier?
4   What is one large cost savings of a tilt-up wall compared to a CIP wall, and what is one large extra cost?
5   Why does the structural engineer 'of record', or the building structural engineer, not provide design for concrete formwork or shoring or tilt-up bracing and others?
6   What is the most expensive item associated with constructing a (A) CIP wall, (B) CIP beam, (C) elevated slab, and/or (D) tilt-up wall?

Exercises

1   Assume a concrete tilt-up building with an outside footprint of 200′ × 250′. Assume that panels are 9 inches thick and no larger than 30 feet wide and all panels are 27 feet tall. The building has two man-doors on each face and eight each 14′ wide by 14′ high truck doors on one face. (A) How many panels will there be? (B) What is the weight of the heaviest panel? (C) Utilizing a suggested estimating rate of 30 minutes per panel for erection, how long will it take to tilt the entire building? (D) With an eight-person erection gang of various crafts, as discussed earlier, how many man-hours will be involved with the erection?

2   Match the labor craft (laborer, electrician, surveyor, cement mason, operating engineer, ironworker, and carpenter) with the following concrete activities. Note some crafts are utilized more than once and others not at all for this exercise.

   a   Crane operation,
   b   Grout placement,
   c   Embedded steel installation,
   d   Patching of rock pockets,
   e   Shoring underside of beams or slabs,
   f   Shoveling concrete,
   g   Welding embeds together, and
   h   Aligning tilt-up panels after erection.

8 Structural steel

Introduction

Chapters 6 and 7 focused on quantity take-off (QTO) measurements for several concrete systems that are often an area of work self-performed by most general contractors (GCs) and comprise a substantial amount of direct labor and corresponding estimating risk. But not all GCs are the same, in that some perform other areas of work on either a sporadic or routine basis depending upon their expertise, geographic location, union preferences or affiliations, owned specialized equipment, availability of quality subcontractors, or market conditions. There are a variety of materials installed by carpenters, which we will cover in detail in the next chapter. Here we are going to focus on the Construction Specification Institute's (CSI) division 05, which is predominantly installed by the ironworker trade. Some of these types of work include:

- Rolled structural steel shapes, including columns and beams,
- Bar joist,
- Miscellaneous steel, and
- Metal decking.

We have grouped all structural steel related work into this chapter. Many of the measuring and counting techniques and procedures presented in the concrete QTO chapters apply to these systems, as well, and others that we will cover in this book. Many GC estimators shy away from estimating structural steel and rough carpentry, but they are not as complicated as appears. Unfortunately, many GC estimators focus on concrete only, as that is what they are most comfortable with. Even if the work described in this chapter is subcontracted out, it helps for the estimator and project manager to know the quantities and pricing that make up a subcontractor's quote. There is more to selecting a best-value subcontractor than just choosing the lowest bid on bid day. We will discuss the bidding process in Chapter 20. Knowledge of the makeup of subcontractor pricing is also a benefit to the GC's jobsite management team when analyzing monthly pay requests and change orders, which are two of the additional types of estimates that we will also introduce in Chapter 23. A new estimator can carve themselves a niche in the construction industry by knowing just a little more about one subject than do others, whether that is mechanical, electrical, and plumbing (MEP) work or structural steel and rough carpentry. In this chapter, as in many others, we use a lot of abbreviations, all of which are also described in the book's front material.

Rolled shapes

In the steel mill or factory, structural shapes are 'rolled' out. Some of the structural shapes common in construction include:

- Wide flanges (WF), formerly 'I' or 'H' beams, also designated as 'W',
- Hollow structure sections (HSS), formerly tube steel (TS),
- Channels,
- Angles, and
- Pipes.

Estimating structural steel is simply a process of counting members, measuring their lengths, noting their weights per linear foot, extending the math, and adding some other elements to that. Structural steel is primarily shown on the structural engineering drawings, or the 'S' series, although some steel may be shown on the architectural drawings and/or the mechanical drawings. Structural steel is specified in CSI division 05 and installed by ironworkers. The process of estimating steel should follow the guidelines discussed previously; start with the work you are doing first and progress through the building, similar to creating the construction schedule.

The steel QTO starts first with the building *columns*. A QTO spreadsheet should be set up by the estimator that lists all of the different column types, whether they are HSS or WF members. Pipes are used for columns, but their connections are difficult and not preferred by contractors. Angles and channels are rarely used for columns. Then, within each type, either HSS or WF, separate the QTO spreadsheet by sizes, for example a vertical column on the spreadsheet for 6 × 6 HSS and another for 8 × 8 HSS. Add a separate column for W30 × 99 columns and another for W12 × 30 columns, and so forth.

The next step is to count each steel column member from the structural drawings. The columns may be shown on the foundation drawing (note columns usually are associated with spot or pad or spread footings or pile-caps). They may also be shown on the first floor structural framing drawing. With each of the counts, it is important to note the length or height of the columns as well. Every math step should be shown on the QTO; do not skip steps. It is easy with Excel to combine many math steps in one cell, but this can be prone to error when cutting and pasting spreadsheet rows and columns. Combining formulas also makes it difficult for the estimator's senior manager to validate the accuracy of the estimate just prior to bid day. So, utilizing our previous examples, the estimator may find:

---

8 each 6 × 6 HSS columns at 15 feet tall = 120 LF
4 each 8 × 8 HSS columns at 15 feet tall = 60 LF
4 each 8 × 8 HSS columns at 20 feet tall = 80 LF
  = 16 EA HSS columns

2 W30 × 99 columns at 20 feet tall = 40 LF
14 W12 × 30 columns at 20 feet tall = 280 LF
2 W12 × 30 columns at 40 feet tall = 80 LF
  = 18 EA WF columns

When taking the column count off of the drawings they should be 'marked up', either with a highlighter or electronically, indicating that each steel column has been accounted for. 'Take-off' means placing all the measurable material results on the quantity take-off sheets, similar to the process introduced in the concrete QTO chapters. Make sure each QTO sheet is properly labeled. The estimator should not group different types of systems together on the same sheet; for example, do not add wood doors to concrete QTO sheets or toilet accessories to the steel sheets.

The next step will be to extend the counts and lengths to a total lineal footage (LF) for each type and size of steel member. At this time, both 8 × 8s can be added together for total LF (60 + 80 = 140 LF), and both W12 × 30s can be added together (280 + 80 = 360 LF). Next, the weight per foot of each of the member sizes must be looked up and multiplied times the total LF to come up with total pounds per member size. Our goal will eventually be to know what the total pounds, converted to tons, is for the project for all of the structural steel. We buy steel by the ton, not per each or per LF or per pound. The estimator does not need to look up the weight of the wide flange members, as the designation already tells us that. A W30 × 99 is approximately 30 inches deep and weighs 99 pounds per linear foot. Weights for a few additional steel members are included in Table 8.1. The estimator can find weights for all structural steel elements in the *Steel Construction Manual* (www.aisc.org/Steel-Construction-Manual) or from a variety of Internet sources. So for our simple example, we would end up with:

120 LF 6 × 6 × 3/8″ HSS @ 27.48#/LF =	3,298#
140 LF 8 × 8 × 1/2″ HSS @ 48.85#/LF =	6,839#
40 LF W30 × 99 @ 99#/LF =	3,960#
360 LF W12 × 30 @ 30#/LF =	10,800#
Total Weight =	24,897#
Plus 15% gussets and plates =	3,735#
Total Pounds =	28,632#
Conversion to tons (T) @ 2000#/T =	**14.3T**

Table 8.1 Structural steel dimensions and weights

Structural Steel Dimensions and Weights

Wide Flanges:	W 8 × 40	40 #/LF, 8.25″ d
	W 24 × 62	62 #/LF, 23.25″ d
	W 36 × 260	260 #/LF, 36.25″ d
Channels:	C 6 × 5/16 web	10.5 #/LF
	C 8 × ¼	11.5 #/LF
	C 12 × ½	30 #/LF
Angles:	3 × 3 × ¼	4.9 #/LF
	4 × 4 × 3/8	9.8 #/LF
	6 × 6 × ½	19.6 #/LF
Hollow Structural Sections:		
	4 ½ × 4 ½ × ¼″ t	13.91 #/LF
	6 × 6 × ½″ t	35.25 #/LF
	8 × 4 × ½″ t	35.25 #/LF
Plates:	3/16″ t	7.66 #/SF
	¼″ t	10.21 #/SF
	½″	20.42 #/SF
Pipes:	4″, Schd 80	14.98 #/LF, 4.5″ OD
	6″, Schd 40	18.97 #/LF, 6 5/8″ OD
	8″, Schd 60	35.64 #/LF, 8 5/8″ OD

We do not add waste or lap with steel, as we expect to have the pieces fabricated to exactly the correct size. But the estimator should add a percentage add-on for gusset plates, base plates, erection saddles, knife plates, web stiffeners and similar appendages that are welded to the steel members by the fabricator in the shop. The more of this we can have prefabricated, the better the labor savings will be in the field. This was part of the constructability discussion in Chapter 5. In our example we have added 15% for gussets and plates. We purchase steel by the ton, so the pounds also have to be divided by 2,000#/ton. The total weight number will now be utilized for several percentage add-ons associated with structural steel, as discussed later in this chapter and in the direct work pricing chapter.

Every steel member typically has two connections, one on each end. But if you calculate two connections per each you would double up on half of them when steel connects with steel. In the case of columns, they have only one end to be connected, the bottom, and the top is accounted for with beams, discussed next. Purchase and installation of anchor bolts (ABs) embedded into wet concrete were included with the concrete QTO. Some estimators choose to quantify ABs with CSI 05 work as that is the vendor which will provide them. Contractors may attempt to have the columns connected to the concrete foundations with para-bolts, which are drilled in and epoxied after the concrete has set. These are more expensive than anchor bolts and are not usually part of the original design and not acceptable to many structural engineers.

Some estimators will take shortcuts with steel by simply allowing 10 pounds per square foot of floor (SFF) for all steel and not performing any QTO, which does not take into consideration the specific project's idiosyncrasies. The next shortcut that some take is to figure labor on the total tonnage for the project. The mistake here is that it takes the ironworkers just as long to install a 6 × 6 HSS column or beam as it does an 8 × 8 HSS column or beam, even though the 8 × 8 weighs more. Many estimating guides allow us to figure labor hours for steel members by the 'each'. In this case we have a total of 16 HSS columns and 18 wide flange columns. It is a good suggestion that the estimator look to their in-house or published databases to see how the work will eventually be priced before they do the QTO and extensions and conversions. This is similar to reading the last chapter of a novel first, to see how the mystery will turn out. Estimating steel elements by the each for labor and by the ton for material purchase is the most accurate method, so the estimator should have both quantities handy.

Horizontal floor and roof steel members have a variety of names including beams, girders (which are large beams), lintels, joist, and purlins, which are beams along the perimeter walls. Let's group all of these into *beams* for this conversation. Also similar to columns, beams are usually wide flange members, HSS, or channels. Angles may be used for lintels and purlins or additional support at blockouts. The QTO for beams is very similar if not identical to the process for the column take-off. Count the 'eaches', measure their lengths, and note their sizes. Then extend out the total length for each size and convert to pounds for each size member. Additional beams are required for blockouts at stairs, mechanical shafts or risers, elevator shafts, and under heavy loads such as mechanical and electrical equipment. The beams should be installed by the each and purchased by the ton, similar to the columns. We will discuss labor productivity and steel material pricing in Chapters 10 and 12.

Beam connections quantified by the estimator are customarily two each, one for each end, but if steel connects to steel, only one is counted. Connections will either be welded or bolted. A structural steel subcontractor, or GC who also specializes in steel erection, will diligently count each bolt and calculate the total welds (in pounds) required in the project. This is a tedious task, as there may be 50 or so different configurations of bolted or welded connections on a typical commercial construction project. A shortcut to this is to count common connections, that is, there are 14 each 2/S601 bolted connection assemblies on the project, and later estimate the labor productivity necessary to assemble that typical connection. A shorter cut to this is to group all welded connections into one pot and all bolted connections into another pot. Then the estimator can seek out the connection for each type that occurs more than the others. For example, 50% of the welded connections on the project are type 16/S601, and the others are split amongst 14 different variations. Then a productivity factor can be figured for that one common configuration and applied to all welded connections, with a similar process for the bolted connections. Connections are not always figured separate from member installation, depending on the database used.

Nelson studs or 'lugs' are a vertical mushroom-looking appendage that is welded through the metal deck onto the top of beams. When the concrete topping slab is placed on the metal deck, this results in a composite deck where the steel and the concrete now work together resisting horizontal shear. Studs are not shown on the drawings but usually included as a note or in a schedule, or in parentheses along with the beam size: W30 × 99 (20) has 20 lugs for that particular beam. Some beams will have a lot of studs, some just a few, and others none at all. The stud spacing will also vary. The estimator must make a count of Nelson studs and note them on the QTO, as they have a significant impact on the amount of ironworker labor necessary to complete the beam and deck installation.

Steel trusses in this context are different from bar joist, discussed next. Bar joist are a manufactured relatively lightweight item, whereas structural steel trusses are comprised of rolled shapes, bolted together either in the shop or the field, and erected all as one assembly. The individual members may be angle iron, channels, tube steel or HSS, or even wide flanges. If the truss is completely assembled in the shop and delivered to the site in one or two pieces, they would be quantified by the truss. If the truss is delivered to the site in many individual parts and pieces, then each piece of steel would need to be quantified separately. Their field assembly would be a very similar process as assembling beams, but the assembly would usually be on the ground at waist height, with the total truss temporarily supported on blocking.

Once the truss is assembled, it would be lifted in place all as one assembly. The installation is then similar to a truss that had been fabricated and shipped to the jobsite whole. This author worked on an aerospace manufacturing facility where the finished building was a 300' × 300' square free space. The trusses were delivered to the jobsite in hundreds of pieces and assembled lying horizontal on the ground to create several 30' deep trusses each 300' long. Then five large cranes tipped the trusses vertical and lifted them to sit atop 80' tall columns. Photographs of this actual project are available on the book's eResource.

Bar joist

Bar joist is an inexpensive and lightweight cousin to steel beams. In this case the structural engineer indicates the desired depth of the joist and the weight it must carry but places the ultimate design responsibility on the supplier, who must submit shop drawings and corresponding calculations. In addition to cost and weight benefits, bar joist also add the flexibility to run small pipes and ducts and conduits at a 90-degree angle through them, whereas this is difficult to do with wide flange beams.

The method to determine the QTO for bar joist can be as simple as measuring the square foot (SF) of area they cover, that is, length times width of the building or portion of building covered by the joist; for example, $30' \times 100' = 3,000$ SF. Some estimating databases will have joist purchased and installed by the square foot of coverage. The problem with this approach is it does not factor the depth of the joist (deeper joist cost more money to purchase) or the weight they must support (stronger joist also cost more money to purchase) or their layout, for example, 30 inches on center versus 48 inches on center. The tighter the layout, the more joists that are required, and that will also cost more money.

An alternative QTO method is to follow our guidelines for columns and beams in Figure 8.1. The estimator first notes the joist sizes, or in this case the depths. There is no need to focus on the weight they support, as this will be up to the fabricator to incorporate. Then measure their length and calculate the total LF of all of the similar depth of joist. Both the labor productivity and material purchase price may be figured on the LF, as will be discussed in the pricing chapter.

An even more accurate way to perform QTO for bar joist, as it relates to labor productivity (at least in this author's opinion), is to count how many joist there are. Sometimes each joist is drawn by the structural engineer, but often a note is just provided indicating they are 2 feet on center. In this case, the estimator measures the width of the area covered, divides by 2 feet to come up with a count of joist, and then adds one more for the end. The quantity of connections will be two per joist, one for each end. The drawings are marked up with a highlighter indicating that each joist, or area of joist, has been accounted for. It takes one crane and two ironworkers (one at each end) to install a joist, regardless if it is 18 inches deep or 36 inches deep and regardless if it is 20 feet long or 40 feet long. The material price is, however, depth and strength related and must follow the LF QTO process, as described earlier. Looking back at our example of 36″ deep joist spanning 30′ at 2′ on center (OC) and covering a 100′ area would require:

$100'/2'$ OC = 50 EA 36″ deep joist, plus one for the end = **51 joist** at 30′ long each
51 EA 36″ deep joist @ 30′ long/EA = **1,530 LF** to purchase

---

Angle iron or steel rod bridging or X-bracing will run perpendicular to the joist and may not all be drawn and are therefore difficult to quantify, especially if threaded rod is used. If possible, a detailed count and weight are calculated. If not, a percentage add-on (10–15%) similar to the previous column example can be applied.

Miscellaneous steel

Often the most difficult element of CSI division 05 to perform QTO on is miscellaneous steel. Any item in construction which has the prefix 'miscellaneous' in it is by definition ambiguous. There are many steel elements in a construction project that fall into this category, even if the building is a concrete or wood structure. The miscellaneous steel may be shown on the structural drawings, but much of it is found on the architectural or civil or mechanical drawings. It is up to the GC's estimator to find it all. There is not a comprehensive list in the documents (some items are called out in the specifications), and

City Construction Company
Quantity take-off sheet

Project: Dunn Lumber   QTO Sheet Number: 25
Location: 3800 Latona, Seattle   Estimator: PJ
Classification: **CSI 05: Structural Steel**   Date: 5/28/2021

Description	No ★	L ★	W	H	Total LF	#/LF	Weight in #	
		Dimensions					**Total**	
Columns: W10 × 49	81	46			3,726	49	182,574	
HSS 5 × 5 × 1/4	15	15			225	22	4,950	
ST Columns:	96	EA					187,524	#
Beams and Girders:								
Level 3: W16 × 31	62	26			1,612	31	49,972	
W24 × 55	137	30			4,110	55	226,050	
W33 × 118	18	30			540	118	63,720	
Level 4: W16 × 31	48	26			1,248	31	38,688	
W24 × 55	106	30			3,180	55	174,900	
W36 × 135	22	30			660	135	89,100	
W40 × 199	17	30			510	199	101,490	
Roof: W12 × 19	48	26			1,248	19	23,712	
W18 × 40	108	30			3,240	40	129,600	
W30 × 90	18	30			540	90	48,600	
W33 × 118	2	60			120	118	14,160	
W36 × 135	19	30			570	135	76,950	
ST Beams and Girders:	605	EA					1,036,942	#
ST Columns and Beams							1,224,466	#
Plus 15% add for gussets and plates and bolts:							183,670	#
Total weight in pounds:							1,408,136	#
Total weight in tons @ 2,000 #/Tons:							**704**	**Tons**
Nelson Studs:	**500**	**EA**				Stud weight w/15% add for beams		
3.5″ d Metal Deck:						147,000	SF	
Add waste @ 10%						16,170		
						163,170	SF	
@ 100SF/Square =						**1,632**	**SQ**	

Notes:

★ Many counts and lengths have been approximated

Many other ancillary items with repeat quantities have not been shown here but will be on recaps

Figure 8.1 Structural steel quantity take-off sheet

nonstructural subcontractors, such as mechanical and electrical, usually exclude it. Some of the items we group under 'miscellaneous' steel include:

- Steel stair flights,
- Handrails,
- Embedded angle or channel,
- Pipe bollards – both inside and outside of the building,
- Grates over trenches or catch basins or decks or air shafts,
- Mechanical and electrical equipment sleepers or curbs,
- Deck and awning supports and metal decking, and
- Ladders, such as an elevator pit or roof ladder.

The best method to perform QTO for miscellaneous steel is to note all of the items and their counts or lengths. Miscellaneous steel may be purchased by the item or each or LF or pound, but rarely by the ton. Installation will also be by the each or LF. Steel stairs would be estimated by the 'flight'. This is another place where it is a good idea to look ahead at the last chapter of the novel, or in this case at the estimating database, to see how the work will ultimately be priced and perform and organize the QTO sheet accordingly.

Structural and miscellaneous steel will also be delivered to the jobsite in different conditions, such as shop-primed, galvanized, or bare steel. Bare steel is likely embedded in concrete or will receive a fireproof coating in the field. Galvanized steel will be left exposed to the weather. Steel that is delivered shop-primed often becomes nicked or damaged in the field and will require touch-up painting before being finish-painted or enclosed in gypsum wallboard (GWB). Field welds and bolts also should be touched up. The GC should include an allowance in the painting subcontractor's scope for touch-up.

Metal decking

Metal decking often receives a concrete topping to create a composite steel and concrete deck. The decking is taken off by the square foot of coverage area. The estimator simply multiplies length times width of the building or area served by the deck. Openings such as stairs or elevators or heating, ventilation, and air conditioning (HVAC) duct shafts are not usually deducted from this SF calculation, as the deck will often be delivered in whole sheets (5′x10′) and the openings cut out later. Even if openings were deducted prior, the cost of working around them is more than straight flat deck, so a deduction in SF would be counterproductive for labor costs. In most cases the structural engineer will choose one size of deck to use throughout the same project, but if multiple depths or gauges of deck were to be used, the QTO sheet would need to be organized around these different sizes. After the deck SF is calculated, it may be converted to squares (SQ) or hundred square feet (CSF). Contractors do not purchase deck by the sheet or pound or ton.

Metal deck that is to receive a concrete topping slab will often be ordered with a shop-fabricated edge angle on the outside building perimeter, and preferably around all interior block-outs. This angle will function as an edge form for the elevated concrete deck; therefore, an additional wood form would not be required. This is another example of constructability the GC will input to the design, either during preconstruction or during buyout of the metal deck supplier. The deck will be puddle-welded to the beams and joist below. An estimator can either count the welds, such as one per 2 feet on center for every joist, or calculate the deck welding on a SF of deck basis, again depending on how the welding operation will be priced.

Metal deck is 'gauge' steel and not a 'rolled' shape. We typically do not include gauge metals with structural steel, but metal decking is the exception because it functions in a structural fashion, especially when combined with beams and concrete to create a structural composite deck. Specialty metals such as stainless steel, aluminum, or copper are also not included in the structural steel QTO. There are many other gauge metals in a construction project that may be installed by trades other than ironworkers and not necessarily specified in CSI 05. Some of them are listed here by CSI division and include:

05: Metal studs and joist, installed by carpenters with gypsum wallboard;
07: Metal wall and roofing panels, installed by sheet metal workers or roofers;
07 and others: Metal flashing, gutters, and downspouts, installed by sheet metal workers or roofers;
07, 10, or 12: Louvers, installed by many trades; and
Old 15, new 23: HVAC ductwork, installed by sheet metal workers, nicknamed 'tinners'.

For a typical structural steel building, the steel QTO would be several sheets long. Each of our structural steel categories (columns, beams, joist, miscellaneous steel, and deck) would all have their own separate QTO sheet, if not many sheets for every category. In addition, each floor and roof would have its own QTO sheet. The steel categories we discussed are likely provided to the GC for installation, or an installation–only subcontractor, by four different suppliers:

- Rolled shapes including columns and beams (likely purchased together),
- Bar joist,
- Miscellaneous steel, and
- Metal decking.

If the main structure of the building was concrete, as discussed in our last chapters, or wood framing, as discussed in the next chapter, there may only be a small amount of steel on the job, and then it would be acceptable to group all of it together. Figure 8.1 is an abridged example structural steel QTO sheet for our case study project.

There are additional structural steel items and equipment we will price, but not necessarily take–off; these items are total tonnage or schedule dependent. Some of the items we will include on our steel pricing recap sheet will be: Shake-out, safety, crane mobilization and rental, operators and riggers, forklifts, man-lifts and scissor lifts, compressors and welders, plumb and align, and others. Prices for these add-ons are included with Chapter 12.

Summary

Rolled shapes of structural steel include wide flange beams and columns, HSS columns, channel beams, pipe columns, and angle iron bridging and lintels. Estimators should have some idea as to how work will be estimated for labor productivity and material purchase and perform their quantity take-off to allow easy transition from the QTO sheets to pricing recapitulation sheets. Steel elements can be taken off by the each, separated by size and category, lengths noted, and converted to pounds and tons. Bar joist can also be quantified by the SF of floor coverage, linear foot, or each. There are many different elements which make up the miscellaneous steel category. Many of these items are shown on the architectural drawings. They are best quantified by the type and each and not customarily converted to pounds or tons. Metal decking is quantified by the square foot of coverage and converted to hundred square feet or squares.

In this chapter we tried to capture many of the structural steel items a GC 'might' install direct, but there are others. These types of work items all have unique quantifying processes, and there are often shortcuts for each. Shortcuts work fine for budget estimating or when there is not sufficient detail in the drawings, but if the detail is available, the estimator should take the time to do a thorough quantity take-off. Shortcuts can result in shortfalls and are risky. Detailed estimates are essential for lump sum bidding and are beneficial for later development of the construction schedule and implementation of the cost control plan, both of which are introduced in Chapter 24.

Review questions

1   How many pounds are there in a ton?
2   Why do we separate different estimating categories by types of craftsmen, for example, carpenters versus ironworkers?
3   Should the rental equipment necessary to erect structural steel be included with the labor cost to install the steel or with the jobsite general conditions, and why there?
4   Wide flange columns and beams have the abbreviation of WF. What are some of the other past and present abbreviations for wide flanges?
5   Why would an angle iron not necessarily make a good column choice?
6   Why are different steel categories (rolled shapes, bar joist, miscellaneous steel, and decking) usually supplied by different vendors?
7   If a GC were to install only one of the four categories of steel, which one would it be, and why that one?
8   Why would a GC prefer para-bolts over anchor bolts?

Exercises

1   Looking back to Chapter 5's discussion of constructability, what suggestions might a GC make to the structural engineer with respect to the structural steel drawings?
2   Figure 8.1 is an abridged QTO for the Dunn Lumber case study project utilized in the book. The actual steel QTO would be several pages long. Compare this QTO to the drawings on the book's eResource: What did we miss?
3   Although our Lee Street Lofts case study included on the book's eResource is primarily a wood-framed building atop cast-in-place concrete walls, there was a substantial amount of structural steel – some of it left exposed as interior architectural features. Prepare a steel QTO for this project.
4   Including allowances for gussets, etc., compute the total tonnage for the following quantities and sizes of rolled steel members. You will have to look up the weights of the steel members for this exercise.

   a   100 LF of 3.5 × 3.5 × 1/4 angle
   b   250 LF of 12 × 3 × 5/16 channel
   c   85 LF of W36 × 135
   d   525 LF of 4 × 4 × 1/4 HSS

5   How many LF of bar joist are there in an area measuring 100′ × 100′ where the joists are 30 inches on center?

6   Building on the joist question in Exercise 5, assume that WF girders sit atop steel columns which are 33′ apart and the joist span from girder to girder. How many joists will be in this 100′ square grid? You may want to draw a sketch for this question.

7   Looking to the list of materials in Exercise 4, and assuming that each steel member was the same length, let's say approximately 30′, and all things being equal (which they never are, but. . .), which element would take the most labor to install?

9 Carpentry

Introduction

Chapters 6 and 7 focused on quantity take-off (QTO) measurements for concrete, and Chapter 8 focused on structural steel. In this chapter we are going to cover several remaining categories of work, many of which may be installed by the general contractor (GC). We have grouped all of these items together in this chapter because they are predominately scopes of work claimed by the carpenters' union and/or usually installed by the carpenter trade. It is customary to think of the carpenter trade performing woodwork specified in Construction Specification Institute (CSI) division 06, but there is also carpenter work in CSI divisions 07, 08, 10, 12, and others. Carpenters are also responsible for CSI 03 concrete formwork, as discussed previously. Many of the measuring and counting techniques and procedures presented in the previous QTO chapters apply to these systems as well and others that we may not have covered. Some scopes of carpentry work include:

- Rough carpentry or framing,
- Siding and roofing,
- Finish carpentry including cabinets and millwork,
- Doors, door frames, and door hardware,
- Punch windows, and
- Specialties including toilet accessories, signage, fire extinguishers, lockers, and others.

Rough carpentry

Steel and wood buildings have significant differences, but many of the QTO processes are similar. We count and measure the materials and extend and convert into purchasable quantities. Carpentry is of course installed by carpenters. Most of this work is grouped into CSI division 06. We separate rough from finish carpentry for a couple of reasons. Some GCs may subcontract out the rough carpentry, or framing, but may perform the finish carpentry themselves; others just the opposite. Although many carpenters will tell you that they can do both rough and finish carpentry, this is not customarily the case. This author was a pretty good rough framing carpenter but was not patient enough to be a good finish carpenter, whereas his father was just the opposite. A rough carpenter uses a sledgehammer and an axe and a nail gun, whereas a finish carpenter uses a nail set and block plane and sandpaper. They belong to the same union and are paid the same wage rate but have different tools in their toolboxes. On a residential project, all of the carpentry may be included on the architectural drawings, whereas commercial projects clearly have the structural framing included with the 'S' or structural drawings, and the finish carpentry is on the 'A' or architectural drawings.

Framing

The very shortcut way to perform a wood framing QTO is simply by the square foot of floor (SFF) or, in the case of an apartment building, by the unit. The second easiest is to run a tape or a wheel along the total wall length and figure wall framing (all types together) by the lineal foot (LF) of wall. Software QTO systems perform the same process. There are multiple problems for the estimator who relies on shortcuts and does not perform detailed QTOs. In this case, some of the variables include

- Engineered lumber versus dimensional lumber,
- 2 × 10 versus 2 × 12 floor framing and joist frequency,
- 2 × 4 versus 2 × 6 wall framing and stud frequency,
- Single, double, or triple top and bottom plates,
- Wall height,
- Double or staggered stud configurations and stud packs,
- Shear walls with multiple layers of plywood sheeting on one or two sides,
- Framing hardware, and others.

Similar to structural steel, a detailed wood framing QTO starts at the bottom of the building and works its way to the top. *Pressure-treated* (PT) plates are applied to the concrete foundation walls. They are measured in LF by sizes, usually 2 × 4 or 2 × 6. If columns or beams are in contact with concrete, they will also be pressure-treated. Pressure-treated lumber should be quantified separate from the other dimensional lumber, as it is generally more expensive to purchase.

Dimensional lumber is 'sticks' cut straight from trees and ranges in size from 2 × 4 and 2 × 6 to 2 × 12 or even 12 × 12. These are 'rough' sizes, before planing. A 2 × 4 actually measures 1.5″ × 3.5″, but when we figure board feet (BF) or thousand board feet (MBF), we figure it on the gross size, 2 × 4. When performing a wood framing QTO it is important to note the type and grade of tree that the lumber came from. For example, a Douglas fir (DF) graded #2 and better (BTR) will be stronger and more expensive (and straighter!) than that cut from a hemlock tree of similar size and grade. If lumber is specified as hem-fir (HF), that means it can be either hemlock or fir, but because hemlock is cheaper than fir, hemlock will be most prevalent in the bundle when received from the lumber supplier. There is no such thing as a hem-fir tree.

Wood *columns and beams* are counted and measured very similar to the structural steel examples in Chapter 8. We count their quantity and note their sizes and lengths. Also similar to steel, estimating columns and beams by the SFF is a shortcut and is risky. The most prudent method is to count each wood framing member, or 'sticks', and add up all of the common sizes and lengths and convert them to board feet and thousand board feet. In Table 9.1 we have provided the reader with conversions for the most common dimensional lumber sizes, from LF to BF. Solid sticks, often cut from old-growth fir trees, include 6 × 6 columns or 6 × 12 beams. A more sustainable way for the designer to specify large sizes such as this today is to use engineered lumber, as discussed later, or 'stud-packs' for smaller columns. For example, an 8 × 8 column could be comprised of five each 2 × 8's nailed together in the field. Each individual 2 × 8 could more easily come from a smaller tree than one 8 × 8.

Wood joist are counted similar to steel bar joist, measuring the area covered and dividing by the frequency of occurrence, whether they are 16″ on-center (OC) or 24″ OC, and always adding an additional joist to the end. The end joist may be known as a *rim joist*.

Table 9.1 Dimensional lumber board foot conversions

Dimensional lumber board foot conversions

Lumber size	Board feet (BF) per lineal foot (LF)★
1 × 2	0.17
1 × 4	0.33
1 × 6	0.50
1 × 12	1.00
2 × 4	0.67
2 × 6	1.00
2 × 8	1.33
2 × 10	1.67
2 × 12	2.00
4 × 4	1.33
4 × 6	2.00
4 × 8	2.67
4 × 10	3.33
4 × 12	4.00
6 × 6	3.00
6 × 8	4.00
6 × 10	5.00
6 × 12	6.00

★ Note: BF/LF can be calculated for any board size by multiplying the two dimensions and then dividing by 12, for example a 2 × 12 joist is 2 × 12/12 = 2 BF/LF and a 10 × 20 beam would be: 10 × 20/12 = 16.67 BF/LF.

Rim joist also run perpendicular to common joist, secure the end of the joist, and sit atop a PT plate. Seldom are the joists all drawn by the structural engineer for all of the floors and roof, as this would cloud up the drawing and it would be very difficult to draw every joist exactly, allowing the contractor to change order for any discrepancies. Also similar to steel, additional wood beams and joist will be necessary at blockouts such as heating, ventilation, and air conditioning (HVAC) duct risers. The total count of joist times the span or length provides the total LF, which is converted to BF and MBF. For example, let's assume again a 100' square area supported by 2 × 8 DF #2 and BTR @ 16" OC. The resultant quantity of joist in LF and MBF would be:

Joist:	100' × 12"/16" OC/EA = 75 rows + 1 @ end = 76 rows @ 100' long
	76 rows @ 100' (don't deduct for the rim joist) = 7,600 LF
Rim joist:	2 EA @ 100' = 200 LF, perpendicular to common joist
Total:	7,600 + 200 = 7,800 LF × 1.10 waste and lap = **8,580 LF**
	8,580 LF × 1.33 BF/LF = 11,411 BF/1,000BF/MBF = **11.4 MBF**
Quantity:	The purchasable quantity will vary depending on layout, but assuming 16' long/EA = 8,580 LF / 16'/EA = **536 EA** of 2 × 8 joist
	(Don't round up, as 10% for waste and lap is already included)

So now, depending upon the method to price both the labor to install the joist and the unit price to purchase the material, the estimator has all of the quantities available necessary to move to the next pricing recapitulation stage, described in Chapters 10 and 11.

Wall framing has a couple of different elements. Most projects have multiple wall types, and the structural drawings will have a wall schedule. Each wall type should be quantified separately. They differ with respect to plate count, stud sizes and layout, plywood sheeting, and framing hardware required. Bottom plates can be one or two boards thick, and top plates are usually two boards thick. Once the total LF of a wall type is known, this dimension can be simply multiplied times the plate quantity to come up with a total LF of plates, which can also later be converted to BF and MBF. Wall studs are not typically shown, similar to joist, either, and it is up to the estimator to figure their count. The different wall types and lengths and heights are known, and the stud frequency, often 16″ OC or 24″ OC, is given. Some walls have double or back to back or staggered studs, and those must be accounted for independently. Our rule of thumb (ROT) for calculating 16″ OC stud count is to figure one stud per LF of wall. This is more than required but will account for added studs at corners and intersecting partitions and cripples at doors and windows. The estimator should figure 1.5 studs per linear foot if they are specified at 2′ OC. The total stud count times the height of the wall (deducting for plate thickness can be done, but this is outside of the 80–20 rule) will produce a total LF of studs and then will convert to BF (.67 BF/LF of 2 × 4 or 1 BF/LF of 2 × 6) using Table 9.1.

If the walls, whether they be structural or non-bearing partition walls, are framed of metal studs and not wood studs, they will be typically installed by the gypsum wall board (GWB) subcontractor. The same holds true if the ceiling joist are metal. Metal gauge studs and joist may be specified in CSI division 05 but will not be supplied by the structural steel fabricator, as discussed in the last chapter. On occasion the gauge material may be specified in 09 along with GWB. Although the carpenter trade also installs the metal studs (and the GWB, acoustical ceilings, plaster, and others), they are installed by different carpenters with a different skill set and a different set of tools. QTO and pricing subcontracted work is discussed in detail in Chapters 14–16.

Wood *door headers* and *window headers* can also be cut from dimensional lumber. Sizes and spans should be separated and counted and then like sizes grouped together to calculate total LF, which can be converted to BF and MBF in the case of dimensional lumber, similar to beams and columns. The thickness of a header usually matches the stud width; for example, 2 × 4 walls utilize 4 × 8 or 4 × 12 headers. Headers today are often fashioned from engineered lumber, which is straighter and does not shrink, as discussed next.

Heavy timbers are those customarily larger than 12 inches square, potentially rough-cut if left exposed, sometimes from old-growth timber or even reclaimed from existing buildings. They are unique structural members, often beams and columns, and ones that an estimator would not want to price from a standard estimating database. They should be counted by the each, with sizes and lengths noted and no additional conversions required.

Blocking and backing may be shown on the drawings, and if so, estimated similar to beams. Some common locations an estimator should look for blocking (or bridging) and backing include rows perpendicular to joist, over or under shear walls, behind toilet accessories and kitchen cabinets, or support for anything else surface mounted to the GWB. But if just an "Add blocking as required" note is on the framing drawings, then it is customary that the 10% we recommend adding to the wood framing QTO will suffice for blocking and backing.

Engineered lumber

Engineered lumber comprises a growing family of wood framing products that are 'sustainable'. Different than dimensional framing lumber, these members are not cut whole from an old-growth tree but rather assembled from smaller trees and glued together to produce a stronger and straighter product. Engineered lumber typically does not shrink or split like dimensional lumber. For ease of matching up with or substituting for dimensional lumber, engineered products are fabricated in the same (final) sizes as are typical framing lumber. For example, Truss Joist International (TJIs) are 9½″ or 11¼″ deep, similar to the planed sizes of a 2 × 10 or 2 × 12 joist. There are many types of engineered lumber products today, including:

- Glue lam beams (GLB),
- Wood 'I' beams or TJIs,
- Parallel strand lumber (PSL) or laminated veneer lumber (LVL),
- Cross-laminated timbers (CLT),
- Oriented strand board (OSB) discussed with sheeting, and
- Trusses: Roof or floor.

The QTO process for these members is similar to dimensional lumber, but it must be kept separate and not grouped together and will often be purchased from alternate suppliers. An adder for 10% lap and waste is usually not necessary for engineered lumber as, similar to structural steel, this material is fabricated and delivered to the jobsite to fit exactly. Engineered lumber is more often purchased by the LF rather than MBF and labor for installation estimated by the LF or each.

Wood *roof trusses* are more common on residential than commercial projects. They can be flat or sloped, such as 5″ rise in 12″ run. These trusses will sit atop load-bearing exterior walls and allow uninterrupted free space between the walls. The trusses are purchased from a supplier and shipped to the jobsite in one piece. The supplier will 'engineer' them and size the wood members and the steel gusset plates. The truck they are shipped on often is equipped with a small crane or boom to allow unloading and placing directly on the walls of the structure. The wood truss take-off is very similar to the steel bar joist take-off. Most published databases such as *Means* have trusses priced by the SFF or SF of roof. This does not account for the frequency of occurrence, i.e. 16″ OC or 2′ OC, nor the framing material, 2 × 4 versus 2 × 6, etc. Many trusses are composed of all wood, but some combine wood with gauge metal for elements in tension. The trusses are estimated by the supplier by the each, and a good way for the GC's estimator is to quantify them by the each. It takes three carpenters just as long to install a 20′ truss as it does a 30′ truss.

Framing hardware

We do not calculate the quantity of nails or screws or glue or caulking required for a typical construction project – they fall outside of our 80–20 rule – but often an allowance of $1,000 or so will be included in the estimate. We do, however, need to calculate the amount and types of framing hardware required. "Simpson" hardware is a common brand of hardware, and all wood framing hardware is often generically thought of to be Simpson hardware. The proper way to perform the QTO for hardware is to categorize each different type and count them from the framing drawings, marking them up in hand with a highlighter as we have discussed previously, or performing the same process electronically with QTO software. As a shortcut they can be all grouped together, similar to structural steel connections discussed in the previous chapter. An even shorter cut is to just allow $/SFF for framing hardware. The problem with either of these last two methods is they are not accurate and are risky. Some of this hardware can cost hundreds of dollars per each and others less than a dollar, so to group them all together would be shortsighted. If the estimator were to be awarded the bid, he or she would still have to come up with definitive counts to be able to place a purchase order with a hardware supplier.

Sheeting

Sheeting for walls and floors and roofs is easy for the estimator to calculate: Simply multiply the length of the area to be covered times the width or height. Similar to metal decking, it is not necessary to deduct for openings unless they were substantial, such as glass curtain wall. Ten percent is added to SF of coverage for waste and divided by 32 SF per sheet to come up with the quantity of either plywood or OSB sheets required. The estimator should keep plywood and OSB separate and all of the different thicknesses of sheeting, such as ½″ and ¾″, separate as well. Sheeting will likely be priced out of a database by the SF, but having the quantity of sheets handy will be helpful during the buyout and construction process. If we had an area of 100′ square requiring ¾″ tongue-and-groove (TNG) plywood, the sheet calculation would be as follows:

100′ × 100′ = 10,000 SF × 1.1 waste = 11,000 SF
11,000 SF/32 SF/sheet = **344 sheets** of 3/4″ TNG plywood

The same rules for combining different structural steel elements apply here as well for the rough carpentry QTO sheets. If this were an all-wood building, there would be many sheets of carpentry QTO, often separated by floor or system, and walls versus floors versus roofs, but if it were primarily a steel or concrete building structure, much of the wood could be captured on just one or two QTO sheets. Figure 9.1 is an abridged example of a rough carpentry QTO for another project we worked on professionally. The commercial case study we chose to include for most of the examples in the book does not have wood framing or significant rough carpentry. We have however included drawings for an executive four-plex townhome project on the book's eResource. That project will require several QTO sheets for each of our carpentry systems discussed in this chapter.

Siding and roofing

There are many different types of siding on a building, including glass and brick and steel and granite, etc. Very few commercial buildings today would have wood siding, and even the houses that do are seldom cedar anymore, but more often a cementitious material such as HardiePlank. Wood siding and cedar wood roofing shingles may be included with wood CSI division 06 or may be with division 07 along with the balance

City Construction Company
Quantity Take-off Sheet

Project: Broadway MXD Apartments   Sheet Number: 1
Location: 1200 Broadway, Seattle   Estimator: LH
Classification: **CSI 06: Rough Framing**   Date: 6/4/2021

Description	No	Dimensions L	W Sizes:	H	Total LF	BF/LF	BF	@1,000BF /MBF
Treated Plate:			2 × 6		1,200	1.00	1,200	1.2
Columns, DF #2 & Btr	40		6 × 6	12	480	3.00	1,440	1.4
Beams:	20	14	6 × 12		280	6.00	1,680	1.7

Joist: 2 × 8 @ 16″ OC spanning 100′ SQ = 100′ × 12″/16″ (+1 @ end) =   76 Rows

Rows:	76	100	2 × 8		7,600			
Rim joist:	2	100	2 × 8		200			
					7,800			
Add 10% for waste and lap					780			
LF × BF/LF / 1,000 BF/MBF =					8,580	1.33	11,411	**11.4**

Joist count @ 16′ span =   8,580   LF / 16 LF/EA =   **536 EA**

Wall Framing: 2 × 4 studs @ 16″ OC, 8′ high, 1,000 LF = 1,000 EA @ 8′ =   8,000   LF
Plates: 2 on top and 1 on bottom = 3 ea @ 1,000 LF =   3,000   LF
                                            11,000   LF
Add 10% for waste: 1,100
@ .67 BF/LF / 1,000 BF/MBF =   12,100   **8.1**

Trusses: 2 × 4 trusses @ 2′ OC, 30′ span/width over 100′ length/area + 1 at end =   **51 EA**

Sheeting: 3/4″ TNG spanning: 100   100   =   10,000   SF
Add 10% for waste: 1,000   SF
ST 3/4″ TNG plywood: 11,000   SF
@ 32 SF/sheet =   **344 Sheets**

Framing Hardware: See separate QTO
Nails: Allow: 2   Box   16d
              3   Box   8d

Figure 9.1 Rough framing quantity take-off sheet

of waterproofing and roofing materials. If a GC were to figure the wood siding to be performed by its own carpenters, it would be quantified by measuring the surface area of all of the walls, length times height, to come up with total square footage. Windows and doors are again seldom deducted from this quantity. In the case of beveled siding, up to

20% would be added to the SF of wall for lap and waste. The siding is then purchased by the linear foot. For example:

---

100 LF × 16′ high = 1,600 SF × 1.2 waste and lap = 1,920 SF
1,920 SF × 2 rows/SF = **3,840 LF** of six-inch beveled siding
Or 8-inch beveled siding = 1,920 SF × 1.5 rows/SF = **2,880 LF**

Regardless if the siding is cedar or HardiePlank, it is usually accompanied with rough-sawn (one face) cedar trim around doors and windows; 1 × 4 and 1 × 6 are often the most common sizes. The material is quantified by the linear foot, and at a minimum 10% is added for waste. It is difficult to find straight cedar trim material without knots. Clear cedar is very expensive. This author's father would buy up to twice the material necessary just to 'cherry-pick' the good 'sticks' from the cedar siding or cedar trim pile; the remainder making good but expensive kindling for the fireplace at home. Cedar siding and trim quantities therefore should include a 20–30% waste factor.

There are many other materials that the siding or building enclosure could be designed from, but these are typically not carpenter or GC installations. Some of them would include masonry (CSI 04), metal siding (CSI 07), glass or curtain-wall (CSI 08), and stucco (CSI 09). These systems are all supplied and installed by subcontractors and will be discussed in an upcoming chapter.

Roofing

Very few if any GCs will install roofing with their own craftsmen, but a few residential contractors will on a smaller scale. Roofing materials used to be all cedar shingles or cedar shakes, and therefore it was also the responsibility of the carpenters to install. Today it is more common to see tile or composition shingles or composite products for houses. Carpenters claim all types of shingles and will also install metal roofing on residential, but sheet metal workers will claim the work on commercial projects. In commercial there is a 'roofing' trade which will essentially take the entire roof, including hot-mopped or single-ply rubber membrane. If an estimator were to quantify the roofing, he or she would measure the square footage looking down on the roof, multiple by a factor from Pythagorean's theory to accommodate for the slope ($a^2 + b^2 = c^2$), factor lap, and convert the total quantified square footage into squares; 100 SF = 1 SQ. An alternate is to measure the length of the roof and multiply times the sloped width or height taken from the exterior elevation drawings. Any underlying waterproofing membrane, such as 30# asphalt, would be quantified the same way. Typical with many QTO processes, as long as the estimator is 'there', it is easy to quantify related materials. In this case, siding and roofing typically have similar quantities to many other division 07 materials, including vapor barriers, waterproofing, and insulation.

Finish carpentry

Finish carpentry, like rough carpentry, is also specified in CSI division 06 but shown on the architectural drawings. Some of these items may be purchased from suppliers and installed by the GC, and some may be packaged together and purchased and installed by subcontractors. Sometimes the GC will purchase the material from one supplier and employ a labor-only subcontractor to install it. Regardless, it is a good idea for the

GC's estimator to figure the quantities in detail to allow him or her to make informed decisions on pricing options on bid day as well as later during buyout. Similar to our discussion in the previous chapter of miscellaneous steel, there are a plethora of finish carpentry materials that could be on a project. We will focus on just a few of the more common, including millwork, cabinetry and countertops, wood flooring, and wood paneling.

Millwork

Millwork is a large group of materials that generally comprise small wood interior trim items. The estimator should categorize them separately by location or use, note the materials (wood and species, versus composite materials, versus plastic), and measure their lengths. Millwork is quantified by the LF and should include an allowance of 10% for waste. If at all possible, it would be beneficial for the GC to purchase all of these materials from the same supplier. Some of the millwork items found in residential or commercial building projects include:

- Base and crown molding,
- Interior door and window casing,
- Wood stair and balcony rails,
- Shelving,
- Backsplash at countertops,
- Stair and balcony wall caps,
- Corner guards,
- Wall bumper or chair rail, and others.

A shortcut for performing the millwork QTO is to group all types of millwork together into one LF total, but variances in material unit prices may cause this to be problematic. Closet and pantry kits include shelves and rods and today are most often metal, covered with vinyl, but still categorized as millwork and installed by carpenters. They are usually quantified by the set or kit or closet.

Cabinetry

Cabinetry exists in residential projects in the kitchen, bathroom (vanities), and the utility room. Commercial projects also have kitchens or coffee or break rooms with cabinets. Commercial projects may also have cabinetry in executive offices, libraries, or conference rooms. Where finishes change, for example from plastic laminate (Plam) to oak to cherry, the cabinet take-off must be kept separate by material type. Many estimators and estimating guides have cabinets quantified by the LF or SF, looking down into a room at the floorplan. These are quick and easy take-offs to perform but have risks associated with them. Lower cabinets are customarily 2 feet deep, and upper cabinets are 1 foot deep. If LF is the basis for the take-off, it will not take into consideration that the lower cabinets are much more expensive to purchase but counted the same as the upper cabinets. Measuring the cabinets by the SF corrects this inconsistency, but does not consider the increased magnitude of labor effort for installing the upper cabinets. They have to be temporarily shored and held in place and leveled before fastening to the wall, whereas the lower cabinets are held in place by gravity and require simple leveling with shim shingles before attaching. If the estimator uses SF as the metric, he or she places twice the labor effort into the lower cabinets. An alternative is to estimate the cabinets by the each or 'box', similar to our discussion with some steel members in the last chapter. It takes just as long to install a 2-foot-wide cabinet as a 3-foot cabinet.

Cabinet tops may be purchased from the same vendor as the cabinets, but depending on materials, they often come from an alternative source. Some countertop materials utilized today include:

- Plam,
- Ceramic tile,
- Soap stone,
- Corian,
- Granite tiles,
- Solid granite, or even concrete slabs.

Carpenters may install the plam tops, but the others are likely subcontracted by firms that employ tile masons. Cabinet tops can be quantified by the SF of top, which should approximately equal the SF of lower cabinets discussed earlier. Cut-outs for sinks and ranges should not be deducted from the SF, similar to our discussion of plywood sheeting earlier in this chapter and metal decking in the last chapter. There is more labor to accommodate these cut-outs than there is a flat uninterrupted counter-top installation. Cabinets or casework specified in CSI division 12, and not 06, are specialized and most often supplied and installed by a subcontractor. These cabinets are common with hospital or laboratory or school projects.

Wood flooring

In most cases wood flooring is supplied and installed by the flooring subcontractor. It is beneficial for the GC to package all of the flooring (and base) together with one firm; that way, if it is on the floor, it is that subcontractor's responsibility, regardless if it is carpet or tile or vinyl or wood. The question of who installs the transition strips is also resolved if packaged under one firm. But because wood flooring is made of wood, it may also be installed by the GC's carpenters. Like many of our items in this chapter, only if the GC was short on work, if there was a small quantity, or if this was one area in which the GC specialized would they choose to do this work. Wood flooring would be taken off by the square foot of floor area, simply measuring length times width. An allowance of 10% for waste would be appropriate here as well. Then a conversion would be necessary for the width of the floor material. If the material were ¾″ thick TNG oak at 3″ wide, then the SFF would be divided by 4 (four boards of 3″ in each foot) to get the LF of material necessary.

Vinyl plank flooring may look like wood but really isn't. It may have a thin particle board backing, but more likely it is made of compressed paper or composite resin materials. Seldom would a GC do this work with its own carpenters but would prefer to subcontract it out with the balance of the other floor systems. This would be quantified by L × W = SFF plus a minimal allowance for waste, say 3–5%.

Wood paneling

Wood paneling was popular in residences in the 1970s but not so much anymore. Paneling can be found in some high-end offices, such as a downtown attorney's office, business reception areas, or elevator lobbies. *Wainscot* is wood wall paneling 36″ or 42″ high located in some of these same upgraded spaces or in hallways. Wood paneling is rarely work installed by the GC's own carpenters and would most likely be subcontracted out. The QTO would be performed similar to the wood floor QTO, except by wall area, not floor area. The total SF would be divided by the material size. If wall paneling were composed of individual boards, it would be similar to our wood floor analysis. If the paneling is in sheets, then the SF would be divided by 32 SF/sheet.

Doors, door frames, and door hardware

Doors, door frames, and door hardware (DFH) are one of the easiest categories to 'take-off', and one often assigned to a junior estimator or project engineer. Even though we promote performing the QTO in a logical sequential fashion as the building would be built, starting with the foundations, it is sometimes difficult for an estimator to get started. Jumping in with the DFH QTO is an easy way to get into the flow of the estimating process. Doors, door frames, and door hardware are specified in CSI division 08 and customarily installed by the GC's carpenters. There are also special gates or garage doors in 08, which may be subcontracted and installed by different trades. But in this section, we are focusing with what in the industry are known as 'man' doors; sorry, ladies!

One mistake inexperienced estimators will make is to simply take the door schedule out of the architectural drawings, even literally cutting it out and taping it to a standard QTO sheet, and performing the 'take-off' right there. That process is very simple and quick and may be accurate, but it may also come up short. The contractor is responsible to provide ALL of the materials shown in the drawings or required in the specifications. Not all drawings are perfect, and they are not all prepared at the same time. It is imperative that the floor plans and exterior elevation drawings be checked for additional doors that may not be on the schedule; the entire contract documents must work together. Even if a door is not on the door schedule but located somewhere else in the documents, it is in the GC's scope and in their contract.

The DFH QTO is a simple process of separating the three major categories of work and then separating them again by material, then again by size. Most projects do not have all of the different categories, and there are significant differences in residential and commercial and industrial work with respect to door assemblies. Here are some of the separations the estimator must consider on his or her QTO sheet:

- First count the *door frames* and separate them by material, such as hollow-metal (HM) (prevalent in commercial) and wood (prevalent in residential). There are also additional subdivisions. Hollow-metal door frames are often solid welded units, but they may be delivered as 'knocked-down' separate pieces that require assembly in the field. Hollow-metal door frames may also require to be grouted with cementitious material in the field, which adds significant labor cost and weight. Wood doors and frames may be pre-hung on residential projects, in that the door and the frame and the hinges are sold as one unit, or they may be separate and require assembly in the field. Wood doors and frames may also be delivered pre-painted or stained, but if not, this work needs to be in the painter's request for quotation.
- There are aluminum door frames as well, often accompanied by aluminum and glass doors, but these are customarily the responsibility of the curtain-wall subcontractor. It would be prudent for the estimator to count them here and include them in the QTO, but when it comes time to price them, he or she should note "by subcontractor" in the pricing column.

- The door frames also need to be separated by size. There are single and double door frames, and although the labor to install them is similar, the cost to purchase is different.
- We classify the doors themselves as 'leafs'. The *door leafs* need to be counted similar to the frames and separated into categories such as:

  - Hollow metal door leaf,
  - Solid core wood door leaf, and
  - Hollow core wood door leaf, and
  - There are also plastic laminate door leafs.

- Many door leafs in commercial construction will have glass panels, or 'relites', in them of various sizes and fire ratings. If relites are included, a separate count with their sizes should be noted. This glass will not be provided by the door supplier or the punch window supplier or the curtain wall subcontractor or the mirror supplier unless the GC specifically requests it and includes it in their bid packages. The GC's carpenters typically install this work. The glass panels will be framed and 'stopped' in with a 'relite kit'. Special counts of all of these are important, as they will affect the direct labor portion of the door estimate as will be discussed in Chapter 12.

The *door hardware* take-off may be as easy as counting one 'set' of hardware for every door leaf. A double door frame requiring two leafs will have two sets of hardware. But the correct way is quite detailed and time-consuming. There are likely 20 to 50 different door hardware set combinations on a large commercial project. Every hardware set type will have hinges and knobs or latches and throws and door stops and kick plates and many other elements that require assembly by the GC in the field. The detailed way to perform a hardware QTO is to note how many door leafs from the DFH schedule apply to each of the different hardware types. Next multiply each of the elements of each set type times the leafs. Then the estimator may combine individual elements from one set type with others when they are the same. For example, hardware type 4 may have the same hinges as type 16, but 4 has a kick plate and 16 doesn't, and 16 has a door closure but 4 does not. Therefore, after extending the hinges times the leaf count, they can be added for these two door types. The result is a very large and very complicated Excel spreadsheet. This is how your door hardware supplier will perform their take-off. This is how one of your contributing authors started his estimating career, including counting Phillips screws.

A shortcut for the estimator is to group all hardware sets together and apply one average price throughout, often with that being the hardware set type which occurs most often. The total hardware set count should approximately equal the total door leaf count – there are two leafs and sets of hardware per double door frame. Grouping all DFH work together as an assembly, regardless of size and material and complexity, to come up with a total count, such as 100 sets or openings at $1,000/set, is an extreme shortcut and one to be utilized by the estimator with schematic budget estimates only and not lump sum bids.

Door hardware is expensive to purchase, and because individual items are relatively small, they sometimes have a way of vanishing from the jobsite before they are installed. The GC will often build a *hardware room* that is locked up to store door hardware until ready for installation. On larger projects, a carpenter will be responsible to 'man' the hardware room. All of the hardware is separated by type and sorted on labeled shelves. When a carpenter is ready to install three sets of hardware type 23, he or she will request an exact list of materials from the hardware room and often is required to sign out for the material.

Assembling and staffing a hardware room, if utilized, will need to be accounted for in the DFH direct work portion of the GC's estimate.

The doors, frames, and hardware may all be supplied by one material supplier to the GC, which makes contracting and hopefully installation smoother. But most often they will be supplied by several different vendors depending upon material and require 'machining' of the frames and door leafs to the hardware. If they are supplied by multiple vendors, the GC may need to make arrangements for the different materials to be shipped from one supplier to another for machining. If the wood doors are pre-stained, they may also make a trip to the painting subcontractor's shop, as well.

Other residential pre-hung doors and frames (usually wood) include bifold doors, sliding barn doors, and pocket doors and are furnished complete with hardware sets. They would be quantified separately by type and by size, as 'sets'.

Some *access doors* may be assigned to subcontractors whose scope of work requires them to be installed. For example, the mechanical, electrical, and plumbing subcontractors may be required by building code to provide access doors to valves, but these may not be shown on the drawings. Access doors that are shown on the bid documents are customarily quantified by the GC by the each, noting different sizes and fire ratings, as with other items, and installed by their carpenters.

Punch windows

Windows are also specified in CSI division 08, and their take-off process is similar to the door take-off. There will be a window schedule, but similar to doors, the estimator should review the floor plans and exterior elevations to develop a complete window count by type. If windows are preassembled with frame and glass 'pre-hung' and ready for installation into a wood or metal stud wall, we classify them as 'punch' windows. These windows will have been specified to come from a manufacturer, and each window will be noted with a size, such as 4-0 × 6-0, and a product number, such as XCY4060D. A window supplier will provide the GC with a lump sum quote for all of these windows, but they will not install them, so therefore the GC's estimator must perform their own take-off, and the contractor will install them with their own carpenters. The window frames may be aluminum or wood or plastic/vinyl, and there will be many different sizes; each of these must be separated on the QTO, again similar to the door take-off. Operable punch windows will come with screens as well, or at least they should, but the estimator should make a note of them. Sliding glass doors are included with this system as well. The GC's punch window QTO can be by the each (it takes two carpenters 30 minutes to install a 3-0 × 3-0 as well as a 5-0 × 5-0 window), or it can be by the square foot of glass. Again, the diligent estimator will have both quantities handy to later apply labor productivity and material pricing to.

Windows can also be classified as *storefront or curtain wall*. These systems will be supplied and installed by a separate subcontractor utilizing the glazier trade rather than the carpenters. We will discuss the GC's subcontractor take-off and pricing procedures in an upcoming chapter. Storefront glass is quantified by the SF, with adders for operable doors. These window systems are often design-build, where the architect has provided some guidelines and details but extensive calculations and shop drawing submittals are necessary from the subcontractor. A mixed-use building may have storefront on the first floor retail level and punch windows on the upper apartment floors, and two different vendors will be involved.

Specialties

It would be impossible to list every architectural 'specialty' item in this chapter. Well, it is possible; it is just that we don't have enough room and likely would miss an item or two, anyhow. What follows are just some of the more common items, listed by CSI order (most from CSI division 10), and how the estimator will quantify them from the architectural drawings. Many will be shown on the floor plans, details, interior elevations, or schedules, and still others are not shown on the drawings at all but will be called out in the specifications. This work is customarily also installed by the GC's carpenters.

- CSI specification section 077000: Roof accessories: It is possible for the GC to group all of the work that occurs on the roof together with the roofing subcontractor, but sometimes, due to specialization or pricing, the GC may choose to install some items themselves. Examples include small self-contained skylights and roof hatches and window-cleaning davits, which would all be quantified by the each.
- 101100: White boards, chalk boards, and tack boards: They can be quantified by the each or the SF.
- 101400: Interior signage: In commercial projects we have door and room and directional signage that may all be grouped into one supply package. They are often wood or plastic and quantified by the each, and also installed by carpenters. This does not include illuminated signs such as 'exit' signs, which will be provided by the electrical subcontractor. Exterior signs are also a separate specialty subcontractor and will not be quantified here or installed by the carpenters.
- 102100: Toilet partitions: Included here only to note they are NOT typically installed by the GC. This is a subcontract item and included in an upcoming chapter. Toilet partitions are typically quantified by the 'stall'. The accessories may be supplied by the same vendor as that which is supplying and installing the partitions.
- 102300: Chair rail, chalk rail, and wall/hallway bumpers: These are manufactured products fabricated from a variety of materials, including wood and plastic and steel, and quantified by the lineal foot.
- 102600: Corner guards: These can be stainless or plastic or wood (and maybe then included with millwork). They have varying sizes, such as 1.5″ or 3″ or 6″, and heights, 3′ or 4′ or 6′. They will be purchased and installed by the each, so taken-off as such.
- 102800: Toilet accessories: Separated by different types (toilet paper dispensers from paper towel dispensers) and counted by the each.
- Mirrors: If they are framed, they are counted with the toilet accessories and ideally supplied by that firm. In this case they will be specified in CSI 10. But if they are unframed, they are usually considered glazing and specified in CSI 08 and may be supplied and installed by the exterior window supplier. Mirrors are quantified by the each if with a frame and by the SF if unframed.
- 104400: Fire extinguishers (FE) and fire extinguisher cabinets (FEC) can come in a variety of sizes and assemblies similar to our door discussion. The FE can be wall-mounted without a cabinet, or it can come with a surface-mounted cabinet, a semi-recessed cabinet, or a fully recessed cabinet. Some FECs are also fire-rated. They must all be quantified separately by the each but are likely purchased from the same vendor. They are not part of the CSI 21 (new division number, previously number 15) fire protection subcontractor's scope, which is a mistake made by many new estimators.

- 105100: Lockers can be supplied and installed by a subcontractor or supplied only for installation by the GC. If GC-installed, they are quantified by the each, with special note if separate sloped tops are required and/or a separate base. Benches are probably provided by the locker vendor or a separate supplier and also quantified by the each and installed by the GC.
- 113000: Kitchen equipment for a commercial project could be substantial if it has a large commercial kitchen or cafeteria. It also could be quite small if it is only for a coffee room or single family residences and therefore installed by the GC's carpenters. Apartment and condominium projects will have potentially hundreds of pieces of equipment, including washers and dryers. Kitchen equipment is quantified by the different types of equipment; for example, ovens are separated from dishwashers. Ideally, all of the equipment may be purchased from the same supplier. Some suppliers will deliver and install the equipment and will therefore receive a subcontract agreement rather than a purchase order. The installation process of kitchen equipment is integral with the electrical and plumbing subcontractors, as they have many connections to make. The cost of their connections will be factored into their separate subcontract agreements and is not part of the GC's take-off.
- 115200: Projection screens: These will be counted by the each. If they are motorized, they may be grouped with the projectors and supplied and installed by a subcontractor.
- 122000: Window blinds: Today, window treatments are customarily provided by a subcontractor, but on residential projects they may be purchased from a supplier and installed by the GC. It takes just as long to install a 3-foot-long blind as a 5-foot-long blind, so counting them by the each but separating by sizes (as that is how they will be purchased) would be appropriate.

Summary

In this chapter we tried to capture many of the remaining direct work carpentry items a GC 'might' perform. Will any GC perform all of these activities? Not likely, but they pick up some on one job and others on another job, contractor specialty and economy dependent. These types of work items all have unique quantifying processes, and there are often shortcuts for each. Shortcuts work fine for budget estimating or when there is not sufficient detail in the drawings, but if the detail is available, the estimator should take the time to do a thorough quantity take-off. Shortcuts can result in estimate shortfalls and are risky.

Were we able to pick up every item? Not likely, but we have picked up the more common items. Even an experienced estimator, when developing his or her initial work breakdown structure, will not pick up every item; new items present themselves as they get further into the documents. If something is discovered later, after the QTO for a system was completed, the new item must still be added somewhere to the estimate. We present a lot of estimating 'rules' in this book, but a couple of the most important ones are:

Include everything somewhere; just don't leave anything out.
Count everything once, but only once.

This section described in detail most of the scopes of work a general contractor may self-perform with their own craftsmen. Many of these scopes may likely also be subcontracted, contractor preference and specialty dependent, but it still behooves the GC's estimator to have a thorough understanding of the quantities. In the next section of the book we price up all of these self-performed scopes and many of the more common subcontracted scopes, including the building envelope, mechanical, electrical, elevator, and civil work.

Review questions

1   How many BF are there in MBF?
2   Why do we separate different estimating categories by types of craftsmen, for example carpenters versus ironworkers versus glaziers?
3   How many total MBF are there in this group of lumber sizes and lengths?

   A   1,000 LF of 2 × 4 studs and plates
   B   500 LF of 2 × 6 studs and plates
   C   820 LF of 2 × 8 joist
   D   160 LF of 4 × 8 door header
   E   60 LF of 4 × 4 posts
   F   40 LF of 6 × 6 treated columns
   G   80 LF of 4 × 12 girder

4   What are three advantages of using engineered lumber over conventional dimensional lumber?
5   How many 2 × 12s would it take to build up a 12 × 12 column cut from individual members or sticks?
6   If the GC is confident it will receive competitive subcontractor bids on items described in this and previous chapters, why is it prudent for the estimator to do a detailed QTO anyhow?
7   Since many GCs employ carpenters, why would they consider subcontracting out any of the work discussed in this chapter? There are many potential answers here.

Exercises

1   (A) List three reasons why a GC may choose to subcontract an item of work that they might normally self-perform or for which they are signatory to that particular trade union, and conversely, (B) list three reasons the GC should do the work with their own forces.
2   Does your answer change in Review Question 6 if it is a slow versus a busy economy, and why?
3   Quantify all of the different types of lumber in Len's Shed. You are going to have to make some assumptions for this exercise, and likely develop a sketch or two, but these are both common estimating activities. Here are the parameters:

   •   12′ square floor plate
   •   8′ high interior space from top of floor to bottom of the truss
   •   One-man door and one 3–0 × 3–0 fixed window
   •   Straight gable roof with 3-tab composition shingles over 30# asphalt paper
   •   HardiePlank 6″ beveled siding over Tyvek over 5/8″ CDX sheeting
   •   Shed sits on two 4 × 8 PT beams which rest on six pre-cast concrete footings
   •   Floor is framed with 2 × 8 joist at 2′ OC and sheeted with ¾″ TNG
   •   Walls are 2 × 4 at 16″ OC with one bottom and two top plates
   •   Pre-fabricated 2 × 4 roof trusses at 2′ OC

- 7/16″ OSB roof sheeting
- 1 × 4 exterior cedar trim
- Include hardware as required
- Everything is nailed with galvanized 8d or 16d or shingle nails

4   How many sheets of 5/8″ plywood wall and floor and roof sheeting are required for a two-story house (18′ high exterior walls) with a footprint of 50 feet across the north elevation, 30 feet across the west, 40 feet across the south elevation, and then an inverted jog of 10′ by 15′ on the southeast corner? There are two exterior man doors, ten windows of varying sizes from 3-0 × 3-0 to 5-0 × 5-0 and two garage doors that are 8′ square each. Assume a 5″ rise in 12″ run (5/12) sloped hip roof. You may want to sketch this structure as well. It is important that the estimator order enough plywood, so the crew does not run short, but not too much, as extra sheets of plywood tend to vanish from the jobsite at the end of the day.

5   If we have a residential kitchen that measures 12′ × 16′ which has full upper and lower cabinets on one 12′ side and one 16′ side, and an 8′x 3′ island in the middle without upper cabinets, how many cabinets will we have, and how many SF of granite countertops? Assume that we have a mix of 1′, 2′, 30″, and 3′ cabinets. Assume that there is a 3′ kitchen sink on the outside wall and a 30″ drop-in range in the island. The sink is centered on a window and does not have an upper-cabinet. Assume that the refrigerator and the double oven are outside of these parameters. The under-counter dishwasher next to the sink will take the place of one 30″ cabinet. Estimating is a mix of science and art, and a little creativity is often needed. You should sketch this one as well.

6   Perform detailed carpentry QTOs for the Lee Street Lofts case study included on the book's eResource. There are a variety of systems available for this exercise including rough carpentry, siding, finish carpentry, DFH, windows, and specialties.

7   In what part of the country do HF trees grow?

8   Prepare a DFH QTO for the Vehicle case study project.

Part III

Pricing for general contractor direct work

10 Direct labor

Introduction

Recapitulation (recap) is the fourth level of the lump sum estimating process shown in Figure 1.1. This involves summarizing and pricing processes that follow the quantity take-off (QTO) work. In Part III, we price all of the direct work that was taken-off by the general contractor's (GC's) estimator, as discussed in the last several chapters. This section of the book includes detailed pricing discussions for direct labor, concrete, steel and carpentry, and construction equipment, in Chapters 10 through 13, respectively.

Quantity totals are used as the basis of the tasks listed on the recap sheets. Quantities were taken-off from the drawings and transferred to and extended on the QTO sheets, and now these tasks are priced and totaled on cost recap sheets. In some cases, the amount of the work requires several recap sheets, and a summary recap is used to consolidate them. The recap process involves transferring numbers from the detailed quantity sheets to more summary sheets. Consequently, errors are likely to occur, and error prevention procedures need to be implemented. An estimator should make it a habit to use these procedures to minimize the potential for making significant errors in the estimate.

Along with error prevention, the estimator should utilize good organizational skills. Attaching the recap to the top of a group of QTO sheets from which data has been transferred completes a package of information that can be easily checked and used later by the project team. If a summary recap is necessary, it should be attached to the corresponding groups of recap/QTO packages used to generate it.

General contracting is a challenge because a contractor bids to perform work for the lowest price with a temporary workforce at a future point in time. The key to being effective is to have key field personnel who can direct and perform the work efficiently and correctly. Direct labor is the general contractor's own labor force that does some portion of the construction work. It also represents the contractor's greatest risk. Labor costs are determined in one of several ways. An estimator must be aware of these methods, because some bids may ask for labor rates, and he or she should know the proper way to determine them. Several of the factors of labor pricing are discussed in this chapter.

Recapitulation sheet

The recap sheet is used to gather the quantified data, apply pricing factors, and perform calculations that result in the various costs that make up a bid. It is also used as a summary recap to consolidate several priced work activities to a single cost item for a particular group of tasks. In Chapter 19, we will see how this number is then entered onto the estimate summary form to complete the estimate. Figure 10.1 is a flow chart that illustrates the recapitulation process.

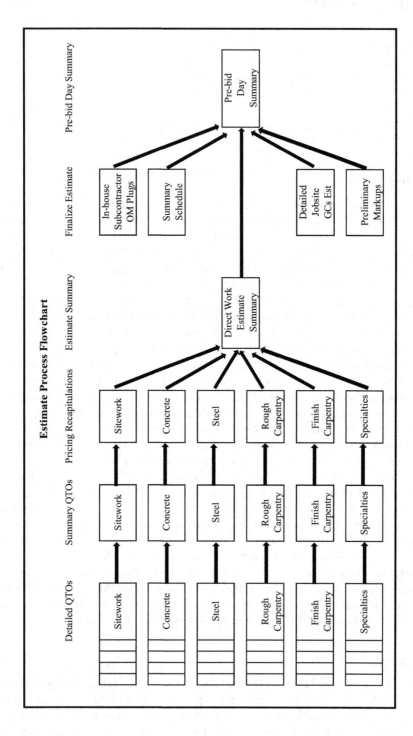

Figure 10.1 Estimate process flowchart

The recap sheet is a relatively standard estimating form. The estimator is encouraged to always fill out the title block information on each estimate worksheet. Starting at the left, the first column on the recap sheet is for a cost code. It can also be used for custom information, such as a drawing or addendum reference. The advantage of the estimator completing the cost codes now is it facilitates a smooth transition into cost control after the project is secured. This column may be used to reference a particular section or detail from the drawings or designate crew type.

The description column on the quantity sheet is used to express the item being quantified. On some recap sheets, the tasks shown will be the same as the column headings on the quantity sheet. For example, on the QTO sheet for the footings (Figure 6.1), the columns were fine grade, formwork, concrete, and reinforcing steel. Excavation and backfill could have also been included on this QTO, but instead the estimator grouped all foundation excavation and backfill on its own QTO sheet. Figure 10.2 shows a completed recap for the interior spot footings. Note that the line item descriptions basically are the same as the column headings on the quantity sheet. Additional pricing line items, or rows, have been added to the recap sheet, as these quantities were similar to other measured and calculated totals. The two columns following the description are the quantity and unit columns. The quantities to be entered are the double underlined totals from the quantity sheet, and the units are always shown. A live standard recap template is included on the book's eResource.

A consistent procedure should be used when transferring information from the quantity sheets to the recap sheet to minimize the chance of errors. When the quantity from the QTO is transferred to the recap sheet, the estimator should make sure that the number has been entered on the recap sheet exactly as it is shown on the quantity sheet, along with the correct units. Common errors include transposing numbers within the quantity being transferred, slipping a digit, or utilizing different units to be priced.

Once the quantity shown on the recap sheet has been verified that it is exactly the same as it is on the quantity sheet, the double underlined total on the quantity sheet is circled. This indicates to others that the checked total has been used on the next sheet in the sequence, in this case the recap. Time constraints as bid day gets closer may affect the thoroughness of checking.

The estimator is encouraged to group only 'like items' on a recap sheet. Figure 10.2 shows the interior spot footings and may eventually include the perimeter footings and grade beams as well. The totals then will reflect the man-hours and the labor and material costs for all of the building footings. The slab-on-grade (SOG), concrete columns, foundation walls, elevated slabs, composite slab, and miscellaneous concrete will each be on separate recap pages. Totals from all of these pages will be consolidated onto a summary recap that will then represent the cost for all cast-in-place (CIP) concrete work. Later, the CIP concrete may be entered as a line item on the estimate summary sheet.

Productivity factors versus labor unit prices

Labor costs can be determined by using either labor unit prices or a combination of productivity factors and wage rates. Labor unit prices are expressed as dollars per unit of the quantity being priced, for example, dollars per square foot, and are subject to revision due to a change in the wage rate, fringe benefits, and/or labor tax rates. Productivity, on the other hand, is expressed as man-hours per unit and is constant over time for a given work task.

Figure 10.2 shows columns for both unit man-hours (UMH) and labor unit prices. Labor costs are calculated by entering the unit man-hours and the current wage rate

City Construction Company
Cost Recapitulation Sheet

Project: Dunn Lumber
Location: Seattle, WA
Arch./Engr.: Flad

Division 3 Concrete

Date: 6/10/21
Estimator: P. Jacobs
Estimate #: 1

Code/ Det	Description	Qty	Unit	UMH	Man Hours	Wage Rate	Unit L Cost	Labor Cost	Material/Equipment Unit Cost	Material/Equipment Cost	Total Cost
	Spot footings:										
23440	Structural excavation	3,168	tcy	0.01	32	35	0.35	1,109	5.00	15,840	16,949
31430	Form footings (4 use)	9,475	sfca	0.09	815	41	3.53	33,409	1.00	9,475	42,884
	Fine grade/Hand Ex/Bf	9,086	sfca	0.02	182	35	0.70	6,360	0.20	1,817	8,177
	Reinforcing steel	22.3	ton	15.00	335	43		14,384	1,200.00	26,760	41,144
	Purchase concrete (1)	957	cy				645		105	100,485	100,485
	Pump concrete	957	cy						15.00	14,355	14,355
	Place concrete	957	cy	0.30	287	35	10.50	10,049	0.05	48	10,096
	Rod off concrete	9,086	sf	0.01	91	34	0.34	3,089	0.00	0	3,089
2315	Backfill	1,924	tcy	0.01	17	35	0.32	606	1.50	2,886	3,492
2020	Dispose of excess excav matl (export off-site)	1,244	tcy	0.01	15	35	0.42	522	15.00	18,660	19,182
	Nails, form oil, etc. (2)	9,475	form $						0.05	474	474
	Reinforcing steel accessories	22.3	tons						100	2,230	2,230

Notes:
(1) includes 5% waste
(2) 5% of form material cost

					Man Hours			Labor Cost		Material/Equipment Cost	Total Cost
Page Totals (to concrete summary)					1,773			$69,528		$193,030	$262,557
Add labor burden @ 55% of labor										$38,240	$38,240
Total Cost											$300,798

Wage Check: $/Hour:	Checks?	Yes	$/Hour:	$39.22
Assembly Check:	Checks?	Yes	$/CY:	$314

Figure 10.2 Spot footing cost recapitulation sheet

without fringe benefits or labor taxes. The unit man-hours are multiplied by the quantity to get the man-hours, which in turn are multiplied by the wage rates, the result being the labor cost. Labor cost can then be divided by the quantity to get the unit labor cost, which can be used as a check against historical information. Using current wage data, this process eliminates the need to make adjustments in labor costs over time. Fringe benefits and labor taxes are accounted for at the end of the estimating process and are discussed later in this chapter.

An estimator should always calculate the unit labor cost and use it as a checking tool. For example, if unit man-hours times the wage rate equals a unit labor cost of $10 per board foot, and $0.10 per board foot is more in line with historical costs, the estimator knows there is an error; in this case a digit (or two) was misplaced.

Productivity

Productivity is the time that it takes a person to do a unit of work. Examples of productivity factors are man-hours per square foot of contact area (SFCA) for concrete forms or man-hours per ton of steel. They are entered in the UMH column on the recap sheet. When multiplied by the quantity, the result is the estimated man-hours to complete the work activity. Most published estimating references show productivity factors and will delineate the crew that does the work and the daily output for the crew. The productivity in man-hours per unit and the crew makeup are important. Daily output is used in planning and scheduling after the project has been won. Published databases such as *RS Means* have productivity factors, as well as *The Guide*, which has agreed to allow students electronic access to their pricing for free at www.bestconstructionsite.com with the username "uw" and the password "class", both without quotation marks.

Some contractors record cost information based on unit labor costs and do not bother with productivity factors. Cost accounting systems in use today can determine historical productivity factors and provide the estimator with a wealth of information. The practice of using productivity factors helps to produce good estimates and reduces the amount of work by eliminating the need to adjust for inflation. Consider the following:

Quantity × Unit labor cost = Total labor cost
Which must be adjusted for inflation for future use, whereas
Quantity × Productivity factor × Wage rate = Total labor cost
For future use, only the wage rate needs to be updated

Adjusting for inflation can be done, but judgment must be exercised to estimate how it has affected construction costs. Conversely, updating a wage rate does not require judgment.

Wage rates

Not all construction participants receive the same compensation or wage, whether in the office or in the field. Every specialized craft is different, and all receive different wage rates. Union craftsmen generally receive a higher wage than do merit-shop craftsmen, primarily because of the time they served in apprenticeships. But even merit-shop crafts receive different wages for different scopes or activities. Some estimators may use a blended wage rate, as discussed later, and although that is easy and often acceptable for

Table 10.1 Craft wage rates

Construction Craft	Base Wage	1.5 Overtime	2 Overtime
Carpenters	$41	$62	$82
Cement Masons	$44	$66	$88
Electricians	$45	$68	$90
Equipment Operators	$40	$60	$80
Ironworkers	$43	$65	$86
Laborers	$35	$53	$70
Plumbers and Pipefitters	$48	$72	$96

budget estimates, it can be risky for detailed competitive bid projects. Every estimator should recognize the scopes of different crafts and their different respective wage rates. Three basic elements that make up the cost of labor are wages, fringe benefits, and payroll taxes. Other elements that may be part of the labor cost are travel, per diem, and overtime.

Wages are paid based on hours worked, with straight time generally being the first eight hours worked per day up to 40 hours per week. Overtime (OT) comes generally in two multiples, time-and-a-half (wages × 1.5) or double time (wages × 2). All time expended over the base wage rate is classified as premium pay in that it represents the premium cost of working a person beyond the agreed-upon normal pay period. Table 10.1 is an example of the wage rates for several union crafts at the location of our case study project. Payroll taxes and labor benefits will be added to this table later in this chapter.

We are using only straight-time wages in this book. This is the approach most estimators take and utilize OT only when needed, often at the discretion of the project superintendent. If a contractor feels that working OT is the only way it can achieve the owner's schedule and avoid liquidated damages, it might be a good option to pass on the bid opportunity. We also will utilize only the base wage rates on the pricing recap sheets, and the labor burden is then calculated from the total direct labor cost and inserted as a line item on the estimate summary page. Some contractors prefer to combine wages and labor burden for a loaded wage, as discussed later.

Labor productivity in UMH and wage rates has been used on Figure 10.2 to determine the labor cost of each line item. The estimator needs to determine which wage rate is the correct one to use for activities and/or composite crews. Some will pick a rate that is representative of all work and use it throughout the entire estimate. A better approach is to determine a composite wage rate based on the type of work being done. An estimator should not use apprentice wage rates for any pricing, unless specifically requested by the project owner and permitted by the union. Foremen are generally paid $1 to $2 per hour more than a journeyman. The lower wage of apprentices balances more or less with the increased wage of the foremen, such that using the journeyman wage as an average is standard in the industry. Where a crew includes a single foreman and a crew of three or four journeyman of the same trade, the composite wage rate is relatively easy to determine. When it comes to a concrete placing crew, several trades are involved, such as carpenters, laborers, cement masons, ironworkers, and maybe an operating engineer and a surveyor, and the calculation is more complex. It is often the easiest and relatively accurate to use the journeyman rate of the one craft that is most dominant for each task, such as:

- Laborers for earthwork, placing concrete, and cleanup activities;
- Carpenters for forming concrete and rough and finish carpentry;
- Ironworkers for installing rebar and erecting structural steel; and
- Cement masons for finishing concrete flatwork.

Carpenters are often the dominant craft employed by a GC, and the journeyman carpenter's wage is a fairly good average to use if the estimator is looking for a placeholder or shortcut in the labor pricing effort – but shortcuts are risky, as discussed elsewhere. This would be acceptable for budget estimating but not bid estimates, which may contain a substantial amount of self-performed work. A check that the estimator should make after pricing the direct labor for a system or assembly is to compute the wage rate for the entire recap sheet. This is done by dividing the total labor cost by the total man-hours. The result should approximate that of the various wage rates used on the sheet. If there is a wide variation, the estimator knows that there is an error in the labor cost calculation and must devote the time to find it.

Labor burden

Direct labor costs contractors much more than just the wages the craftsmen and administrative supervisors receive on their paycheck. There are several add-ons to the cost of labor, whether that be indirect labor, including the project manager and superintendent, or direct craft labor such as carpenters and laborers. Contractors pay an additional markup or percentage add-on on top of all of the wages they pay; this is known as labor burden. Labor burden is not a fee or profit markup, but it is a direct cost of doing work. The amount of this markup is not established at the jobsite level by project managers or superintendents; rather, labor burden is determined at the chief financial officer and chief executive officer (CEO) level. The burden is usually a journal-entry charge from the contractor's home office accounting department to the jobsite and is not accompanied with a separate invoice. Project managers, superintendents, and estimators do not typically have any direct input to the amount of labor burden estimated for or charged to a project. Labor burden has two major components, labor taxes and labor benefits, as reflected in the following equation:

$$\text{Labor burden} = \text{Labor taxes} + \text{Labor benefits}$$

The combination of labor taxes, or payroll taxes, and labor benefits, or fringe benefits, is commonly referred to as labor burden. This burden can vary widely from one trade to another, and the estimator can spend an inordinate amount of time calculating them for each crew. It is common to determine an average labor burden based on the overall makeup of all tradesmen employed by the company. This average can then be used for most estimating purposes.

Required labor taxes

Some members of the built environment community will refer to all of labor burden as either labor taxes or labor benefits, but they are distinct and have different costs and rates for different types of labor. Labor taxes are also known as payroll taxes. Labor taxes are government determined and have at least four major elements that contractors are required to pay; this includes:

- *Social security* was created by the Federal Insurance Compensation Act (FICA). The employer pays one half or approximately 6.2% up to the first $133,000 of wages (changes yearly) for FICA contributions and the employee pays the other half as a withholding from their weekly check, for a total tax of 12.4%.
- *Medicare* is also a joint contribution from the employer and the employee similar to FICA and amounts to approximately 1.45% each for a total of 2.9%.
- *Unemployment tax* has two elements, federal and state:
  - Federal unemployment tax costs 6% on the first $7,000 of income, but much less if there is a state unemployment tax; and
  - State unemployment tax varies by state. The amount of unemployment tax percentage paid by any company is proportional to the amount of unemployment claims they experience from personnel lay-offs.

- *Workers' compensation insurance* markups, or *workers' comp*, vary considerably due to a variety of factors, including the safety record of the contractor and its associated experience modification rate (EMR). In addition, the potential safety risk of the labor craft and differences between indirect and direct labor will impact workers' comp. The baseline EMR is 1.0. Contractors which have a higher incident rate of safety accidents have an EMR rate greater than 1.0, and those with fewer accidents a rate below 1.0. Some crafts are more prone to accidents, and they will have a higher workers' compensation rate. Indirect salaried employees have a much lower chance of a safety accident and therefore have a much lower workers' comp rate than a direct work craftsman.

Labor benefits

Labor benefits are also known as fringe benefits; they are determined by the contractor and include a variety of items. These are not 'taxes' per se but are voluntary contributions. Labor benefits are a significant portion of the total labor burden that is attached to wages. If contractors are signatory to labor unions, they will likely pay higher labor benefits than do contractors that employ merit-shop labor. Some items that may be included with labor benefits include:

- Health insurance,
- Dental insurance,
- Eye insurance,
- Disability insurance,
- Life insurance,
- Union dues,
- Pensions and retirement,
- Use of company cars and cell phones,
- Vacation,
- Sick leave,
- Bonus,

- Education and training,
- Safety add-on,
- Supervision add-on, and
- Small tools add-on.

Some labor benefits, such as medical insurance, are a shared cost between the employee and the employer. Depending upon contract terms and the definition of reimbursable costs on open-book projects, some contractors may include more or less of these potential labor benefits in their labor burden rates. Benefits are paid based on each hour worked and are not subject to the effects of premium time.

Combined labor burden

There are significantly different labor burden rates for direct craft labor (carpenter and electrician) than for indirect labor (project manager and superintendent and CEO). In addition, different crafts or trades have different rates depending on the type of work and associated safety risk; ironworkers and electricians are subject to more safety incidents than painters and landscapers. There are a variety of union issues that also affect labor burden; for example, some trades do not provide their own tools – the contractor does. Subcontractors are responsible to pay their own labor burden; the GC does not get involved with subcontractor burden. Table 10.2 includes several trades' wage rates and their associated labor taxes and labor burdens, which add up to a combined loaded wage rate.

It would be very cumbersome for a construction company to invoice or journal entry each jobsite's different labor burden rates for each type of craft or administrative labor category. Most contractors will develop a 'blended' burden rate at the beginning of the year, which is based on labor mixes from the prior year. A blended burden rate for a general contractor that is signatory to the carpenter, laborer, ironworker, and cement finisher unions might be 55% or higher. Contractors that utilize merit-shop labor will have a lower total labor burden percentage markup due to a mix of fewer labor benefits that were listed in Table 10.2, maybe 30%. Usually direct and indirect labor is kept separate, as the burden rate of indirect labor is much less than direct (maybe 35%), but even these may be blended on certain projects, again depending on contract terms. In addition, some contractors will invoice their clients on open-book projects a 'loaded' wage rate,

Table 10.2 Craft labor burden

Construction Craft	Base Wage	Labor Taxes	Labor Benefits	Total Burden	Loaded Wage
Carpenters	$41	$8	$13	$21	$62
Cement Masons	$34	$6	$11	$17	$51
Electricians	$45	$9	$14	$23	$68
Equipment Operators	$40	$10	$12	$22	$62
Ironworkers	$43	$12	$19	$31	$74
Laborers	$35	$8	$11	$19	$54
Plumbers and Pipefitters	$48	$11	$17	$28	$76

which includes the base rate plus a blended burden rate. Contractors may choose to do this because a loaded wage reflects the total cost of an hour of direct labor, but it is difficult to substantiate during an audit of an open-book project. The loaded wage rate for laborers in Table 10.2 would therefore be $54 per hour. Some contractors include other markups with labor burden such as liability insurance, but these are volume-dependent and not labor-dependent and would not be accurate on projects with a different mix of direct labor versus subcontracted labor.

Construction projects that receive federal funding require contractors to pay direct craft employees a 'prevailing wage rate', which is the wage rate most common to the area being worked plus labor benefits. This is also known as the Davis-Bacon wage rate. The prevailing wage rate for carpenters in Table 10.2 would therefore be $54 per hour. It is up to the contractor whether they want to contribute additional labor benefits beyond the prevailing wage rates. Labor taxes are mandatory percentage add-ons whether the labor wages are union, merit shop, or prevailing wages.

Labor burden is charged on a contractor's own direct and indirect labor only. Labor burden is not added to material costs, equipment rental, or subcontractors. Subcontractors are required to factor their own labor burden within their bid prices to the general contractor.

Occasionally the estimator will encounter a project in a remote location. The workforce has to travel to the site and stay there for certain periods of time. The travel and per diem costs have to be included in the bid. These, like the labor burden, are shown as a line item on the bid summary.

Summary

The direct work pricing recap process is the listing and pricing of the items that have been quantified. Some work activities, such as CIP concrete, require summary recap pages, which consolidate the totals of several recaps into a single total. This total is then entered as a single line item on the bid summary. During the recap process, quantities are transferred from the QTO sheets to the recap sheet. This operation can be a source of errors, and the estimator needs to be diligent in following procedures for reducing them. Circling the totals on the QTO after they have been transferred eases the checking process later on.

There are two ways to use the wage rates in determining labor costs. The estimator can develop an average wage for all work of a class of activities and use it throughout those activities, or he or she can develop a composite wage for crews used on particular activities. While the latter may be slightly more accurate, the average wage method is sufficient for most estimates and is easier to use.

Payroll taxes are government-mandated taxes and are based on the amount of total wages paid. They consist of FICA, state unemployment insurance, federal unemployment insurance, and workers' compensation insurance. Labor taxes are calculated as a percentage of the actual wages paid. It is important for the estimator to know how the fringe benefits and payroll taxes are calculated, because they will affect the labor burden percentage used later on the bid summary sheet.

Contractors have some flexibility in their calculation of total labor burden and how it is used in estimating and job cost accounting. This flexibility is sometimes utilized differently on closed book competitively bid lump sum projects compared to open-book negotiated projects. In this book we are recommending that the estimator only use bare wage rates on the pricing recap sheets, and the labor burden is added on the estimate summary page as a percentage markup to labor. Labor burden is discussed again in Chapter 18, with other markups.

Review questions

1 What is the purpose of a recapitulation sheet?
2 What is a productivity factor, and what are its units?
3 What can a wage rate check indicate?
4 What is one of the best sources of information for pricing and productivity factors?
5 Which craft in Table 10.2 has the highest: (A) bare wage rate, (B) total labor burden, (C) percentage markup for labor burden, and/or (D) loaded wage rate?
6 Which person costs the contractor a higher percentage of labor burden, the CEO or the carpenter foreman?
7 What is the difference between labor taxes and labor benefits?
8 What is the difference between a loaded wage rate and a blended wage rate?

Exercises

1 Assuming five craftsmen for each of the crafts in Table 10.2, (A) what would the total blended labor burden percentage markup be? (B) What would happen to that rate if you eliminated the craft with the highest burden percentage and doubled the manpower from the craft with the lowest burden percentage? (C) What would happen to the original rate if you eliminated the lowest burdened craft and doubled the manpower from the highest craft?
2 If contractors charge a blended labor burden rate to a client and alter their craft mix as discussed in Exercise 1, what would that potentially do to their profit?
3 If a contractor charges the blended labor burden rate you calculated from Exercise 1.A to a client but applies it to both direct and indirect labor, how would this affect their profit?
4 What customarily happens to labor burden rates journal-entered to a jobsite after employees top $133,000 in wages paid in one calendar year? Is this ethical? How does this affect project profits? How does this affect corporate profits?

11 Concrete pricing

Introduction

In the last chapter, we discussed a contractor's direct labor productivity and wage rates. In this chapter, we price the material costs of all of the cast-in-place (CIP) and pre-cast concrete systems that were quantified in Chapters 6 and 7. The best source of material pricing for any estimate is current market pricing. Second would be a contractor's in-house historical database. The third choice would be to utilize material unit pricing from published databases that combine averages for all types of construction and in all geographic locations.

In this chapter, we build on the foundation and slab-on-grade quantities developed in Chapter 6 and discuss all of the other typical CIP concrete systems an estimator may come across, many of which are included in the Dunn case study project. In addition, we price the tilt-up quantity take-off (QTO) from Chapter 7 and develop a comprehensive pricing recapitulation (recap) sheet for that unique method of construction. After all of the pricing is complete, the estimator may want to combine CIP concrete systems onto one summary recap and another for pre-cast systems before forwarding to the estimate summary sheet. Pricing for structural steel and carpentry and other GC direct work is discussed in the next chapter.

Material pricing

Material pricing can be straightforward when properly done on the pricing recap sheet. The material unit price is entered in the column headed 'Unit material cost' (Unit M Cost). It is then multiplied by the quantity to get the total material cost in dollars. An estimator must ensure that the units of the unit price are the same as those of the quantity. It seems self-explanatory, but multiplying 1,000 square feet (SF) of plywood (versus sheets) times $30 per sheet of plywood (versus SF) will result in a significant estimating error – but these types of errors happen.

Unit prices can be obtained in several ways. One way is to use published estimating reference data. Most estimating references are published annually, and their prices have been updated during the previous year. They are representative of a nationwide average and may not reflect the prices of a particular location. The estimating student will typically use this approach, as they do not have access to other prices. One such published database is *The Guide*, which has agreed to allow students electronic access to their pricing for free at www.bestconstructionsite.com with the username "uw" and the password "class", both without quotation marks.

A second source is from historical data maintained by the contractor. Data may come from cost reports of previous jobs or from recent estimates. The age and verifiability of these prices must be kept in mind if the estimator decides to use them. Prices that are a year or more old should be, at a minimum, adjusted for inflation. The third and best way to obtain market unit prices is to solicit them from vendors. The estimator should describe to a supplier what the job is and when it is expected to be performed. As a result, the vendors will usually honor their pricing throughout the duration of the work.

An estimator must understand that some material prices can be solicited during the estimating process, but others will not be received until bid day. Vendors that are major material suppliers, such as concrete reinforcing steel (rebar) and structural steel fabricators, will submit their prices as complete bids on bid day. Those materials that are priced by the unit, such as concrete, lumber, gravel, etc., can be priced by calling the vendors during the course of the estimate preparation. In some cases, bid prices also may be received from them on bid day, which may require the estimator to make a last-minute adjustment to the overall bid.

Concrete foundations

Concrete is priced by the cubic yard delivered to the jobsite. The vendor needs to know the concrete specifications and the approximate quantity of each type being used. For example, the case study specifications call for 3,000 pounds per square inch (psi) concrete for footings and slab-on-grade (SOG), 4,000 psi for foundation walls and elevated slabs, and 6,000 psi for columns. Exterior paving is to be air entrained to 6%, and maximum slump for all concrete is 4 inches. The project manual should be checked to see if superplasticizers or other additives that aid in the placing, finishing, or curing can be used. The estimator should send the specifications to concrete suppliers and then follow up for pricing.

Concrete formwork is typically the most labor intensive and therefore deserves attention from the estimator. There are different ways to build forms depending on the type and size of the concrete elements being constructed. General construction firms generally have historical records that provide data on the cost of the material per square foot of contact area (SFCA). This cost includes all lumber and sheet material, bracing, kickers, and form ties. A concrete subcontractor would get into more detail. Items such as nails and form oil may be priced separately as a percentage add-on.

The forming is typically the most expensive of all concrete tasks for many elements, such as walls, elevated slabs, and elevated beams. A contractor, in an effort to reduce this cost, will buy form material that can be used more than once. For example, forming material will be purchased for only one-fourth of a long wall section. After the first placement, the forms will be stripped, cleaned, patched, and reused. Since the forms do not have to be built from scratch, a major cost reduction is realized in both material and labor. It is common to use forms up to four times, and on occasion five or six times. When pricing form material, an estimator must have a good idea of how many times the forms will be used.

Using unit costs for multiple uses of foundation forms will provide reasonable savings over single use. The same is true for many elevated structures, such as beams, columns, and structural slabs. But the estimator is usually not the 'builder', and he or she should discuss the planned means and methods, including quantity of concrete pours and form reuse, with the project superintendent. A good rule of thumb is if a contractor can get more than four uses out of the same form material, they are making money on that aspect.

Prefabricated metal forms can be rented for some common shapes, such as columns and beams. On large concrete construction projects, form costs can be reduced by having custom metal forms made and using them an indefinite number of times. Prefabricated forms are usually made of steel and are strong enough to minimize any extra bracing that might be required for wood forms. If prefabricated forms are more economical, the estimator needs to obtain a quote from a form-rental supplier.

Occasionally the specifications will prescribe how long forms are to be left in place. Most superintendents remove forms as quickly as possible so they can be reconditioned and set up for the next concrete placement. The estimator must be aware of anticipated form use and concrete curing requirements before pricing them.

Other material prices that need to be determined are fasteners, bonding agents and curing compounds, rebar chairs and tie wire, lifting inserts, and any other similar items. For nails and bonding agents, the estimator should use a percentage of the total form material cost, showing it as a line item on the recap sheet. Curing compound is applied only to flat slab surfaces, elevated slabs, and tilt-up panels. It is priced by the area of coverage. Reinforcement steel tie wire and chairs are classed as rebar accessories and are also calculated as a percentage of the rebar costs or an allowance of $/ton of rebar.

One other item the estimator needs to consider is any hoisting equipment that may be required to lift forms into place. Some situations will allow the forms to be handled by a jobsite forklift. Multi-story projects will require a crane to lift the forms and set them. If a tower crane is to be used, it will be priced separately and does not need to be included in the concrete work. But if a 20-ton crane is to be brought onto the jobsite just for handling the forms, its cost needs to be included in the concrete section of the estimate.

Applying pricing factors

In illustrating how the pricing recap sheet is used, reference will be made to the quantity sheets developed in Chapters 6 and 7. Referring ahead to Figure 11.1, the application of the pricing factors is shown in the Unit M Cost column. The estimator should make sure that the units of the unit prices match those units shown after the quantity unit column from the QTO sheets.

The unit prices and quantities are extended (multiplied) to determine material cost of the line item. Note that the figures to the right of the decimal point are insignificant and not shown. There are some spaces in the Unit M Cost column for which no cost is entered. A zero or dash may be entered so that others will know that there is no material rather than wondering if a cost was accidentally omitted.

Two line items are shown that were not quantified on the quantity sheets. These are accessory items that are priced by calculating them as a percentage of some other cost. Nails and form oil are noted as 5% of the form material costs. The second accessory line item is miscellaneous material needed for installing the reinforcing steel. This cost covers the purchase of the tie wire, chairs, dobies, or other supports specified to give the proper spacing to the bars. An allowance for these is calculated as $100 per ton of reinforcing steel material. The actual price for the reinforcing steel will not be known until bid day, but it has been quantified by weight; thus, a simple extension will determine a cost. Unless there is a wide variation in the weight shown on a supplier's bid, an adjustment in the calculation of the reinforcing steel and accessories on bid day will not be necessary. This then becomes part of the buyout process discussed in Chapter 24. Additional line items appear on the pricing recap, as their quantities were similar to other measured quantities on the

City Construction Company
Cost Recapitulation Sheet

Project: Dunn Lumber
Location: Seattle, WA
Arch./Engr.: Flad

Date: 6/10/21
Estimator: P. Jacobs
Estimate #: 1

Division 3 Concrete

Code/ Det	Description	Qty	Unit	UMH	Man Hours	Wage Rate	Unit L Cost	Labor Cost	Material/Equipment		Total Cost
									Unit Cost	Cost	
	Slab on grade:										
23440	Fine grade	56,160	sf	0.002	112	35	0.07	3,931	0.05	2,808	6,739
	Visqueen	61,776	sf	0.002	124	35	0.07	4,324	0.03	1,853	6,178
31430	Edge form	960	LF	0.06	58	41	2.46	2,362	1.00	960	3,322
	Blockouts	78	EA	1.00	78	41	41	3,198	50	3,900	7,098
	Expansion joint	2,400	LF	0.06	144	41	2.46	5,904	1.00	2,400	8,304
	Construction joints	600	LF	0.06	36	41	2.46	1,476	1.00	600	2,076
	Slab rebar	25.5	tons	12.00	306	43	516	13,158	1,200	30,600	43,758
	Purchase concrete (1)	721	cy						105	75,705	75,705
	Pump concrete	721	cy						15	10,815	10,815
	Place concrete	721	cy	0.40	288	35	14	10,094	0.00	0	10,094
	Finish SOG	56,160	sf	0.01	562	34	0.34	19,094	0.00	0	19,094
	Sawcut control joints	4,000	LF	0.01	40	35	0.35	1,400	0.30	1,200	2,600
	Protect and cure	56,160	sf	0.001	56	35	0.04	1,966	0.05	2,808	4,774
	Nails, form oil, etc. (2)	7,860	form $						0.05	393	393
	Reinforcing steel accessories	25.5	tons						100	2,550	2,550
	Page Totals (to concrete summary)				1,804			$66,907		$136,592	$203,499
	Add labor burden @ 55% of labor									$36,799	$36,799
	Total Cost										$240,298
	Wage Check?	Yes							$/SFS:	$4.28	$/Hour: $37.10
	Assembly Check?	Yes									$/CY: $333

Notes:
(1) includes 5% waste
(2) 5% of form material cost

Figure 11.1 Slab-on-grade cost recapitulation sheet

QTO sheet. Figure 11.1 is a pricing recap for the SOG for our case study project. Pricing recap for the spot footings was included with Chapter 10. The other self-performed systems or assemblies will have a similar pricing format and process.

Concrete pumping was not a separate QTO item but is essential for most commercial construction. An average per yard pump cost of, say, $15/CY is easy to plug into the recap sheet, but as we indicate throughout this book, estimating shortcuts can be risky. The backup to the $15 is explained in detail in Chapter 13.

Cast-in-place concrete systems

Most of the processes we have discussed for extending quantities and adding labor productivity rates, wage rates, and market material unit prices for other CIP systems are similar to those utilized for the foundations and SOG. In this section we briefly describe many other CIP systems and the cost differences they may have that the estimator should pay attention to. In many cases different concrete systems have similar elements, such as:

- Structural excavation and backfill (for foundation systems),
- Fine grade or hand excavation,
- Formwork including layout,
- Reinforcement steel,
- Embeds, including anchor bolts,
- Purchase and pump and place concrete,
- Finish: Rod-off for foundations and trowel finish for slabs,
- Protect and cure, and
- Patch and sack exposed or finish surfaces.

Pile caps and grade beams are very similar to spread footings and continuous footings, as discussed previously. Because they often sit atop piling and support shear and bearing walls above, these concrete systems likely are specified with higher strength concrete than typical foundations, and they will have more rebar. Their sections or depths may also be deeper. But the process to perform the QTO and extend the quantities on the recap sheets is similar − it is just that the quantities may be greater and the unit prices may be slightly more.

As indicated, the strength of redi-mix concrete in psi varies with different concrete systems. *Columns* typically have higher strength concrete, in this case 5,000 psi, and the purchase cost will be greater. The estimator should plug in as accurate of material pricing in the direct work as possible, and if significantly different supply prices per cubic yard (CY) are received on bid day, be able to make a pricing adjustment, or at a minimum be able to analyze the differences during buyout. A separate recap for purchasing concrete for our case study may look like the following example. If concrete prices varied by $10/CY on bid day, there would be a substantial change in estimated cost to purchase concrete.

Concrete redi-mix supply
2,497 CY of 3,000 psi concrete @ $105/CY = $262,185
7,963 CY of 4,000 psi concrete @ $110/CY = $875,930
267 CY of 6,000 psi concrete @ $120/CY = $32,040
Total material cost included with direct work estimate: $1,170,155

Patching and sacking of finished concrete elements may be defined in the specifications, but it is often a judgment or allowance made by the estimator. Patching and sacking is a broad term which means to touch up the concrete, often with a concentrated cement and sand and water mix, either by troweling over the finished surface or applying it with a sponge. Gunny sacks were at one time the tool of choice for 'sacking'. If patch and sack is required, the estimator will estimate sacking using $/SF of surface area. Concrete elements typically considered for patching and sacking are CIP columns, *foundation walls*, and tilt-up walls, which are discussed later. Concrete that will be covered with other building elements, such as studs and sheetrock or suspended ceilings, are not typically touched up after stripping the forms. Exposed architectural concrete will require snap tie holes to be filled, concrete fins left between joints in forms need to be ground, and rock pockets must be filled. Significant rock pockets may require review by the structural engineer, even if the concrete is later covered with other building elements. The estimator may assume a percentage requiring touch-up, say 50%, or may conservatively estimate 100% of the surface area of all concrete left exposed should be patched and sacked. Bid jobs are likely the former allowance and negotiated projects the latter. The estimator for City Construction Company approached material cost patch and sack as indicated here. Labor was plugged at 0.025MH/SF and figured with the cement mason's wage rate.

Patch and sack material recap
20,000 SF of columns @ 50% @ $0.50/SFC = $5,000
106,000 SF of walls @ 50% @ $0.50/SFW = $26,500
Total material cost: $31,500

As mentioned with QTO, elevated concrete structures must be 'shored' in addition to formed until the concrete has cured to its design strength. This includes *beams, drop caps,* and *elevated CIP slabs.* After the forms are dropped from the concrete, often with assistance of small forklifts, the slabs will be 're-shored' with fewer shoring poles laid-out on a predetermined grid. Although this is somewhat a means and methods process, the GC will typically employ a structural engineer to calculate shoring requirements. Shoring is a cost that must be included with the recap sheets, which is usually calculated on a $/SF of form area basis.

Composite concrete *slabs on metal deck* differ from elevated slabs because the metal deck is essentially the temporary form for the wet concrete but remains with the concrete. The metal deck will also require shoring and re-shoring, similar to the previous process. Structural steel beams that support the metal deck often have lugs that are welded to the beams and penetrate through the deck and are embedded in the concrete. The supply cost of the deck is typically $/SF or $/hundred SF and is included with CSI division 05 pricing recaps.

Supply and installation and tensioning of *post-tension cables* in elevated concrete slabs are often a subcontract specialty. A plug estimate may be included by the GC on a $/pound basis within the direct work or carried on the subcontractor list in hopes of receiving quotes on bid day. *Shotcrete* is also a subcontract specialty and will be priced based on $/SF of shoring wall. The GC may be involved in supportive activities of the shotcrete subcontractor with scaffold or cleanup and, if so, will include these costs in its direct work estimate.

There are many other *miscellaneous concrete systems* that will likely warrant their own QTOs and recaps and a summary recap, as discussed later. Some of this concrete may be shown on design drawings other than the structural drawings, such as civil, architectural, and mechanical and electrical. But regardless of where the work is shown, it is still in the GC's scope and contract with the project owner. All of these systems have similar makeup (forms, rebar, embeds, concrete, finish) and will be estimated similar, but because their quantities may be less, the unit pricing may be higher. Some of the miscellaneous concrete systems include:

- Cast-in-place concrete stairs,
- Curbs for mechanical and electrical equipment,
- Housekeeping pads for mechanical and electrical equipment,
- Site concrete including curbs and walks and pavement, and
- Architectural concrete.

Early in this book we discussed that the estimating team should decide what work they intend to construct with in-house craftsmen and what work they will subcontract out. There is much more estimating and cost control risk associated with self-performed work over subcontracted work, and the estimator will spend his or her time appropriately. If a contractor has an option to self-perform or subcontract out, they should estimate the work both ways to give them flexibility on bid day. But this approach requires a lot of work and accuracy up front and careful attention to complete scope analysis on bid day. The conservative approach would be for the estimator to figure it all self-performed, receive subcontractor quotes on bid day but place them aside, and if successful with the bid go back into the estimate and develop a complete assembly price that can be compared with the subcontractor quote. If it makes sense considering cost, schedule, quality, and safety, then during buyout the GC may choose to award the work to a subcontractor. An example of this analysis is included in the box that follows. Some of the concrete work that may be self-performed or subcontracted includes:

- Structural excavation and backfill using equipment such as a backhoe,
- Formwork and shoring for elevated beams and slabs,
- Rebar placement,
- Post-tension cable installation and stressing,
- Slab finishing, and
- Supply and installation of complete concrete systems such as beams and elevated slabs.

Form elevated CIP slabs

(1) Direct work:
    Labor: 112,346 SF @ 0.2 MH/SF @ $41/HR @ 1.55 labor burden = $1,427,918
    Material: 112,345 SF @ $1.00/SF = $112,345
    Equipment: 2 hard-tired forklifts @ 2 months each @ $1,500/mo. = $6,000
    Total direct estimate: $1,546,263, or $13.76/SFS, or
(2) Subcontractor pre-bid day unit price budget of $15/SFS, or
(3) Subcontractor lump sum quote received 30 minutes before the bid was due: $1,350,000

Pre-cast concrete tilt-up walls

The pricing recapitulation of concrete tilt-up panels will be similar to that of other concrete elements except that it will also include the hoisting, shimming, bracing, and grouting of the panels. Figure 11.2 is an expanded pricing recap sheet for the panel samples

City Construction Company
Cost Recapitulation Sheet

Project: The Training Center
Location: Kent, WA
Arch./Engr.: LM
Division 3 Concrete

Date: August 18, 2021
Estimator: P. Jacobs
Estimate #: TTC-1

Code	Description	Qty	Unit	UMH	Man Hours	Wage Rate	Unit L Cost	Labor Cost	Material/Equipment Unit Cost	Material/Equipment Cost	Total Cost
	Tilt-up Wall Panels (30 EA):										
	Prepare pouring slab [1]	25,220	sf	0.00	101	35	0.14	3,531	0.05	1,261	4,792
	Form wall panels	6430	lf	0.06	386	41	2.46	15,818	1.50	9,645	25,463
	Reinforcing steel	33	tons	0.03	1	43	1.16	38	1,200	39,600	39,638
	Embeds	240	ea	0.25	60	41	10.25	2,460	25.00	6,000	8,460
	Hoisting anchors	240	ea	0.25	60	41	10.25	2,460	22.25	5,340	7,800
	Purchase concrete	581	cy						120	69,720	69,720
	Pump concrete	581	cy						15.00	8,715	8,715
	Place concrete	581	cy	0.50	291	35	17.50	10,168			10,168
	Finish panels (steel trowel)	25,220	sf	0.02	378	34	0.51	12,862			12,862
	Curing compound	25,220	sf	0.00	42	35	0.06	1,483	0.15	3,783	5,266
	Hoist and set panels	30	ea	1.00	30	40	40	1,200			1,200
	Braces	90	ea/mo						12.00	1,080	1,080
	Shims	2670	lb						1.65	4,406	4,406
	Grout	30	cf	3.00	90	34	102	3,060	125	3,762	6,822
	250-ton crane:	15	hr						480	7,200	7,200
	in & out & mobilization	1	LS						10,500	10,500	10,500
	Detail lift points	30	pnls						75	2,250	2,250
	Weld joints: L, M, & EQ	247	lf	0.20	21	43	3.58	885	5.13	1,267	2,152
	Patch and sack panels	30	ea	0.75	23	34	25.50	765	70	2,100	2,865
	(1) Building SOG, includes cleaning, layout, and bond breaker										
					1,482			**$54,729**		**$176,629**	**$231,358**

Assembly Check: /sf of panel =   Checks? Yes $/SFP: $9.17
Wage Check: $/HR =   Checks? Yes $/HR: $36.93

Figure 11.2 Tilt-up wall panel cost recapitulation sheet

quantified in Figure 7.3. The same checking and circling procedure is used on the tilt-up quantity sheets as was done for the spread footings and SOG.

Published references provide pricing data on tilt-up wall panels. Some will give a detailed account of the items that make up the construction costs, and others will only give an average price per square foot of panel. If the estimator lacks a good sample estimate to use as a model for his or her work, it is worthwhile to find an estimating reference that provides details of the various costs. These details can then be used as a guide to obtain local pricing.

Many of the tasks required to construct tilt-up panels are similar to a slab-on-grade, and applying bonding agent is similar to a sprayed membrane curing compound. Edge forms are similar to SOG forms, and placing and finishing of the concrete are also similar to a SOG. Reinforcing steel is a little heavier, but labor productivity is based on weight, so this task is easily priced. This leaves only the inserts, hoisting, and setting the panels as the unknowns. Grouting is given in most estimating references. Backfilling in preparation for the closure strip will have been accounted for in the SOG section of the estimate. This analogy will aid in developing most of the costs of the tilt-up panels, and consulting with a knowledgeable superintendent or ironworker foreman will provide the rest.

Note that Figure 11.2 shows a detailed listing of the work required. The first item, 'Prepare pouring slab', has been noted to include the application of a bond breaker to the slab. This is done after the forms are in place and is the only material cost in the original slab preparation process. The form material was quantified in lineal feet similar to that of a slab-on-grade.

The number of hoisting anchors and their placement was determined at the time of the panel detailing. The estimator can determine how many will be needed and consult with a superintendent or ironworker foreman to determine what type of anchor is preferred. The estimator will also want to have a catalog of concrete accessories that shows hoisting anchors and their load data. He or she should also have some basic knowledge of structural properties.

To determine the number of lifting inserts in a sample panel, the weight of the panel is calculated by multiplying the concrete volume in cubic feet by a unit weight of 150 pounds per cubic foot. This results in a panel weight of approximately 85,950 pounds. Typical vendor data recommends that even though the sizing information incorporates a 2:1 safety factor, 33% should be added to allow for the extra tension required to break the bond between the panel and the forming slab. Adding 33% to the calculated weight results in a total weight of 114,305 pounds. An insert is selected and its capacity is divided into the total weight. Using a 16,000 lb. capacity face tension insert:

114,305 pounds ÷ 16,000 pounds per insert = 7.14 inserts required/panel

---

The estimator should use eight inserts for this panel. It is a good idea to have the vendor's sales engineer also verify these calculations and use the results only for pricing purposes. The final panel detailing may cause a variation in the number and type of inserts, but the cost difference is generally insignificant.

There are some items on the tilt-up recap that are unique to this particular type of construction. These include braces, shims, and grout. When a panel is installed, it needs to be leveled and set at the proper elevation. Tops of footings are designed to be slightly low so the panel can be set on shim packs. This does two things: (1) allows adjustment so the panel can be set to the proper elevation and (2) provides a space in which grout will be inserted so the panel will have full bearing on the footing. Once the panels are in place, they need to be braced until the structural steel and/or roof system is completed. A concrete-accessories vendor will provide shim packs, grout, and rent braces to the job. This vendor is a good source of pricing information. From a labor standpoint, the installation of the shims and all of the work involved with the braces is included in the panel hoisting and setting cost. Grouting is a separate operation and needs to be priced accordingly.

Near the end of the listing on the recap are two line items for the crane that will be used to hoist the panels. The weight of the heaviest panel is calculated including an allowance for breaking the initial tension bond between the tilt-up and the SOG. The estimator must either have someone who is knowledgeable about cranes and their operation select the crane required or must learn to use crane rating tables that are available from vendors who rent cranes. Crane capacity drops off very fast from their nominal rating the further the boom reaches out from the cab of the crane.

Some estimators may show equipment such as a tilt-up crane in the general conditions section of an estimate, but we recommend that wherever possible, equipment should be estimated and charged to the work. This is the approach behind activity-based costing, which is a function of lean construction. The crane for hoisting the panels is used only for that purpose and should be included with the recap that shows the tilt-up work. By contrast, if the building were constructed with a stucco finish or a brick veneer, the tilt-up crane would not be required. Also, the estimator must include the transportation to and from the site. There will also be a similar cost at the end of the work to prepare the crane for transportation back to its base. Some large cranes require an 'assist crane' to assemble the tilt-up crane on-site. The elements associated with tilt-up crane costs are discussed in detail in Chapter 13.

Two miscellaneous items need to be added to the tilt-up recap sheet. The first is the detailing of the lift points, which must be done by a qualified structural engineer. This cost is the responsibility of the general contractor and is to be included in the estimate. The estimator needs to find out whom his or her company regularly employs for panel detailing and have them provide an estimated cost for the work.

Another potential cost item is patching of the lifting and bracing points after the panels have been set and the structural steel erection has been completed. The patching is a cement-based material, and the work usually is done by the GC's own forces. Estimators will use a historical percentage for the patching work, but a conservative approach is to use 100% of the tilt-up surface area that will remain exposed when the project is complete. This rule can be applied to patch and sack of other exposed CIP elements such as walls and columns and beams.

When finished, the estimator must do the operational and accounting checks as shown for footing and SOG recaps. The assemblies check is done on a square foot basis for a given thickness of wall. The overall total of $9.17/SF is reasonable, but if excessively low or high, the estimator would need to go back and look for adjustments in the material and labor columns.

Summary recapitulation sheets

A summary recap is an extra sheet on which to consolidate several related recaps, as might be the case for the CIP concrete. The commercial estimate summary sheet that will be introduced in Chapter 19 is limited in the number of line items that can be used for direct work. Using a summary recap on some work allows more concise information to be entered onto the bid summary.

CIP concrete frequently is recapped on several sheets. There may be one for footings and foundations, one for floor slabs, one for CIP walls, and one for miscellaneous concrete, among others. The totals of these are listed as line items on a summary recap sheet and then summed to a bottom line total for all CIP concrete activities. These totals are then ready to be transferred to the bid summary as a single line item. Figure 11.3 represents a summary concrete recap as it might appear for the case study. Tilt-up panels are not part of the commercial case study project but would not have been included on the CIP concrete summary recap because of the significant hoisting and setting operations that are unique to that work.

It should be noted that the summary recap does not contain all of the information that is on a recap. It is only necessary to show the man-hours, labor cost, material cost, and the total cost. But total quantities of a system, such as cubic yards, are useful to include as well as $/hour and $/assembly checks. After checking for transposition errors, range summations, and an accounting crosscheck, the totals on the summary recap are double underlined to show that they are ready to be transferred to the bid summary.

Not all work activities will use a summary recap. Only those that require several recaps for sub-activities, such as the CIP concrete, structural steel, and possibly the rough

---

City Construction Company
Cost Recapitulation Summary Sheet

Project: Dunn Lumber   Date: June 10, 2021
Location: Seattle, WA   Estimator: P. Jacobs
Arch./Engr.: Flad   Estimate #: 1
CIP Concrete

Description	Qty	Unit	Man Hours	Labor Cost	Material Cost	Total Cost
Interior Spot Footings	957	CY	1,773	$69,528	$193,030	$262,558
Continuous Footings	102	CY	764	$31,325	$23,760	$55,085
Grade Beams	117	CY	665	$27,273	$28,000	$55,273
SOG	721	CY	1,804	$66,907	$136,592	$203,499
Columns	267	CY	2,700	$102,597	$78,000	$180,597
Fnd'n & Shear Walls	2,324	CY	18,900	$718,200	$554,750	$1,272,950
Elevated Slabs	4,156	CY	19,100	$783,117	$1,294,000	$2,077,117
Composite Decks	1,283	CY	6,415	$256,600	$320,750	$577,350
Miscellaneous Concrete	200	CY	1,100	$41,800	$53,600	$95,400
Site Concrete	600	CY	3,060	$119,340	$120,600	$239,940
Total CIP Concrete:	**10,727**	**CY**	**56,281**	**$2,216,687**	**$2,803,082**	**$5,019,769**

Wage Check $/Hour: Checks?   Yes $/HR: $39.39
Assembly Check $/CY: Checks?   High, but decks and shear Wls $/CY: $467.96

Figure 11.3 CIP concrete cost recapitulation summary sheet

carpentry, will have them. Smaller items such as door installation, toilet accessories, and minor amounts of finish carpentry will have their own single pricing recap sheets.

Summary

Pricing of materials should be a straightforward process. Data may be available from past projects the construction company has done. Published data also is available. Vendors provide the most current unit prices for many materials. Labor pricing can be more difficult, especially if unit labor costs are used. These have to be adjusted for inflation. Productivity factors, which are man-hours per unit, are easier to use. They seldom change for a work activity unless there is a new and innovative way of working more efficiently. Determining labor costs using productivity factors only necessitates the use of current wage rates, which are commonly available. This eliminates the need to apply inflation judgments to the costing process. Labor productivity and wage rates were discussed in the previous chapter. The estimator may then calculate the unit labor cost, if so desired, to use as a check against other similar projects. In the next chapter we will price the materials for the balance of GC self-performed work including structural steel, carpentry, doors/frames/hardware, and architectural specialties.

Review questions

1   When is a summary recap page used?
2   How is a crosscheck performed?
3   What can an assemblies check indicate?
4   If the estimator (not you) made the mistake of using the wrong units on the pricing recap, as in the example given at the beginning of this chapter, how much would have been included for the plywood? How much should have been included?
5   Why should CIP concrete systems be priced separate from pre–cast or tilt-up concrete?
6   What is an advantage of using UMH over labor unit costs?
7   Should a GC issue a purchase order or a subcontract agreement for each of the following different concrete scopes:

   a   Form purchase
   b   Form rental
   c   Rebar supply
   d   Rebar installation
   e   Concrete pump
   f   Concrete purchase
   g   Slab finishing
   h   Tilt-up crane

Exercises

1   Create a pricing recap sheet for the concrete walls quantified in Exercise 1 of Chapter 7. Using a current published database, price the material. Price the labor costs using the productivity factors, also from the reference, but with current wage rates for the trades in your state. How do the labor costs calculated from your local wage rates differ from those in the database or those used in this book? What is the cause for the differences? Hint: What is included in the wages, and are they union or merit-shop?

2 Combine the wage rates of one each laborer, carpenter, cement mason, operating engineer, and ironworker from Chapter 10 and calculate an average wage rate. Why would a GC (A) use the average wage for all its work, or, conversely, (B) not use an average wage rate? (C) Which craft wage rate is closest to your average?

3 Looking back at the CIP elevated slab formwork example at the end of the CIP concrete systems section, which of the three estimating options would be the best choice: (A) in a budget estimate, (B) in a GMP, (C) in a pre-bid day summary estimate, (D) on bid day, and/or (E) during buyout?

4 If concrete redi-mix supply prices came in on bid day of $95/CY and $102/CY and $107/CY for 3,000, 4,000, and 6,000 psi mixes, respectively, what would be a potential bid adjustment the estimator might make? Refer to the quantities in the example shown in the CIP concrete systems section. Provide an argument why he or she should or should not (choose one position) make this adjustment on bid day – there are a couple of reasons for both.

5 What are the assembly costs/CY of each of the concrete systems in Figure 11.3?

6 What is the cost/CY of our tilt-up example shown in Figure 11.2?

7 In Chapter 6 we discussed both detailed and rule-of-thumb methods for quantifying rebar weights. We have not given you rebar solutions for each CIP concrete system. Using 100#/CY as an average and the summary recap Figure 11.3, what is the approximate total tonnage for our case study rebar? Using rebar unit prices from figures in these last two chapters, what is our total anticipated rebar supply cost? We will be able to use these results on bid day if we receive rebar supply quotes.

8 Why do the dollar totals in the self-performed option of boxed-in elevated CIP slab example included in this chapter not exactly match the values in Figure 11.3?

9 Price up the concrete substructure system(s) you quantified in Exercise 6.4. How does the total system $/CY compare with a comparable system from the Dunn commercial project? Why might they be different?

12 Pricing balance of general contractor self-performed work

Introduction

The last chapter priced all of the concrete work. Our coverage for this chapter is the balance of the direct work a typical commercial general contractor (GC) may perform, which includes Construction Specification Institute (CSI) division 05, structural steel, and all of the carpentry work in CSI divisions 06, 08, and 10. Much of this work was quantified in Chapters 8 and 9. General contractors will choose to perform different aspects of work with their own direct forces or to subcontract work out for a variety of reasons, including:

- Union affiliations,
- Specialization,
- Equipment and tools,
- Economy,
- Client or contract requirement, or
- Magnitude (quantity) of that type of work.

In this chapter we will again assume that the contractor is estimating the work utilizing labor productivity (hours per measured unit) and not labor unit pricing. Labor is priced with bare wages, not loaded, and we recommend adding labor burden to the end of the estimate on the estimate summary page, as described in Chapter 19. The contractor will also be purchasing materials from independent specialized material suppliers and fabricators, for which they will have executed either long-form or short-form purchase orders. Many materials will be priced utilizing database material unit prices and not plugs or lump sum allowances. Other materials will receive competitive supplier quotations on bid day. It is imperative for the estimator to understand in what units materials will be priced so that during the quantity take-off (QTO) process he or she measured and counted and converted quantities into units that allow a smooth and error-free pricing process. If the QTO was developed detailed enough, and measured quantities were converted to purchasable units, the pricing is very straightforward. Many of the labor productivity rates and material unit prices used in this book were provided by *The Guide*, which has made electronic access free to students. Other rates and prices were provided by our industry partners.

Structural steel pricing

Structural steel columns and beams: Installation labor of many elements of structural steel is figured by the unit man-hours (UMH), and material is purchased by the ton, as shown in the next boxed-in example. Note that for simplicity we have not defined every abbreviation used within these estimating formulas and calculations. All of them, and others, are spelled out in the front material of the book.

---

15 HSS columns @ 1.5 MH/EA @ $43/HR = $968 labor cost
And 2.5 tons @ $2,600/T = $6,500 material cost

20 W33 × 118 beams @ 2 MH/EA @ $43/HR = $1,720 labor cost
And 39 tons @ $2,600/T = $101,400 material cost

500 Nelson studs @ .33 MH/EA @ $43/HR = $7,095 labor cost
Material cost was accounted for with the gusset add-on to beams

---

Estimating references commonly provide productivity in terms of man-hours per ton of steel. If the estimator has not quantified the steel by weight, this productivity factor is of no use. A more accurate way to perform the labor estimate is to know the crew size and capabilities. For most projects, structural steel will be erected with a five-person crew commonly called the raising gang. The raising gang will be supported by an operating engineer operating the crane and an ironworker (IW) rigger on the ground hooking steel up to the crane. This crew should be capable of installing 40 pieces of steel a day for most of the structural members. Once the steel is plumbed and aligned, a two-person IW bolt crew will install the remaining bolts. This is done on a mechanical lift, and a good crew should be able to install and tighten a bolt every two minutes. This time includes required inspections by the building department. A surveyor or two will also assist with the steel alignment.

Structural steel material prices are furnished as competitive bids by fabrication shops. It is important for the estimator to know what is included and what is not. Bolts have been discussed in Chapters 6 and 8, but as a reminder, the fabricator furnishes all steel-to-steel connecting bolts and all anchor bolts placed in concrete. The general contractor will furnish all other bolts. Anchor bolts and other embedded steel will be installed by the carpenters and steel-to-steel bolts installed by the IWs.

Trusses: If steel trusses are purchased and shipped to the job in parts and pieces, they will be field assembled and priced similar to beams, discussed earlier. If prefabricated and pre-assembled, they are purchased by the each, for example:

---

10 trusses @ 40MH/EA @ $43/MH = $17,200 labor cost
10 trusses @ $10,000/EA = $100,000 material cost

Bar joist: A fabricator will design the bar joist and enter into a lump sum purchase order with the GC. The estimator will develop their price based upon the depth and strength and lineal footage of joist. The installer may use a MH/SF or MH/LF productivity factor, but the most efficient is a MH/EA calculation, as follows:

---

51 Joist @ 0.5MH/EA @ $43/HR = $1,097 labor cost
1,530 LF @ $30/LF = $45,900 material cost

Miscellaneous steel: Because there are so many different sizes and shapes and types of potential miscellaneous steel items, it is difficult to provide rules of thumb for all of them.

Some are purchased by the pound, such as embedded steel; others by the each, such as bollards; and others by the linear foot, such as handrails, and metal stairs are estimated by the flight. In all cases, a supplier quotation should be provided for the material and the contractor's own in-house historical database used for labor productivity. If the IW crew is still on-site when the miscellaneous steel arrives, they may assemble parts and pieces of it, but often a carpenter will assist with installation later in the project.

Metal decking: Installation labor and material purchase are generally calculated by the squares (SQ) which is equal to 100 square feet (SF) of deck, although some estimators may price it by the SF. An example would include:

1,632 SQ @ 2.25 MH/SQ @ $43/MH = $157,896 labor cost
And 1,632 SQ @ $180/SQ = $293,760 material cost

Joist and decking are installed relatively quickly. Estimating references often show joist installed by the lineal foot. The estimator should talk to an ironworker foreman and determine the time it will take to install the joist and related bracing. Bridging for the joist is listed and priced as a separate line item. Decking is priced by the square foot (or 100 SF or SQ) for slab form and by the SQ for roofs, and productivity includes its welding to the support structure. Shear studs, or Nelson studs, involve the labor to weld them and special welding equipment.

Equipment such as cranes, man-lifts, scissor-lifts, and welding machines will be figured on total structural steel duration. Additional costs associated with structural steel erection include shake-out (receiving materials, unloading from the truck, and assembly on the ground for easy installation), plumb and align, and safety. These costs are predominantly labor-based and are estimated using either an assumed crew size and duration of activity, or duration of the project, or as a percentage add-on to total tons of steel on the project.

All structural steel needs to have hoisting equipment for handling. This can be anything from a forklift to a heavy lift crane or traveling derrick. Equipment selection and costing can challenge the estimator. He or she should consult with a superintendent and an ironworker foreman to determine the type and size of equipment that is best suited for the project. It is also important to discuss operating personnel and mobilization costs. Additional consultation with a company that rents hoisting equipment will provide rental rates, travel times and costs, and the fuel and maintenance requirements. Equipment costs are expanded on in the next chapter.

As discussed earlier, pricing recapitulation (recap) sheets for structural and miscellaneous steel erection contains tasks that are not necessarily shown as columns on the quantity sheets. Many scopes that require pricing have similar quantities as other measured items, and some are too small to quantify but still require allowances or plug estimates. One small steel item that is often overlooked is shim material. Shims are small pieces of steel plate that are used under column base plates to set them to the proper elevation. The fabricator does not furnish shims, and the estimator must determine their cost and include them as a line item on the pricing recap. Some companies have historical data that allows a percentage of the steel cost to be used to cover items such as shims.

Often, an estimator will find cement-based items on the structural steel drawings such as grout under the column base plates and fill for stair treads and landings. Both of these should be quantified and recapped as miscellaneous concrete. This work will be accomplished by laborers and cement masons and not ironworkers.

We have brought many of the steel quantities developed for our case study in Chapter 8 forward and added labor productivity rates, labor wage rates, and material and equipment unit pricing and assembled them on Figure 12.1, which is an abridged pricing recap sheet.

City Construction Company
Pricing Recap Sheet

Project: Dunn Lumber
Location: 3400 Latona, Seattle, WA
Architect: Flad

Date: 6/10/2021
Estimator: PJ

CSI 05: Structural Steel

Crew No.	Description	Qty	Unit	Labor					Material/Equip		
				UMH	Man Hours	Wage Rate	Unit L Cost	Labor Cost	Unit M Cost	Material Cost	Total Cost
	Columns:										
	Purchase HSS & WF columns	108	Ton					0	2,600	280,800	280,800
IW	Install HSS 5 × 5	15	EA	1.1	17	43	47	710		0	710
IW	Install W10 × 49's	81	EA	3	243	43	129	10,449		0	10,449
	Beams:										
	Purchase WF beams	596	Ton					0	2,600	1,549,600	1,549,600
IW	Install beams	605	EA	2	1,210	43	86	52,030		0	52,030
IW	Nelson studs	500	EA	0.33	165	43	14	7,095		included	7,095
IW	Metal deck	1,632	SQ	2.25	3,672	43	97	157,896	180	293,760	451,656
IW/OE	Shakeout	704	Ton	1.5	1,056	43	65	45,408	50	35,200	80,608
IW/SU	Plumb and align	6	Days	24	144	43	1,032	6,192		0	6,192
IW	Safety, lanyards, cables, harness	704	Ton	1	704	43	43	30,272	75	52,800	83,072
	Crane:										
	Mob and de-mob	2	EA					0	5,000	10,000	10,000
	Rental	8	Wks					0	5,000	40,000	40,000
OE	Operator	8	Wks	40	320	40	1,600	12,800		0	12,800
	Welding machine	2	Mos					0	500	1,000	1,000
	Scissor lift	2	Mos					0	750	1,500	1,500
	TOTALS				7,531			**$322,852**	Checks v/	**$2,264,660**	**$2,587,512**
	Wage check: Total labor $/Total MHs = Average $/HR:							$42.87		Checks v/	
	Assembly check: Total $/Total tons:										$3,675
	Assembly check: Total $/SF of structure:										$15.85

Figure 12.1 Structural steel pricing recap sheet

Totals are calculated at the bottom of each pricing recap page for labor hours, total labor cost, total material and equipment cost, and the far right-hand total cost column. Each of the total columns should then also be added horizontally across to verify the far right-hand column is correct. With Excel it is easy to cut and paste and insert rows and columns and cells, and sometimes automatic formulas may be impacted. A manual cross-check of pricing recap totals is a good practice.

An average wage check also should be made on each pricing sheet for each system or assembly. The estimator should divide the total labor dollars by the total labor hours. The average should be within the range of different wage rates used on each sheet. Major estimate errors come from slipping a digit, not from a miscount of anchor bolts.

An assembly check could be made on rolled shapes based on tons of steel. But if this were a small steel project, and bar joist and metal deck and miscellaneous steel were all added to the same recap sheet, $/ton would not be an appropriate check. A more common assembly check would therefore be to unitize the man-hours and total cost in terms of square foot of the steel structure. This can be checked against other square foot cost data in either published or historical references.

After validation, the totals at the bottom of each of the columns are double-underlined, indicating they are ready to be forwarded to the next summary level in the estimating process. Once forwarded, each of these totals is then circled. After all of the hours and costs have been forwarded from an individual recap sheet, the estimator may place an "X" in the lower right-hand corner of that sheet, indicating he or she is done with it.

A large structural steel building would require multiple CSI division 05 pricing recaps separated by area or floor or phase or member types, such as one for columns and another for beams. But if this were a wood or concrete structure, steel may be recapped all on one or two pages. These cost organization features then all feed into an efficient cost control system, as introduced in Chapter 24.

Rough carpentry pricing

The process of applying labor productivity factors and wage rates and material unit prices to rough carpentry quantities developed in our previous carpentry QTO (Chapter 9) is similar to structural steel.

Columns and beams and joist: Labor is figured for rough carpentry members by multiplying the converted quantity by the contractor's in-house UMH productivity rate times the bare wage rates. Material purchase cost is estimated by the thousand board feet (MBF) multiplied times database or market or quoted unit prices, for example:

1.4 MBF of <u>columns</u> @ 41 MH/MBF @ $41/HR = $2,411 labor cost
1.4 MBF @ $2,400/MBF = $3,360 material cost

100 LF of <u>GLB</u> @ 0.1 MH/LF @ $41/MH = $410 labor cost
100 LF @ $15/LF = $1,500 material cost

536 <u>joist</u> @ 0.5 MH/EA @ $41/MH = $10,988 labor cost
11.4 MBF @ $1,500/MBF = $17,100 material cost

Wall framing: Headers would be estimated similar to beams or joist. Studs and plates can be estimated by the LF of wall or the MBF of material, for example:

18.1 MBF of <u>studs</u> @ 19 MH/MBF @ $41/MH = $14,100 labor cost
18.1 MBF @ $2,000/MBF = $36,200 material cost

Trusses are estimated similar to steel bar joist. They are engineered by the supplier, and the GC enters into a purchase order for a lump sum. The supplier would have developed their own supply price by the each. Some GC estimators will figure labor productivity on the LF of truss or SF of roof, but the most accurate way to figure labor cost is by the each, for example:

---

51 trusses @ 1 MH/EA @ $41/MH = $2,091 labor cost
51 trusses @ $250/EA = $12,750 material cost

Framing hardware: These should be separated by type; the following example applies to Simpson beam-to-column supports:

---

50 EA @ 0.2 MH/EA @ $41/HR = $410 labor cost
50 EA @ $100/EA = $5,000 material cost

Sheeting: Wall and floor and roof sheeting should all be separated and estimated independently, as well as separating plywood from oriented strand board and varying thicknesses of sheeting. Some estimators and databases will figure plywood by the square foot, but it is supplied by the sheet and the carpenters install it by the sheet. For our example, we estimate 3/4" tongue-and-groove plywood floor sheeting as follows:

---

344 sheets @ 0.5 MH/sheet @ $41/HR = $7,052 labor cost
And 344 sheets @ $30/sheet = $10,324 material cost

Siding and trim and roofing: Whether the estimator takes the siding off by the square foot of wall or converts to LF, the process of multiplying the measured quantity times a historical UMH and bare wage rates plus published or market material prices is the same, for example:

3,840 LF 6" beveled siding @ 0.1 MH/LF @ $41/MH = $15,744 labor cost
And 3,840 LF @ $3.75/LF = $14,400 material cost

420 LF cedar 1 × 4 trim @ 0.1 MH/LF @ $41/MH = $1,722 labor cost
And 420 LF @ $2/LF = $840 material cost

25 SQ cedar shakes @ 10 MH/SQ @ $41/MH = $10,250 labor cost
And 25 SQ @ $300/SQ = $7,500 material cost

Rough carpentry *equipment* such as a crane or compressor or nail gun is based on activity duration times market rental pricing. The quantities from the rough carpentry Figure 9.1 QTO have been brought forward and extended with productivity rates, wage rates, and material pricing and assembled in an abridged rough carpentry pricing recapitulation sheet, Figure 12.2.

City Construction Company
Pricing Recap Sheet

Project: Broadway MXD Apartments
Location: 1200 Broadway, Seattle, WA
Architect: The Design Center

Date: 6/10/2021
Estimator: LH

CSI 06: Rough Carpentry

Crew No.	Description	Qty	Unit	Labor					Material/Equip		
				UMH	Man Hours	Wage Rate	Unit L Cost	Labor Cost	Unit Cost	Material Cost	Total Cost
CA	Treated 2 × 6 plates	1.2	MBF	32	38	41	1,312	1,574	1,950	2,340	3,914
CA	Columns: 6 × 6	1.4	MBF	42	59	41	1,722	2,411	2,400	3,360	5,771
	Beams:										
CA	Dimensional 6 × 12	1.7	MBF	27	46	41	1,107	1,882	2,000	3,400	5,282
CA	GLB 5 1/8″ × 15″	100	LF	0.1	10	41	4	410	15	1,500	1,910
	Joist: 2 × 8:										
CA	Buy	11.4	MBF					0	1,500	17,100	17,100
CA	Install	536	EA	0.5	268	41	21	10,988	0	0	10,988
CA	Wall framing: 2 × 6	8.1	MBF	19	154	41	779	6,310	2,000	16,200	22,510
CA	Headers: 4 × 12	1.5	MBF	17	26	41	697	1,046	1,550	2,325	3,371
CA	Trusses:	51	EA	1	51	41	41	2,091	250	12,750	14,841
CA	Framing Hardware:	50	EA	0.2	10	41	8	410	100	5,000	5,410
CA	Plywood: 3/4″ TNG	344	Shts	0.5	172	41	21	7,052	30	10,320	17,372
	Nails: 16d Galv (50#/box)	2	Box					in	80	160	160
	8d Galv	3	Box					in	90	270	270
	Compressor	1	Mos					0	350	350	350
	Nail guns (2 @ 1 mo/ea)	2	U.Mos					0	100	200	200
	Boom Truck	1	WK					0	500	500	500
	Total MBF	25	MBF								
	TOTALS				834			**$34,174**		**$75,775**	**$109,949**

Wage Check: Total labor $/Total MH = Average $/HR = $41.00   Checks v/
Assembly Check: Total $/Total MBF = $/MBF =   Checks v/ $4,346

Figure 12.2 Rough carpentry pricing recap sheet

Lumber suppliers do not typically furnish prices on bid day. They prefer to have the GC's estimator determine the approximate quantity of lumber and sheet goods that are required, and then they will furnish unit prices over the telephone. Most lumber suppliers price lumber according to each size. Sheets of plywood are also priced according to type and size. The estimator should talk to a lumber supplier prior to estimating the work to learn how the pricing will be received. He or she can then set up the estimating sheets accordingly. Residential GCs will send the framing drawings to the lumber supplier, and the supplier may perform its own take-off and provide a lump sum bid, but there can be complications if the supplier underestimated a quantity and the framing crew runs short of material out on the jobsite.

Finish carpentry pricing

There are potentially as many different types of finish carpentry items as there are miscellaneous steel items. Here are some pricing examples for wood base, lower kitchen cabinets, and wood flooring:

---

450 LF oak <u>base</u> @ 0.1 MH/LF @ $41/MH = $1,845 labor cost
And 450 LF @ $1.75/LF = $788 material cost

24 oak kitchen <u>cabinets</u> @ 1.5 MH/EA @ $41/MH = $1,476 labor cost
And 24 @ $400/EA = $9,600 material cost

1,250 LF oak <u>flooring</u> @ 0.15 MH/LF @ $41/MH = $7,688 labor cost
And 1,250 LF @ $3.50/LF = $4,375 material cost

All of the finish carpentry items would be priced on a recap sheet separate from the rough carpentry recap, and the totals from each of the columns forwarded to the estimate summary page. These two types of CSI division 06 work are performed by different crews at different times. Independent recap sheets facilitate an easier transition to cost control, as we will discuss later.

Door and window pricing

Several examples follow as a sampling of pricing for some of the major items self-performed by a GC in door and window CSI division 08.

---

100 single hollow-metal (HM) <u>door frames</u> (DF) @ 1 MH/EA @ $41/MH = $4,100 labor
100 single HM DF @ $300/EA = $30,000 material cost

50 double HM DF @ 1.5 MH/EA @ $41/MH = $3,075 labor cost
50 double HM DF @ $300/EA = $15,000 material cost

150 wood <u>door leafs</u> @ 1 MH/EA @ $41/MH = $6,150 labor cost
150 wood door leafs @ 520/leaf = $78,000 material cost

25 door <u>relites</u> @ 0.5 MH/EA @ $41/MH = $513 labor cost
25 relite kits @ $150/EA = $3,750 material cost

150 door <u>hardware</u> sets @ 5 MH/set @ $41/MH = $30,750 labor cost
150 sets @ $1,100/set = $52,500 material cost

1 <u>hardware room</u> @ LS $10,000 labor cost
1 hardware room @ LS $5,000 material cost

35 EA <u>access doors</u> @ 1 MH/EA @ $41/MH = $1,435 labor cost
35 EA @ $250/EA = $8,750 material cost

750 <u>punch windows</u> @ 1.6 MH/EA @ $41/MH = $49,856 labor cost
750 EA @ $400/EA = $300,000 material cost

A GC may self-perform single, or 'punch' window installation with its own carpenter crew. This is especially true if there are only a few windows on a commercial project or a residential project. Commercial storefront glazing is different than punch windows, as the frames and glass and hardware are engineered by a design-build subcontractor and assembled in the field by the glazier trade. A GC plug estimate for storefront window wall system is discussed in Chapter 14.

CSI division 08 would likely have at least two separate pricing recaps, one for doors, frames, and hardware and the other for glass and glazing. They are provided by different suppliers and installed at different times. The totals from these recap pages would also be forwarded to the estimate summary page.

Specialty pricing

The pricing process for most of the CSI divisions 10 through 12 specialties is similar. The items were counted and taken-off by the each, and most will be installed by the GC's carpenters. Some items may be provided by similar suppliers, but it is possible to end up with ten or so specialty item suppliers – hence the term, 'specialties'. Some pricing examples include:

75 interior <u>Plam signs</u> @ 0.25 MH/EA @ $41/MH = $769 labor cost
75 signs @ $50/EA = $ 3,750 material cost
17 semi-recessed <u>paper towel dispensers</u> @ 1 HR/EA @ $41/EA = $697 labor cost
12 dispensers @ $150/EA = $2,250 material cost
117 recessed <u>fire extinguisher cabinets</u> (FECs) @ 2.5 MH/EA @ $41/
   MH = $11,993 LC
117 FECs @ $175/EA = $20,475 material cost

Most of the specialty items would be grouped to one pricing recap page or separated by major CSI division. If this were a large apartment or hotel, then there would be one sheet dedicated to just toilet accessories and another for kitchen appliances.

Summary

The key to an efficient pricing process is a thorough detailed quantity take-off process with quantities converted to units that will reflect pricing. The experienced estimator knows how items will be priced. The new estimator should look ahead at databases to facilitate this important transition.

The resulting bottom line of each pricing recapitulation sheet needs to be carried forward to the estimate summary page, as described in Chapter 19. If there are several pricing pages for large systems such as concrete, steel, or carpentry, they may need to be gathered to a separate summary recap page before forwarding to the summary page. All of the columns should be totaled and cross-checked on each page before forwarding. These columns include the labor hours, labor cost, material and equipment cost, and total cost. The estimator should add rows across and compare to totals from the vertical columns. Average labor wage checks and system checks such as total dollars per cubic yard of concrete or per ton of steel or per apartment unit are relevant to perform on the pricing recap sheets as well. Similar to the QTO process, when pricing is forwarded to the next summary level, totals are circled, and when every double-underlined total has been forwarded from a recap, an 'X' is placed in the lower right corner, indicating the estimator is done with that sheet.

It would be impossible for us to have provided pricing examples for all work items that may be found in any type of project. We have hit on a few of the more common ones. If an estimator cannot find an example that exactly applies to their work item, they should look around in any estimating database and find something similar or ask a coworker. Superintendents are great sources for labor productivity. Although superintendents may not think in terms of $/SF or UMH, they do think in terms of a four-person crew working two weeks to accomplish a task, which totals 320 man-hours. How does this total feel? The estimator must compare the superintendent's plan to the total estimated cost or hours for that scope of work. We call this a 'gut-check'.

Review questions

1   Why should estimators utilize standard estimating forms?
2   What is the downside of using labor unit rates, such as $3.55/SF, versus labor productivity rates, such as 0.1 MH/SF?
3   How would mistakes from the QTO process discussed in the four earlier QTO chapters affect the pricing processes in these four pricing chapters?
4   What happens if the estimator's hours and the superintendent's hours do not agree?
5   In addition to IWs, what other crafts may be involved with steel erection?
6   In addition to carpenters and laborers, what crafts may be involved with concrete system construction?
7   What is wrong with a residential wood framing subcontractor figuring a labor estimate for a new house on a $/SFF basis? They often do that.

Exercises

1   Prepare a pricing recapitulation page for the lumber quantities from Chapter 9 Review Question 3. Utilize published database productivity rates and material unit prices. Utilize the carpenter wage rate presented in Chapter 10.
2   Assume the Exercise 1 project will take five days to install. What will your crew size be? What equipment is necessary, and how much will that cost? What other items might the estimator add to this work package so that the foreman has a complete estimate and bill of materials? Expand the Exercise 1 pricing recap to include all of these items.
3   Utilizing a standard database and making whatever assumptions are necessary (yes, estimators do this), price up Chapter 9's Exercises 3, 4, or 5. Utilize standard pricing recap forms.

4   What are the average wage rates in Exercise 3? What are the systems costs? Are those reasonable? How might you verify?

5   Some estimators used 'loaded' wage rates on their pricing recap sheets; others use bare wage rates. What are the advantages and disadvantages of each approach?

6   Prepare an argument why an estimator would choose to (A) use one standard average wage rate for all work and, conversely, (B) use different wage rates for different crafts.

7   Price up one of the carpentry systems that was quantified in a Chapter 9 exercise.

13 Equipment

Introduction

The estimator needs to think where the cost of certain equipment will go in the estimate. Some equipment is unique to certain activities, and its cost should be included in the direct work part of the estimate. A concrete pump, for example, is used only for concrete work and is a direct cost to those activities. The cost of a crane to hoist tilt-up wall panels should be shown on the pricing recapitulation (recap) sheet for the panels. The cost of a crane brought on site specifically for hoisting structural steel should be shown on the recap sheet for the steel, because once this work is done, the crane will be removed from the site. After completion of direct labor, material, and equipment pricing, the estimating team will organize and summarize all of the recap sheets in preparation of posting the information to the estimate summary page. Proper placement of equipment costs will have an effect on the checking procedures as the estimate is completed.

Equipment that is used throughout the job is considered a jobsite overhead cost or jobsite general conditions cost and will be discussed in Chapter 17. For example, a tower crane or jobsite forklift is used for many work activities for most of the duration of a project. Its cost is difficult to apportion to specific tasks and is thus considered part of the jobsite general conditions. Estimating, and later cost coding, of the equipment used for the work is known as activity-based costing (ABC), which is an element of lean construction techniques. Because tower cranes and personnel and material hoists are such significant costs and have an impact on several parts of the construction process, we describe techniques for estimating their costs in this chapter.

The major topics covered in this chapter are:

- Activity-based costing;
- Sources of equipment pricing;
- Concrete equipment, including concrete pumping and tilt-up crane;
- Hoisting for steel erection and wood framing;
- Tower cranes;
- Material and personnel hoists; and
- Pricing recap completion.

Use of market databases to estimate equipment cost is not recommended. Databases may be used to estimate equipment productivity, but published rental rates or internal company rental rates are the best sources of equipment cost data.

Activity-based costing

Activity-based costing is not a process to reduce costs; it is a means for identifying indirect costs that can be applied to the cost of direct construction activities. This allows contractors to understand the true cost of direct work and enable them to focus on improving costs where needed most. Applying project indirect costs to construction activities also allows a contractor to improve their estimating capabilities. Controlling overhead costs has always been of interest to contractors, and the first step in ABC is to understand these costs and track them. This is achieved by establishing cost codes to link project indirect costs to specific work activities.

Activity-based costing is an approach to the costing and monitoring the total cost of individual construction activities. This involves tracing resource utilization and assigning all project costs. This results in better as-built cost data that can be used in estimating future projects.

Sources of construction equipment

Contractor-owned equipment

Few construction companies own any equipment. The exception would be heavy civil, marine, and industrial contractors. Most commercial and residential contractors either establish separate equipment companies or rent from outside sources. Equipment may also be provided and operated by subcontractors as part of their scopes of work. The equipment manager for company-owned equipment will establish internal rental rates based on the estimated annual use of each piece of equipment. These rental rates are charged to individual projects whenever the equipment is used.

Contractor equipment companies

Many construction companies choose not to own equipment themselves but rather set up separate equipment companies that own the equipment. The equity ownership of these companies is usually a select group of individuals within the construction company, often the corporate officers. These equipment companies charge rental for equipment use on projects at rates similar to rental charges from equipment rental companies. Often there are two rental rates: An operating rate and a standby rate. The standby rate is charged when equipment is on a project site but not being operated.

Equipment rental companies

Another primary source of construction equipment is an equipment rental company. Rental rates typically are quoted by the day, week, and month. Equipment rental is often accompanied with mobilization or delivery charges and demobilization or pickup charges. Some pieces of equipment have meters installed that measure hours used or miles driven, and if the equipment is used overtime or on weekends, the equipment rental company may require additional rent, as indicated in the forklift lease agreement shown in Figure 13.1.

Highline Equipment Company   Date: August 3, 2021
2785 East Marginal Way   Lease No: 19–07–42
Seattle, WA 98212
Tel: 206–856–1257

Equipment Lease Agreement

Highline Equipment Company (hereinafter "Highline") leases to:

Name: City Construction Company (hereinafter "Lessee")
Mailing address: 1449 Columbia Avenue, Seattle, WA 98202

The following equipment: one (1) Case Model 588H Forklift, serial no: 98567 (hereinafter "equipment") for use at the following project location: 3800 Latona Avenue, Seattle, WA 98105.

Rental Price: $3,800 per month, excluding sales, use, personal property taxes; licenses; and fees which are Lessee's responsibility. Lessee shall pay rental on the first of each month in advance at Highline's office. The rental rate is based on usage up to 184 hours per month. Upon completion of rental, the total hours used will be divided by the months of rental and $90 per hour will be charged for hourly usage exceeding the aggregate hours.

Lessee responsible to lubricate, oil, maintain, operate, adjust, and inspect the equipment in accordance with manufacturer's specifications, and provide records of all maintenance performed to Highline.

Approximate delivery date is August 17, 2021. Rent shall begin upon initial shipment of equipment from the designated delivery yard to Lessee, but no later than the Guaranteed Rental Start Date of August 24, 2021. The Guaranteed Rental Period is fifteen (15) months.

Inbound freight: Deliver yard: Outbound freight: Return yard:
$1,200   Highline $1,200   Highline

Lessee has read all terms of this lease and agrees to said terms in their entirety.

LESSEE: City Construction Company   Highline Equipment Company
By: Rhonda Martin, Project Manager   James Smith, President
By: *Rhonda Martin*   By: *James Smith*
Date: August 3, 2021   Date: August 3, 2021

Figure 13.1 Equipment lease agreement

Subcontractors

Subcontractors typically provide any equipment needed to perform their scopes of work. Often the general contractor (GC) will provide hoisting support rather than requiring subcontractors to provide their own cranes or other hoisting equipment. Any equipment used on a project where the equipment supplier also provides an operator will be obtained by a subcontract issued to the supplier. The subcontract defines the scope of services to be provided, establishes the duration of the agreement, and delineates insurance requirements.

Concrete equipment

Estimating the cost of concrete pumping includes several variables. Many estimators will price pumping at a historical rate such as $15.00 per cubic yard (CY) for all concrete. This is satisfactory as long as the estimator understands what makes up the pumping cost and what affects the rate. A concrete pump has a minimum charge for just showing up at the jobsite plus a rate per CY for pumping plus an hourly component. Travel costs will vary for different projects depending on the anticipated travel time from the concrete pump supplier's yard. For most quantities to be pumped at a given time, the rate used by the estimator is usually fairly accurate. If a small quantity is to be pumped, the pumping cost will be much higher. This is caused by applying the minimum pumping charge to the small quantity of concrete. Large placements will have a lower pumping unit cost. If the pumping costs are too high, the estimator may need to explore another more economical means of placing the concrete such as waiting until a larger volume of concrete is needed.

Concrete pumps come in various sizes, and the estimator must anticipate a placement rate and the distance that the concrete must be pumped. Pumps that are too large are more expensive, while those that are too small will affect the placing labor cost. A small pump can cause additional costs by keeping a concrete truck on site too long. A concrete pump supplier's quote for the case study project is shown in Figure 13.2.

Acme Concrete Pumping
12590 33rd Avenue Northeast
Seattle, WA 98121
(206) 362–8650

Date: June 24, 2021
Company: City Construction Company   Salesman: Jack Brown
Project: Dunn Lumber   Effective Date: July 1, 2021
Contact: Rhonda Martin   Ending Date: August 31, 2022

Acme Concrete Pumping is pleased to quote the following equipment:

	Price per hour	Price per yard	Travel
Line Pump	$192.00	$4.20	$192.00
17 meter / 54 foot Boom Pump	$198.00	$4.20	$198.00
32 meter / 105 foot Boom Pump	$198.00	$4.50	$198.00
36 meter / 120 foot Boom Pump	$222.00	$4.50	$222.00
39 meter / 130 foot Boom Pump	$234.00	$4.50	$234.00
42 meter / 140 foot Boom Pump	$264.00	$4.80	$264.00
47 meter / 154 foot Boom Pump	$282.00	$4.80	$282.00
58 meter / 190 foot Boom Pump	$354.00	$5.40	$354.00
61 meter / 200 foot Boom Pump	$402.00	$5.70	$354.00

Conditions:

- Steel plates or equivalent crane mat required on all working outriggers of 50 meter or larger concrete pumps.
- Billing begins at start of set-up and continues until cleanup is completed.
- Overtime is applicable Monday–Friday at an additional $60.00 per hour on pours over 8 hours.
- Overtime Saturday is an extra $60.00 per hour; Sunday and holiday is an extra $120 per hour.
- There is a (4) four hour minimum charge for any pump.

This quote is for the above named project only and valid for (30) thirty days from the above date, after which time it may be withdrawn without notice.

Contractor: City Construction Company   Acme Concrete Pumping
Accepted By: *Rhonda Martin*   Accepted By: *Jack Brown*
Date: June 24, 2021   Date: June 24, 2021

Figure 13.2 Concrete pumping quote

The following example illustrates the use of the quote shown in Figure 13.2.

A 105-foot boom pump is to be used to pump 100 CY of concrete in 4 hours (HR). What is the estimated cost per CY to pump the concrete?

Concrete pump travel cost: $198
Rental cost per hour: 4 hours @ $198/HR = $792
Pump cost per cubic yard: 100 CY @ $4.50/CY = $450
Total cost per this pour: $1,440
Cost per cubic yard: $1,440/100 CY = $14.40/CY

Note that the pumping cost at $14.40 per CY is less than the estimator's historical rate of $15.00 per CY but is applicable for this placement.

The concrete pump company is really a service provider or subcontractor and not a supplier because the pump is operated. The pump supplier should therefore receive a subcontract agreement from the GC and will in turn be required to provide a certificate of liability insurance before arriving on the jobsite. Another item to consider is the time a concrete truck is at the jobsite. The price of the concrete typically includes transportation to and from the site and 20 minutes for standby and placement time. Time is money, and the superintendent must move the concrete placement along to avoid extra costs for the truck and pump.

Other equipment used for concrete construction includes forklifts, cranes, and mobile man-lifts. Forklifts are used to offload delivery trucks and to move materials around the jobsite. Since they are used for multiple tasks, they usually are included in the jobsite general conditions estimate and not charged to specific concrete construction tasks. Cranes are used to hoist formwork and reinforcing steel as well as lift smaller pre-cast concrete panels. Tower cranes often are used to hoist concrete formwork and reinforcing steel if available on the jobsite. If a tower crane is not available, a mobile crane may be required. If the mobile crane is only used to support concrete construction, its cost would be included in the direct cost for the concrete work. If the crane is required for multiple tasks on the jobsite, its cost may be prorated across each task or included in the jobsite general conditions cost estimate.

Concrete wall panels for a tilt-up structure typically are cast on a concrete floor slab. When properly cured, a crane is used to hoist the panels into place. Once elevated, the panels must be braced and connected with adjoining panels. Because the safe lifting capacity of a crane is reduced as the boom is lowered, the crane may need to be relocated multiple times around the walls being constructed with the tilt-up panels. Mobile cranes with hydraulic booms are often used to set the concrete panels. Since the crane is to be used for a single purpose, its cost would be included in the direct cost of wall erection. To lift the pre-cast panels, the GC would issue a hoisting subcontract to a crane company for the equipment and labor needed to operate the crane.

Let's look at an example.

A contractor has selected a 250-ton self-propelled hydraulic crane. Equipment cost for the crane is $485 per hour. Mobilization and demobilization costs are estimated to be $10,500. Each panel will need to be raised and held in location until metal supports are installed and the connecting brackets welded. After conducting a time analysis, the contractor estimates that the crane will be able to erect two panels per HR. A total of 30 panels are required for the project, so 15 HR will be required for panel erection.

250-ton tilt-up crane mobilization: $10,500
Crane rental per hour: $485/HR
30 panels @ 30 minutes/panel = 15 HR
Crane rental cost: 15 HR @ $480/HR = <u>$7,200</u>
Total tilt-up crane cost: $17,700
Crane unit cost per panel: $590/EA

Structural steel and wood framing hoisting

The estimator needs to understand the characteristics of hoisting equipment when sizing it for a particular job. A forklift has a rated capacity based on the load being a certain distance from the backstop and with the forks positioned near the ground. Once the load is elevated or a boom is extended, the lift's capacity is reduced due to an increased tendency of tipping. This is basically true for cranes also. An accurate determination of the weight of a load and equipment rating tables will provide information on sizing the equipment to the job. The estimator should always discuss hoisting with a superintendent so that properly sized equipment is priced.

A forklift should be operated by a full-time operating engineer, if a union project, or a certified carpenter foreman. No one on the jobsite should be allowed to jump on the forklift unless they are properly trained and licensed.

Small cranes can be rented without an operator by using a rental agreement similar to the lease agreement in Figure 13.1, in which case the GC will need to supply a certified operator. If the crane is rented with an operator, a subcontract agreement and liability insurance will be required, as previously discussed for concrete pumps. Larger cranes will also require a full time rigger or bellman to connect loads to the crane. If open lattice boom cranes are required to support steel erection, there will be increased mobilization and demobilization costs. This is because the booms must be detached from the crane for transportation and then installed and rigged with cable once the crane arrives at the jobsite.

Tower cranes

Tower crane estimates are typically a substantial part of the overall estimate, with costs generally being in the tens of thousands of dollars per month. Estimates for tower cranes can generally be put into three categories – primarily due to how they are determined: Foundations and infrastructure can be estimated based on quantities and unit prices; erection and dismantling are typically bid out or done on a lump sum or price per each basis; and operations and maintenance estimates are based on the amount of time (months) that the crane will be on the jobsite.

Estimate preparation typically is started with a boilerplate template from the contractor's database, where line items can act as questions to be answered in regard to the crane's requirements. The estimator should review these line items, as well as bring up other questions based on project and site requirements, with the project team to determine appropriate line items to price in the tower crane estimate. Because of the multiple factors involved in determining location, duration, and size of crane, the estimate should not be completed until some planning analysis is performed. This section is not meant to be a detailed 'how-to' for preparation of a tower crane estimate but more of an overview of the sections and processes. We assume the reader has some prior estimating knowledge.

Once decisions have been made on location, type of crane, and the length of time it will be needed on the project, it is time to prepare a cost estimate for the work. If the project is a lump sum bid, the GC may plug in a generic estimate and then modify it if they are successful in obtaining the project, and a hoisting plan is developed. Most GCs use a generic or boilerplate estimate to build off of that asks all the questions and allows for associated dollars to be attached for different situations that may or may not apply for a particular installation. The estimate may be developed by a staff estimator, but our survey found that most project managers responsible for administering the crane contract will develop the estimate with input from their operations officer, the project superintendent, and a hoisting superintendent. The following subsections correspond to the crane estimate template sections and are organized as the crane would go together; first the foundations, then crane erection, followed by operations, and finally disassembly.

Foundations and infrastructure

The foundations required to support the loads induced by the tower crane can be substantial. Typically, the most economical way to construct these foundations is to integrate with the building foundations to lessen both the initial installation cost as well as potential demolition costs if the pad is constructed outside the building footprint.

In regard to estimating the footing, the first step (like anything in construction) is to determine the design. This is done by a licensed structural engineer who is hired by the GC. The structural engineer will take into consideration the foundation requirements for the crane, the contractor's preferred location, and the building foundation system when designing the footing. Once the design is complete, the contractor should be able to estimate the cost similar to any other building element. Similar to any other foundation, the estimator would figure excavation and backfill, formwork, concrete, and rebar costs. In addition, choices of anchor bolts or a sacrificial lower steel section of the crane that are both embedded into the concrete need to be considered. Tower crane foundations are estimated similar to spread footings, as discussed in Chapters 6 and 10.

Other costs to consider when evaluating the infrastructure required are power requirements for the crane and how that will be provided. Generally, this can be determined by giving an electrical subcontractor the tower crane requirements and some parameters on electrical equipment locations and service. With this information, they should be able to give the estimator a bid for the work.

Lastly, and often forgotten when evaluating the infrastructure costs of installing the crane, is how it affects other building elements – especially when installed within the footprint. Some of the items to consider are:

- If the crane is installed in an elevator shaft, what are the costs to expedite the work that was held off in the shaft because it had to wait until the crane was removed?
- Are there 'pour-backs' required in concrete decks after the crane is removed?
- Is there other work that needs to be expedited because it was held up while the crane was in place?
- Are horizontal braces required between the crane mast and the building?

These are important questions to consider and address when preparing the tower crane estimate. As mentioned in the overview, these questions are generally included in the template provided by the GC for crane estimates to help determine costs when necessary.

Crane erection

It should be noted that most cities have crane erection companies that will construct and dismantle tower cranes. When deciding how to erect the crane, the contractor should compare the cost of managing the work themselves (if they are capable) with the cost of using a crane erection company, as well as the risks involved, when making their decision. The following paragraphs describe the costs involved with erecting the crane regardless of who does the work.

Tower cranes are engineered to be built relatively quickly, like an erector set, but they cannot go up themselves. Typically, erecting a tower crane requires the use of another mobile crane or crawler crane to hoist the pieces into place. Generally, it can be assumed that the tower crane can be built in one or two days, and the rental costs for the crane and their hoisting crew can be established by requesting a quotation from a mobile crane company. Similar to determining foundations, the location and hoisting requirements for the mobile crane must be known before the price can be established. This second crane is known as an assist crane.

In addition to the mobile crane costs, there needs to be an erection crew who can guide and fasten into place the pieces being hoisted by the mobile crane company. The erection crew size can vary from four to eight ironworkers depending on the size and complexity of the crane installation.

Lastly are the incidentals or other factors that can greatly affect the cost of erection. Some of these considerations are:

- Does the work have to be completed on the weekend?
- If an urban site, are there street use fees to be paid for building the crane on public property?
- Are there power or trolley lines to be temporarily (or permanently) moved in order to construct the crane?
- Will the crane swing over a private neighbor's property, and can a temporary easement be gained to allow that swing?
- What types of inspections are required?

These questions, as well as others, need to be considered and addressed before completing the estimate.

Operations and maintenance

There are four fundamental costs to operating and maintaining the crane: Rental, crew (operator and rigger), utility costs, and maintenance. Tower crane rental is generally similar to renting any other piece of equipment – although more expensive. General contractors may have their own cranes and rent to the project, or they may rent through a third party supplier. An extensive analysis of the advantages and disadvantages of crane ownership along with other major pieces of equipment is an interesting study but beyond the scope of this book. Generally, the rental agreement includes a mobilization cost, demobilization cost, supplier monthly rental fee, and some unit prices for maintenance of the crane. The crane supplier handles shipping and factors those costs in their bid. Like most subcontractor and vendor work, this can be bid out competitively between different vendors, and the most economical selection can be made for the project. In many areas, only one supply source of tower cranes is available.

Operator costs are generally hired direct by the general contractor or through a third party subcontractor who provides an operator labor pool. However, when estimating operator costs, it is important to include time and monies for climbing the crane (most union agreements call for the shift to start when they start up the ladder), operator inspections, and overtime. Generally, it is safe to assume a 50-hour week for operator costs, but the estimator should evaluate schedule, hoisting requirements, and other factors with the project team before determining the hours required by the operator on a given project.

Rigging costs are sometimes included in the crane estimate or in other places in the overall project estimate. This depends on how the crane is going to be used, subcontractor agreements, and how the general contractor likes to present their estimates. Rigging costs can be quite expensive. The rigger and the operator need to be well coordinated. Every time there is a pick, the rigger is responsible for guiding the operator, making appropriate and safe connections to whatever is being hoisted, and then guiding the load into place. Some contractors prefer to have a dedicated person do this task; others let it be handled by a variety of trained riggers. From an estimating perspective, the important thing to remember is that this person(s) needs to be accounted for somewhere in the estimate. It is also safe to assume that the number of hours required will be similar to those of the operator.

Utility costs could be estimated based on the power usage of the crane, operating time, and local utility rates, but most contractors just plug an allowance based on experience. Sometimes the electrical costs are shifted to the owner, who assumes the risk but also saves on tax and overhead costs. Regardless of how the costs are paid for, the cost can be tens of thousands of dollars and needs to be accounted for in someone's budget.

Lastly are the maintenance costs of keeping the crane in good working order. Typically, the rental agreement will allow for routine inspections and service or if there is a major issue with the crane but will not include costs for service due to wear and tear to the equipment. Generally, it is not appropriate to include significant costs for repairs due to wear and tear, but it is reasonable to include costs for things that would be expected to need replacement/repair during the course of the project. For example, think of the tires and brakes on your car. Again, this is something that the estimator should evaluate with the project team when reviewing project requirements and the rental agreement when determining appropriate amounts to include in the estimate. Major service or repair items not covered under the rental agreement should be clearly excluded from the contractor's contract estimate to the owner if this is a negotiated project. Initial periodic inspection costs should also be included with the operational costs.

Because the tower crane is expected to be on the project for several months, it is important to thoroughly evaluate the monthly operation and maintenance costs when estimating the project. Underestimating the time the crane is required on a project by even one month could cost the project tens of thousands of dollars.

Dismantling

The process for dismantling or disassembly of the tower crane is similar to erection but is oftentimes more difficult due to the fact that there is now a building structure to maneuver around. It may be simple to assume the same costs for dismantling as the erection, but factors that may have not been required during erection, such as street use fees, may now come into play. Some other more complicated options for disassembly in tall high-rise buildings include a helicopter (which is very dangerous and requires all sorts of other permits) and the use of a jib crane mounted on top of the building that disassembles and lowers the crane to the ground. It is important for the estimator to evaluate these conditions with the project team before completing the estimate. A tower crane estimate for our case study is shown in Figure 13.3.

City Construction Company
Cost Recapitulation Sheet

Project: Dunn Lumber
Location: Seattle, WA
Arch./Engr.: Flad

Date: 6/10/21
Estimator: P. Jacobs
Estimate #: 1

Tower Crane

Code/ Det	Description	Qty	Unit	UMH	Man Hours	Wage Rate	Labor Cost	Matl/Equip/Sub Unit Cost	Matl/Equip/Sub Cost	Total Cost
	Base equipment charges:									
	City permit	1	allow					10,000	10,000	10,000
	Monthly rental	12	mo					15,000	180,000	180,000
	Excess rent @ 40 HR/mo	480	HR					95	45,600	45,600
	Monthly repair agreement	12	mo					1,980	23,760	23,760
	Technical support: Supplier	6	days					3,500	21,000	21,000
	Retorque tower secction	1	LS					1,600	1,600	1,600
	Freight in	1	LS					9,750	9,750	9,750
	Freight out	1	LS					7,000	7,000	7,000
	Structural cost:									
	Foundation	1	LS	400	400	40	16,000	14,000	14,000	30,000
	Foundation anchorbolts	1	LS					13,500	13,500	13,500
	Erection/dismantle:									
	Erection (subcontractor)	1	LS					72,900	72,900	72,900
	Traffic control	2	LS	30	60	35	2,100	0	0	2,100
	Street use permit	2	LS					950	1,900	1,900
	Dismantle (subcontractor)	1	LS					77,300	77,300	77,300
	Operation cost:									
	Rigging supplemental steel	1	allow					6,000	6,000	6,000
	Temporary power	12	mo					1,500	18,000	18,000
	Safety/small tools	1	allow	100	100	40	4,000	4,000	4,000	8,000
	Tower Crane Insurance	940,000	ST$					1%	9,400	9,400
	Inspections: Supplier	12	mo					2,000	24,000	24,000
	Communications	12	mo					250	3,000	3,000
	Operator straight time	52	wks	40	2,080	50	104,000	0	0	104,000
	Operator OT	52	wks	20	1,040	75	78,000	0	0	78,000
	Rigger/bellman straight time	52	wks	40	2,080	56	116,480	0	0	116,480
	Rigger/bellman OT	52	wks	20	1,040	84	87,360	0	0	87,360
	Page Totals (to general conditions)				6,800		$407,940		$542,710	$950,650
	Add labor burden @ 55% of labor								$224,367	$224,367
	Total Cost									$1,175,017
	Wage Check:	Checks? High, but premium for TC operation plus OT							$/Hour:	$59.99
	Monthly Check:	Checks? Yes							$/MO:	$97,918

Figure 13.3 Tower crane cost recapitulation sheet

Personnel and material hoists

Personnel and material hoists, formerly 'manlifts', are also known as temporary construction elevators. They are typically required on buildings taller than four stories. State labor laws and union agreements will have an impact if a GC is required to provide these hoists, but often it is a practical time and cost efficiency decision made by the GC. The process to prepare a hoist estimate is very similar to the tower crane estimate discussed in the previous section. An example estimate for a personnel and material hoist is shown in Figure 13.4.

Exterior areas of the building skin will need to be left out to allow access to each floor. These areas will later be in-filled, but the work is completed out of sequence and subcontractors will need to remobilize. In addition, platforms will need to be built at each floor to allow safe access in and out. A rigger is not required, but a full-time elevator operating engineer is required.

City Construction Company
Cost Recapitulation Sheet

Project: Broadway MXD Apartments
Location:1200 Broadway, Seattle, WA
Architect: The Design Center
Personnel/Material Hoist

Date: 7/13/21
Estimator: LH
Estimate #: 1

Code/ Det	Description	Qty	Unit	UMH	Man Hours	Wage Rate	Labor Cost	Matl/Equip/Subs Unit Cost	Matl/Equip/Subs Cost	Total Cost
	Personnel/material hoist	12	mo					12,000	144,000	144,000
	Hoist premium cost for overtime	12	mo					1,900	22,800	22,800
	Erection	1	EA					15,000	15,000	15,000
	Dismantle	1	EA					20,000	20,000	20,000
	Hoist technician – 2 × 2 days	4	days					1,400	5,600	5,600
	Jump/climb hoist – 2 times	2	EA					2,500	5,000	5,000
	Upper floor connection	3	EA					750	2,250	2,250
	Floor gates rental – 12 mo	168	mo					125	21,000	21,000
	Floor gates install	14	EA					380	5,320	5,320
	Floor gates remove	14	EA					125	1,750	1,750
	Freight	2	EA					3,500	7,000	7,000
	Temporary power	12	mo					1,400	16,800	16,800
	Foundation	1	LS	75	75	40	3,000	15,000	15,000	18,000
	Anchor bolts	1	LS	20	20	40	800	9,000	9,000	9,800
	Operator – straight time	52	wk	40	2,080	44	91,520			91,520
	Operator – overtime	52	wk	10	520	66	34,320			34,320
	Radio communications	12	mo					850	10,200	10,200
	Page Totals (to general conditions)				2,695		$129,640		$300,720	$430,360
	Add labor burden @ 55% of labor								$71,302	$71,302
		Total Cost								$501,662
	Wage Check:	Checks? High, but premium for OT							$/Hour:	$48.10
	Monthly Check:	Checks? Yes							$/mo:	$41,805

Figure 13.4 Personnel/material hoist estimate

Pricing recap completion

When all labor, material, and equipment pricing recap entries have been completed and the extensions made, the man-hour, labor cost, material cost, equipment cost, and total cost columns are summed to totals at the bottom of the page. Two checks should be made at this point. First, the page should be scanned to make sure it is complete and that every line item has the entries and appropriate calculations. The summation ranges should also be checked. The format of the page is such that the line items are totaled to the total cost column, which is then summed to a total at the bottom of the page. A crosscheck now should be made. This check adds the totals of the labor, material, and equipment cost columns to make sure that the result equals the sum of the total cost column. If it does not, there is an error on the recap sheet.

When completed, the estimator needs to verify if the system or assembly cost is competitive. A quick assemblies check, as shown at the bottom of the pricing recap in Figure 11.1, will provide a unit cost that can be used as a check against other estimates or the contractor's database. An assemblies check unitizes certain figures to a common quantity on the recap sheet. In this case, a cost per man-hour check and the total cost per cubic yard placed are calculated. The first figure results in an average $37.10 per man-hour to construct the spread footings. The second figure is the total cost of $333 per cubic yard. If the estimator has done several estimates, he or she will attain a level of comfort with these figures.

Summary

The estimator needs to ensure that the cost of all equipment used on a project is properly accounted for. Equipment that is used for a single activity should be charged to that activity, while equipment used for multiple activities should be included in the jobsite general conditions estimate. Use of published databases for equipment pricing is not recommended because rental rates will vary based on the source of the equipment. ABC is used to allocate project indirect costs to the cost of direct construction activities.

Pricing for concrete pumps is commonly needed for many construction projects. The cost of a concrete pump will vary based on the size of the pump, the quantity of concrete to be pumped, and the length of time required to pump the concrete. Tower cranes typically are used for extended periods of time and support multiple construction activities. Therefore, the tower crane cost is included in the jobsite general conditions estimate. Tower crane cost estimates have four components: foundation and infrastructure, crane erection, operation and maintenance, and crane disassembly. There are many other types of equipment used on construction projects, and similar approaches should be used in estimating their cost.

Review questions

1   What are four sources that might be used to obtain construction equipment needed for a construction project?
2   What type of cost information can be obtained from an equipment lease agreement?
3   A contractor needs a concrete pump for a project that can pump concrete a distance of at least 150 feet. The concrete placement will last six hours and involve the placement of 150 CY of concrete. What is the estimated pumping cost in $/CY for this concrete placement?
4   Where is the rental cost shown for equipment that is specific to an activity of work such as concrete tilt-up or structural steel?
5   What is the difference between the rigger and the operator? What labor craft is the rigger usually, and why that choice?
6   With whom should the estimator discuss the tower crane estimate before completing it?
7   Why are certain pieces of construction equipment included with jobsite general conditions and not attributed to specific scopes of work?

Exercises

1   Without looking ahead to our general conditions chapter, what equipment not addressed in this chapter might also be estimated and cost-coded to jobsite general conditions?

2   Make a list of ten pieces of equipment that might be used on a construction project and identify how each might be procured. Are they subject to ABC or costed to jobsite general conditions? Will they be obtained with a subcontract agreement or a lease agreement and operated by the GC's craftsmen?

3   Even if it is not a union project such that a full-time operating engineer was required to run the forklift, why is that still a good idea?

4   What is the cost of a 20 FT × 20 FT × 8 FT concrete tower crane footing based on the following parameters? The process to be used is similar to the process described in Chapter 6 for quantity take-off of a spot footing and the pricing process described in Chapter 11. Make sure you perform a $/MH and $/CY check when you are done. Make whatever assumptions are necessary.

   a   Formwork material cost at $2/SFCA and 0.5 MH/SFCA
   b   Concrete purchase including pumping at $150/CY
   c   Rebar tonnages at 200 LBs/CY
   d   Rebar purchase and install costs at $1,000/ton
   e   Rebar template at $100 for material plus one-hour labor
   f   Eight tower crane anchor bolts at $5,000
   g   Overhead and profit at 20%
   h   Excavation and backfill at $1,550

5   What portion of the tower crane estimate is based on time?

6   Why would a GC choose not to own some of the major pieces of equipment discussed in this chapter?

7   A photograph of a tower crane is included in Chapter 16. What type of tower crane is that: Horizontal jib (also known as a hammer-head) tower crane, self-erecting tower crane, or luffing tower crane?

Part IV

General contractor estimates for subcontracted work

14 Envelope and finish subcontracted scopes

Introduction

In the last two parts we focused on the quantity measurement and pricing practices that general contractors (GCs) undertake for the cost estimation of direct work performed by their own craftsmen. Now, with this chapter, we shift our attention to the companies that generally perform 80 to 90% of the work on a typical commercial project, the subcontractors. Subcontractors may also be known as specialty contractors or trade contractors, and today many negotiated GCs prefer 'trade partners'. Much of the work required to complete a project involves subcontractors and major material suppliers. Some GCs estimate this work in detail and plan, upon winning the project, to solicit subcontractor and supplier quotations afterward. This is risky and may not always prove to be the best method of doing business. A more common practice is for general contractors to solicit quotations from subcontractors and suppliers while preparing their bids. The advantage is that the GC, upon being awarded the contract, can immediately enter into agreements with the subcontractors and material suppliers knowing their final prices. The general contractors in this scenario, however, do not know the subcontractors' and suppliers' prices until near the time that bids must be submitted to the owner. To protect themselves, the contractors should prepare order of magnitude (OM) estimates for the subcontractors' scope of work and the major suppliers' materials. That process is the focus of this and the next two chapters.

The sixth step in the lump sum estimating process shown in Figure 1.1 provides a way to determine an approximate estimate of the cost of the overall project. In this step the GC estimator creates OM or 'plug' estimates for the work that will be done by subcontractors and material furnished by major suppliers. These are allowances or placeholders until competitive market subcontractor quotes are received. Order of magnitude estimates are a reasonable determination of the expected range of subcontractors' or major suppliers' quotations. The method for developing these estimates is not as detailed as was used for self-performed direct work but is based on historical or published reference data. Since these estimates are based on averages for several projects, they do not consider some of the unique aspects of the project being bid. Order of magnitude estimates for major scopes should not be relied upon to complete the contractor's bid to the owner. Some of the smaller OM estimates, however, can be used in cases where no competitive vendor quotations are received. Order of magnitude estimates are useful in determining the pre-bid summary estimate so that various markups can be determined. The pre-bid summary estimate will be discussed in Chapter 19.

In this chapter we focus on the subcontractors covered by Construction Specification Institute (CSI) divisions 04, 07, 08, 09, and 12. This will cover most of the subcontractors involved in the building envelope, which may also be known as the building enclosure or 'skin'. The major subcontractors here include waterproofing, weather protection, curtain wall, roofing and others. In addition, this chapter will cover many of the subcontractors involved in building finishes, which includes drywall, painting, floor covering, and others. Although we presented the substructure and superstructure as areas estimated direct by the GC's estimator, some of the work in those categories is also subcontracted and is covered in this chapter as well. In the next two chapters we will expand our subcontractor coverage to some of the most significant team members, including mechanical and electrical and site work subcontractors.

Subcontractor work

In Chapter 1, the estimator determined what portions of the project will be performed by subcontractors and what materials would be provided by major suppliers or fabricators. When listing subcontractors on the project item list, the estimator must be aware that some direct work sections will also be listed. For example, the general contractor may install the reinforcing steel direct, but a fabricator supplies it. The same usually is true for the structural steel. On bid day, the general contractor may receive bids not only for supply but also for reinforcing steel installation and structural steel erection.

General contractors typically perform work direct if that work is accomplished by their own craftsmen as a normal course of business. Commercial GCs likely employ the following labor crafts to install these associated scopes of work:

- Carpenters install concrete formwork, rough carpentry or wood framing, finish carpentry and millwork, doors/frames/hardware, punch window installation, and specialties.
- Laborers shovel and rake earthwork, perform general cleanup, install below grade vapor barriers and waterproofing, assemble perforated pipe drain systems, and place concrete.
- Ironworkers install rebar in concrete forms and erect structural and miscellaneous steel.
- Cement masons finish concrete slabs and patch and sack vertical concrete surfaces.
- Operating engineers run heavy equipment and also perform surveying.
- Teamsters primarily drive trucks.

General contractors typically do not employ crafts such as glaziers, plumbers, sheet metal workers, or electricians; therefore, they almost always subcontract that work out to specialty contractors. The work normally or potentially installed with a GC's own workforce receives the most attention from the estimator and was the focus of our last several chapters. Figure 14.1 is a partial pre-OM subcontractor list for the case study project.

When completing the subcontractor list, the estimator uses the work breakdown structure as a guide. The estimator should populate a preliminary subcontractor list with in-house developed 'plug' estimates and not wait until bid day. This list is used more than once in the bid completion process. Each time that it is used, it should be prominently identified, such as "OM Estimates", "First Run", or "Final Run", so that the wrong version will not be used in the final bid. The completed pre-bid day subcontractor OM list for our case study is included at the end of Chapter 16.

City Construction Company
Subcontractor OM Plug Estimates
(Partial Template)

Project: Bin	Spec	Dunn Lumber Description	Estimator: Paul Jacobs Est. #: OM GC OM Plug	Comments
1	24100	Demolition	$	Buildings & Site
2	31113	Form Elevated Slabs	$	
	32100	Reinforcement Steel	$	w/Directs
	33000	Pump Concrete	$	w/Directs
	33500	Finish Slabs	$	w/Directs
3	39000	Shotcrete	$	
4	42200	Interior Garage CMU	$	
	51000	Misc. and Structural Steel	$	w/Directs
16.1	54100	Exterior Metal Stud Framing	$	Sub w/GWB
5	64100	Casework	$	
6.1	71400	Waterproofing	$	w/Traffic

Figure 14.1 Subcontractor OM plug estimates

As a reminder, not all materials are on this list. Only those that are procured from major suppliers, such as reinforcing steel and structural steel, are listed. Those direct work items for which the estimator expects to receive subcontract bids should be broken out of the direct work estimate and listed on the subcontractor list. This is done by taking the total cost for the work, including labor taxes, and treating it as a vendor bid, with the general contractor being the vendor. This price is then entered as one of the prices on the bid evaluation worksheet on bid day and noted as "GC Direct" or in our case "CCC Direct". The estimator must make sure that the cost of these work items is eliminated from the direct work portion of the estimate. If the subcontractor's quotation is lower than the estimated cost of self-performing the work, the general contractor may choose to execute a contract with the subcontractor.

The procedure for generating an OM estimate is to perform a simple quantification of the work to be done and then apply assembly unit prices. Sources of unit price data include published estimating references, historical databases, and the estimator's personal notebook. Subcontracted unit prices from *RS Means* can be derived from the far right-hand pricing column for most materials and systems, which includes labor, material, equipment, and subcontractor markups. *The Guide* has separate sections within each major CSI division – one for direct labor, one for direct material, and one for systems in place – that can also be utilized for subcontractor pricing plug estimates.

Select subcontractors may also provide the estimator with a pre-bid day budget or unit price to use as a placeholder for the GC until their official bid is submitted. Prices for OM work do not need to be broken down into labor and material. Most references will show an overall unit price that includes a reasonable overhead and profit for the subcontractor. If productivity information is available, it can be entered on the recap sheet and extended to provide the total man-hours for the work. This information will be helpful when generating the project summary schedule discussed in Chapter 17. Creating OM estimates for the work that is to be subcontracted on the case study project involves the use of several techniques, as will be discussed in the next sections.

After the bid has been tendered, the GC estimator should compare the OM estimates with the vendor quote. If there is a wide variation, it is useful to determine the reason. Performing this post-bid analysis will increase an estimator's skill in developing more realistic OM estimates in the future. This chapter will describe the basic procedures a general contractor estimator will utilize to prepare OM estimates. Order of magnitude estimates for building envelope and interior finish subcontractors will be discussed in detail along with specific examples for our case study project.

Building substructure and superstructure

Masonry (CSI division 04) is one of the major work items in the substructure or superstructure that is done by a subcontractor. Most estimating references have good information for preparing masonry subcontractor OM estimates, and often company historical data can be used. Units generally are square foot measure or piece count. The GC estimator measures height times width and typically only deducts for major openings, such as curtain wall. The resultant square footage is then multiplied by an all-in subcontract unit price such as:

1,610 SF of CMU wall @ $20/SFW = $32,200 of CMU

We utilize a lot of abbreviations in these boxed-in examples and equations but have not spelled out each for simplicity. All of the abbreviations used in this book are included in the front material.

Most of the work in division 05 involves materials that are furnished by steel fabricators and installed by the general contractor or a specialty contractor that employs ironworkers. The major variation is cold-formed metal framing, which usually is furnished and installed by a drywall subcontractor as part of the gypsum wallboard (GWB) systems. There may be a few other minor work items that will also be subcontracted from division 05, and a thorough review by the estimator will pick these up.

Division 06 covers wood and plastics. Rough carpentry is commonly used for the superstructure of various housing projects, but its use in commercial and institutional buildings is limited. It is common to find rough carpentry for accessory items, such as canopies, soffits, and walls in commercial projects. If subcontracted, an OM estimate would be similar to that for direct work, and the estimator should use reference data for unit prices or contact lumber suppliers to determine current prices. Rough framing is generally a subcontracted item for residential construction. Our case study includes a variety of enclosure or siding materials, including some wood siding. The GC has chosen to subcontract all that work, and an OM estimate will be included as follows:

18,500 SF of wood siding @ $25/SFW = $462,500

Building envelope

Division 07 covers those items that provide thermal and moisture protection for the building structure. The building envelope includes all of the materials that surround the building and protect it from the elements. The main items are foundation moisture protection, waterproofing, insulation, roof systems, fire stopping, and caulking and sealing. Most quantities are either square foot or lineal foot, and dimensions previously determined in other parts of the estimate can be used for these as well. Several pre-bid day subcontractor OM estimates have been prepared by City Construction's estimator for the division 07 work for our case study project and are reflected in this section.

Roofing

Roofing is a fairly simply process of measuring the length of the building times the width. If there is a significant slope, such as is the case with residential projects, then the slope needs to be accounted for in an $a^2 + b^2 = c^2$ formula. Some roofing continues up the inside of parapet walls and also needs to be added in. Included with the roofing scope often are cant strips, curbs, roof accessories, flashing, and rigid insulation. Traffic coating may also be performed by roofers, and because of our case study's garage ramps, this is also a significant cost. A roofing subcontractor will quantify and estimate all this work separately, but the GC's estimator will factor a combined systems estimate such as:

Roofing: 42,000 SF of roof @ \$13/SFR = \$546,000
Traffic coating: 20,000 SF of deck @ \$20/SFD = \$400,000

Siding and waterproofing

There may be a multitude of different types of siding and specialty subcontractors involved with a commercial building's envelope. Wood siding will be installed by a different trade – carpenters – and was discussed earlier. The siding subcontractor, if there is only one, may also install the metal studs, insulation, and waterproof barriers and is often responsible for their own scaffold systems. Siding and waterproofing plug OM estimates for our project are reflected as follows:

Metal siding: 40,000 SF @ \$22.50/SF = \$900,000
Waterproofing: 38,462 SF @ \$16/SF = \$615,392

Fireproofing

Fireproofing is extremely difficult to take-off exactly because it generally covers most of the structural steel members and metal decking, but there are exceptions to this rule. Typically, the fireproofing subcontractor will need to generate shop drawings, often shared with the city, to confirm what areas need to be coated and in what thickness. The easy way for a GC's estimator to develop a plug estimate is by the underside square foot of the floor structure to be treated and is reflected in the following:

163,200 SF of steel structure @ \$2.70/SF = \$440,640

Insulation

Insulation is also specified in CSI division 07 and can be estimated here along with siding and roofing, or can be estimated along with finishes, discussed later. It is common for the drywall subcontractor also to be the insulation installer, as the two scopes of work are typically installed together. Insulation should be separated by type (rigid, batt, spray-on) and by the insulation value, or 'R' value (R19 versus R30, etc.). An insulation QTO and pricing recap should also be separated by thermal or acoustical. An insulation subcontractor would definitely make these distinctions, and more. But a GC estimator preparing a plug estimate may group all insulation together, as reflected at the end of this section. Many quantities will have similar measurements, such as the roofing and siding may match the thermal insulation, and the drywall, discussed later, may include the acoustical insulation.

Additional division 07 work that appears on all projects may be difficult to quantify; it is estimated as a lump sum allowance and will be replaced out with market-priced subcontractor or supplier prices on bid day. Examples of this include caulking and fire stopping. Many GCs will try to buy these scopes out with other subcontractors such as the waterproofing or drywall subcontract or shift responsibility to the mechanical and electrical subcontractors to caulk and fire stop their own work.

Insulation: 133,000 SF @ $2/SF = $266,000
Caulking: Allow one lump sum $25,000
Fire stopping: Allow one lump sum $12,000

Doors and windows

Doors and windows are included with CSI division 08. As explained in the direct work QTO for self-performed work, the GC may install single or *punch windows* purchased from a supplier with its own carpenters but would need to hire a glazing subcontractor that would design the *storefront window systems* and install it with their own glaziers. Storefront is estimated on a square foot of wall (SFW) for the GC, and an OM installed price is applied. Skylights may be priced on a square foot of glass or by the each. Exterior glass doors will be plugged on a $/leaf basis, similar to other doors, except these will be furnished and installed by the storefront subcontractor.

Doors, door frames, and hardware (DFH) are furnished by major suppliers and often installed by the general contractor. The estimator already has prepared a QTO and an installation estimate when estimating the direct work. Depending on the material for the doors and frames, there may be one supplier who bids all doors, door frames, and hardware, or there may be up to six vendors supplying different portions, such that one bids wood door leafs and another that bids hollow metal (HM) door leafs and frames, and others. The OM material estimate is prepared based on the direct work QTO and will be priced in units similar to the following:

- Doors: $/door leaf, for example $520 for solid core wood door and $700 for HM door leafs. Double doors will count as two leafs.
- Door frames: $/frame or opening, for example $300 for a single HM door frame and $500 for a double HM door frame. Frames that require grouting will be accomplished

in the field by the GC's laborers or cement masons. Grouting should be included on the GC's 08 direct work estimate recap.

- Door hardware: $/set such that each leaf has one set of hardware, for example $1,100 for commercial door hardware sets for projects that have panic hardware, such as our case study, can be plugged as an OM estimate until competitive bids are received on bid day. Nonpublic commercial door hardware sets may cost in the $300 to $500 range per set to purchase.

Overhead doors are also supplied and installed by a subcontractor and will be estimated on a $/each basis but noting the differences in sizes and materials. Door operators may be supplied by the door subcontractor or may be separate and would also be plugged with a $/each allowance. Our case study project includes two different categories of these specialty doors for the garage and for the retail spaces. Access doors are quantified by the each and may be included with division 08 or division 10 specialties, or the GC may assign them to the mechanical and electrical subcontractors whose work they affect.

The estimator for our case study project has grouped all of the CSI division 08 subcontractors and suppliers together onto one pricing recap sheet, as reflected in Figure 14.2. Each of these specialties will be separated out into individual bins for bid day, as discussed later in this book. The same cost recap sheet as was used for the direct work is the best format for preparing subcontractor OM estimates or may be compacted for simplicity and eliminate the labor columns.

City Construction Company
Cost Recapitulation Sheet

Project: Dunn Lumber
Location: Seattle, WA
Arch./Engr.: Flad
Date: 6/10/21
Estimator: P. Jacobs
Estimate #: 1

CSI Division 08: Doors and Windows

Code/ Det	Description	Qty	Unit	UMH	Man Hours	Wage Rate	Unit L Cost	Labor Cost	Subcontractor/Material Unit Cost	Cost	Total Cost
8.01	Single HM Door Frames	170	EA	1.0	170	41	41	6,970	300	51,000	57,970
8.02	Double HM Door Frames	50	EA	1.5	75	41	61.50	3,075	500	25,000	28,075
8.03	HM Door Leafs	70	Leafs	1.0	70	41	41	2,870	700	49,000	51,870
8.04	Wood Door Leafs	150	Leafs	1.0	150	41	41	6,150	520	78,000	84,150
8.05	Door Hardware	220	Sets	5.0	1,100	41	205	45,100	1,100	242,000	287,100
8.06	Relights	25	EA	0.5	13	41	20.50	513	150	3,750	4,263
8.07	Grout HM Door Frames	220	EA	2.0	440	34	68	14,960	50	11,000	25,960
8.08	Door and Hardware Room	1	LS	244	244	41	10,000	10,000	5,000	5,000	15,000
8.09	Access Doors	35.0	EA	1.0	35	41	41	1,435	35	1,225	2,660
	Subtotal:				2,297			$91,073		465,975	557,048
8.11	Large Garage Security Doors	5	EA				Sub	0	47,800	239,000	239,000
8.12	Retail Roll-up Glass Doors	7	EA				Sub	0	20,000	140,000	140,000
8.21	Punch Windows	750	EA	1.60	1,200	41	65.60	49,200	400	300,000	349,200
8.22	Storefront Glass System	34,000	SFG				Sub	0	100	3,400,000	3,400,000
	Page Div 08 Totals				3,497			$140,273		$4,544,975	$4,685,248
	Not transferred direct to bid summary as some are direct and some are materials only and some are subcontracted scopes										
	Wage Check: $/Hour	Checks? Yes								$/Hour:	$40.12

Figure 14.2 Doors and windows cost recapitulation sheet

Finishes

Division 09 covers finishes throughout the building, which is typically all subcontracted work. Most of this work involves dimensions and quantities that have been determined for other items. Floors, walls, ceilings, tile, painting, and wall coverings as well as floor coverings are shown on the architect's finish schedule. Most of the finishes are defined in the drawings and specifications and can be quantified readily. There is almost an infinite amount of finish specialty materials, with new ones developed daily. We have included examples of some of the more common ones that appear in most commercial construction projects. In projects that are budgeted early, before a room finish schedule has been established, the estimator will simply use an allowance or plug in dollars per square foot of floor ($/SFF). In most cases, the unit prices can be determined from published estimating databases or historical records. In this section, we have presented the finishes scopes in the relative order in which the work will be installed, not necessarily by strict CSI order.

When preparing the OM estimate for *walls and partitions*, an estimator should be aware of different construction methods and how they affect cost data. Two-sided partitions will be priced per square foot for a complete wall system, which includes framing, wallboard on both sides, and taping. Perimeter walls will be estimated as one-sided.

The experienced estimator will do two things to enhance his or her ability to efficiently prepare OM estimates. The first is to use assemblies or systems pricing whenever possible and reliable unit prices that include all related costs. Second, the estimator will make sure the quality level of the system is understood and representative unit pricing is applied. For example, an interior partition will include the framing, wallboard on both sides, and one of four quality levels of taping finish. Also, adjustments may have to be made for things such as $^5/_8$-inch wallboard in lieu of ½-inch or fire-rated or waterproof board instead of standard board. Some wall assemblies also include insulation and potentially paint.

When taking off the quantities for OM estimates, an estimator often encounters fractional dimensions. These should be rounded up to the nearest reasonable unit (inch or foot). The estimator, in the interest of improving efficiency, may want to make a template QTO that lists all of the rooms and their dimensions. Copies can then be used for the various finishes, as well as walls, floor coverings, or ceilings.

Columns on the *floor covering* subcontractor QTO worksheet are set up for all different types of floor covering and base. Base is typically quantified in lineal feet and most flooring in square feet, although carpet is often later converted to square yards. The floor covering template with the room dimensions can be used for other finishes as well. *Ceilings* are nearly always the same as the floor area, and columns are set up for the different types that may be used, such as acoustical ceiling tile (ACT) and drywall. Acoustical or wood or GWB ceilings are priced by the $/SF. Another QTO sheet will be used to quantify the walls by type and finish, and yet another for paint and wall covering. *Paint* is estimated by the $/SF of wall and $/SF of ceiling in the case of GWB ceilings.

Subcontracted quantities are transferred to the pricing recap sheet using the same error control procedures that are used in direct work estimating, discussed in previous chapters. For OM estimates, several changes are made to the recap sheet. The material unit cost and cost columns are changed to read S/C (subcontract) unit cost and subcontractor cost. Since these are subcontractor and major supplier OM estimate sheets, the hours, labor, and total cost columns are not used and have been eliminated for simplicity from

City Construction Company
Cost Recapitulation Sheet

Project: Carpenters Training Center   Date: 6/10/21
Location: Kent, WA   Estimator: R. Martin
Arch: LMA   Estimate #: 1
CSI 09: Floor coverings

Code	Description	Qty	Unit	Sub U. Cost	Sub OM
9.6.1	Sheet Vinyl	102	SY	45	$4,590
9.6.2	Carpet	959	SY	50	$47,950
9.6.2.1	Carpet Pad	959	SY		included
9.6.3	VCT	8,755	SF	3.00	$26,265
9.6.4	Rubber Base	6,120	LF	2.00	$12,240
9.6.5	Floor Preparation	18,304	SFF		included
	Total floor coverings:				**$91,045**

Figure 14.3 Floor covering subcontractor OM cost recapitulation sheet

the following example. Figure 14.3 shows an OM subcontractor recap for all of the floor covering scopes for another project one of your authors managed.

Most projects will have some architectural wood, such as moldings, base, and other types of trim. Using the room finish templates discussed earlier, the estimator should be able to quantify the items and complete the OM estimate. Cabinetry and casework will be furnished and installed by subcontractors on commercial projects, but the GC may install small installations with its own carpenters. The estimator should use published or company data sources as a guide for quantifying and pricing.

Specialties, equipment, and furnishings

CSI divisions 10, 11, and 12 contain many items that may be specified for a given project. The estimator needs to be sure how they are specified in relation to the general contractor's direct work estimate. Some will be purchased and installed by the general contractor, some furnished by the contractor for installation by a subcontractor, some furnished by a supplier to be installed by the general contractor, and some furnished and installed under a separate subcontract agreement. There are no required rules or procedures for specialty OM estimates, and the estimator should consult references to determine the best way to determine unit costs. Many items are customarily purchased from a supplier and installed by the GC's carpenters, and preparation of their OM estimates is straightforward. Others, such as projection or sound equipment, are more complex and may be subcontracted work. Some specialty scopes and their unit of measure for applying OM unit prices include:

- Many items that have already been quantified on the direct work QTOs and are purchased by the each, such as toilet accessories, fire extinguishers, signs, corner guards, white boards, and others.

- Toilet partitions are included in CSI division 10 and will be supplied and installed by a subcontractor. A GC's estimator will plug them on a $/stall basis. The same supplier who furnishes the toilet accessories (with a purchase order) may also provide the toilet partitions (with a subcontract agreement).
- Dock levelers are included in CSI division 11 and priced as $/ea. This work will be performed by a specialty contractor. Steel embeds associated with dock levelers should be included with the concrete estimate and are likely furnished by the miscellaneous steel supplier. Power to the dock leveler is in the electrical subcontractor's scope.
- Window blinds or curtains and other window treatment are included with CSI division 12, along with furnishings. Window treatment may be estimated by the GC's estimator on a $/EA or $/SF basis, size and material dependent. If more than a few blinds are required, this work will also be performed by a specialty contractor. Furnishings are typically not in the GC's scope.

Summary

Order of magnitude estimates should be developed by the GC's estimator for all subcontracted work items to have a means of preparing a pre-bid day summary estimate and to have a basis for comparing subcontractor quotations on bid day. For most of the project scope, OM estimates can be developed by making simplified QTOs and applying either assembly costs or unit prices per square foot of floor area. When taking off building dimensions, an estimator can set up one QTO sheet for the flooring, one for the ceiling, and one for the framed walls, using the same data. The best source of information is to use unit prices, first from subcontractors and then, if not available, from historical references based on the type of project. Mechanical and electrical OM estimates are more difficult to determine and are discussed in the next chapter.

Review questions

1 What work should OM estimates be done for?
2 What are two purposes for developing OM estimates?
3 Where are the results of the OM estimates listed?
4 What type of source material is good to use for pricing OM estimates?
5 If subcontractor prices are received for work usually done as direct work, how should these be handled?
6 How can the estimator make his or her work easier when preparing OM estimates?
7 Why is the resilient base estimated separately from the resilient flooring?
8 If a GC is certain of receiving competitive subcontractor bids on bid day, why should it bother preparing OM estimates? There are several good reasons.
9 What is the difference between a subcontractor and a supplier? Note that just answering "labor" is not correct, as suppliers often fabricate materials and deliver them to the jobsite with their own labor.

Exercises

1 Prepare an OM estimate for a system or assembly from one of our case study projects that was not resolved in this chapter, such as the Vehicle project's metal siding and roofing scopes.

2   If you prepared an OM estimate for ACT for $100,000 on your own project and one of the following happens on bid day, what do you do?

   a   Not one subcontractor bid is received,
   b   One subcontractor bid of $90,000 is received,
   c   One subcontractor bid of $110,000 is received,
   d   Two subcontractor bids are received, one at $50,000 and another at $95,000, or
   e   Two subcontractor bids are received, one at $105,000 and one at $150,000.

3   The Lee Street case study project did not include a formal finishes schedule in the contract documents, as the builder was also the developer and the finishes evolved as the building was constructed. Utilize the photographs and drawings from the book's eResource and prepare a CSI 08 or 09 OM recap for one of these executive townhomes.

15 Mechanical, electrical, and conveyance systems

Introduction

Major portions or elements of the completed building 'function' or run in that they are operating 'systems'. This includes the mechanical, electrical, plumbing (MEP), fire protection, and elevator systems that are the focus of this chapter. Because their costs are so significant, development of accurate order of magnitude (OM) budgets by the general contractor's (GC's) estimator is important. The more accurate the total estimate is leading into bid day, the more efficient will be the bid day process.

Most of the systems in this chapter are typically procured and designed in one of two ways. The conventional method is for the building architect to employ mechanical and electrical engineers and fully design the systems in preparation of a lump sum bid. Architects also engage an elevator consultant to assist with the elevator design early in the schematic design phase. The other common method is for all or some of these systems to be design-build (DB) by the subcontractors that will perform the work. Their prices can be negotiated or can be bid. In the case of DB subcontractors, the GC's estimator would have little to no design drawings to assist with his or her OM estimate. The third approach is a hybrid of these two known as bridging or design-assist. In this case the design consultants prepare preliminary schematic drawings and specifications or design criteria from which the subcontractors can prepare estimates and complete the design themselves.

Some categories of subcontracted work such as mechanical and electrical represent significant portions of the project cost and are based on information that may not be completely detailed in the bid documents. It is difficult for an estimator to anticipate all of the work because such things as routing of pipes, ducts, and wiring are often left up to the subcontractor to determine. This is part of their means and methods. The GC estimator therefore often relies on rough OM estimating techniques for these systems.

These systems all represent significant costs in any typical commercial construction project, and the more accurate an OM estimate the GC can prepare, the smoother the bid day operation will be. The best method to prepare these high-value OMs is to call the subcontractors early, shortly after the bid documents have arrived and the GC has made a decision to pursue the project. If subcontractors can provide early 'budget' pricing for the GC, such as "the fire protection for this project will likely be in the $3/SF to $3.25/SF range", this would be very beneficial and preferred over using a database. These costs can vary significantly depending on the type of work. Unfortunately, some GC estimators will spend an inordinate amount of time only quantifying scopes they are familiar with, such as counting concrete formwork snap ties or anchor bolts, but may miss the MEP scopes by millions of dollars. All of the plug estimates discussed in this chapter include the subcontractor's labor and associated labor burden, materials, equipment, second- and third-tier subcontractors, startup and testing, and add-on markups.

Conveying equipment

Construction Specification Institute (CSI) division 14 includes a variety of complicated 'conveyance' systems, including elevators, escalators, and horizontal people-movers. Internal crane systems for industrial projects may be part of division 14 or included with 13. Division 14 scopes are not typically included with MEP, but we have added it to this chapter because the equipment operates, is typically long lead, and costs significant amounts of money. The GC's estimator will approach OM work for this division similar to that of the MEP systems. Because elevators 'move' or function as do other systems discussed in this chapter, the beginning estimator mistakenly views this as mechanical work. Elevators, for example, are not designed by the mechanical engineer but typically by a separate elevator consultant working direct for the architect. There is substantial published estimating data on elevators, but the estimator has to be careful to fully understand unique specifications. Our case study has six elevator cabs with varying stops. The other conveyance systems can be quite customized, and rule-of-thumb unit pricing can be risky. The estimator should discuss these scopes with a potential bidding subcontractor.

Traction elevators used in high-rise buildings have more extensive controls due to their coordinated operation with other units and for some of the life safety issues. The estimator needs to gain some understanding of what is specified in order to find the costs in published or historical reference data. General contractor estimators typically price elevator OMs by the floor or 'stop' and separate by hydraulic versus traction and passenger versus freight elevators. A two-story building will have an elevator with two 'stops'. Standard elevator cab finishes are not what many architects prefer, so an allowance of $10,000 per elevator cab for finish upgrades is customary and has been called out in the case study specifications. Hydraulic elevators will require a recessed cast-in-place concrete elevator pit to be constructed by the GC. The elevator's plunger or piston will recess into a casing that must be drilled and installed by a separate subcontractor – such as the piling or shoring subcontractor. The casing will be installed in a 'jack hole', which, along with the pit, is one of the first things to be constructed on the project because it is typically one of the deepest building elements.

Escalators are also part of conveyance systems and included in CSI division 14. They are very expensive and would be plugged with an allowance based on distance or height covered, such as $250,000 per escalator.

An effective way to develop an OM estimate for special items and equipment such as a *bridge crane* is to have one or more subcontractors look at the requirements and ask them for help in determining the costs. These are likely DB systems that will be required on industrial projects more than commercial. One of your authors recently installed one on such a project, and it is included with the guaranteed maximum price estimate in Chapter 23. Many times these specialty contractors will provide a budget price based on their experience and may guide the GC estimator on related costs and qualifications. If possible, more than one supplier should be solicited for all these scopes so a competitive quotation will be received.

Mechanical

Mechanical systems include fire protection, plumbing, heating, ventilation, and air conditioning (HVAC), and controls. This work was formerly all grouped in CSI division 15 but is now separated into divisions 21, 22, 23, and 25, respectively. Division 24 has been reserved for future use, along with several other new division numbers. Secondary support systems and subcontractors include mechanical insulation, labeling, balancing, and potentially commissioning. Most commercial buildings have all of these elements. Fire protection sprinklers are not customarily included in single family residences. The mechanical work may all be performed by one contractor in-house direct, or the HVAC subcontractor may contract with the other entities as second- or third-tier subcontractors, or the GC may take some of the systems under separate subcontracts. Regardless of how the mechanical system ends up being packaged, it behooves the GC's estimator to develop separate OM estimates for each portion and be prepared on bid day to analyze both separate and combination bids.

HVAC estimates are very dependent on the type of system specified. The MEP delivery system for the case study project is design-assist, and only performance criteria is included with the bid documents. The estimator thus can use only basic unit price information to develop the OM estimate. The easiest method again is for the estimator to call a subcontractor and ask for an early budget allowance. Short of that, the estimator will likely plug a dollar per square foot of floor ($/SFF) allowance in. The HVAC subcontractor typically performs the mechanical controls direct or employs the services of a specialty third-tier subcontractor.

Another method a GC estimator may use to develop mechanical plug estimates is to perform a quantity take-off (QTO) of the basic system assemblies. Items that can be measured or counted and then labor and material unit pricing applied include:

- Ductwork: Separate by systems (supply, return, and exhaust), duct material (spiral, galvanized, stainless), size and shape of duct (round, square, rectangular); measure lineal footage of each; and convert to pounds of ductwork.
- Diffusers and grilles: Count the devices and separate by sizes.
- Equipment: Sometimes $/ton of cooling is a metric for performing rough HVAC plug estimates. A mechanical subcontractor would perform all of this detail, and more, to prepare its in-house estimate.

The mechanical contractor is often the largest subcontractor on the project. On mechanical-intense work such as biotechnology laboratories, electronic clean rooms, or hospitals, the mechanical subcontractor may account for 50% of the overall cost. The success of a construction project for a GC will rely on all of its subcontractors, especially the mechanical subcontractor, as will the success of a bid or proposal.

Plumbing

Plumbing is also designed by the same mechanical engineer that designed the HVAC. The plumbing scope includes the service water and sewer systems within building lines, along with roof drainage if handled internally and not part of the roofing subcontractor's scope. Process piping associated with heating and cooling will be performed by pipe fitters, which are from the same union as plumbers but will typically be employed by the HVAC subcontractor.

A GC can prepare a plumbing estimate by using a $/SFF allowance or perform a fixture count, such as toilets and sinks, and utilize a $/fixture system plug. When doing a fixture count, some confusion can occur about what is and is not a fixture. Most fixtures have a supply and drain connection, and some have a hot water connection. Water heaters have a water supply and possibly a fuel connection and an exhaust (unless electric). Water heaters can be considered the same as a fixture as far as installation and plumbing rough-in are concerned. An OM estimate on a $2,500/fixture basis includes purchasing and installing the fixture itself, all supply and waste piping, and associated hangers, pipe insulation, labeling, and testing. All fixtures can be grouped together to come up with a total count, or they can be separated by types. The latter would be recommended for apartment projects, which have a substantial quantity of plumbing fixtures.

The plumbing drawings for the case study show that there are 130 plumbing fixtures requiring piping connections including floor drains, roof drains, mechanical condensate drains, and four water heaters. The building floor area is 265,000 SF including the garage, offices, retail space, and restaurants. Dividing this area by 130 fixtures equals approximately 2,000 square feet of floor per fixture. The estimator may then look for a unit cost of a representative building. But because this is a mixed-use building with such diverse areas and considering the restaurant space is 'shelled', pending a tenant lease agreement, representative $/SFF would be difficult. This applies to electrical, discussed later, as well. A more accurate estimating approach is to figure types and sizes and lengths of piping, including associated fittings. This is difficult because often the small bore piping, less than 1.5 inches in diameter, is not drawn, and routing is left to the subcontractor as part of its means and methods. A plumbing subcontractor may take this approach, but a GC estimator will fall back on a $/SFF OM without additional design detail.

The plumbing estimate does not typically include the outside service piping. That work will be designed by a different engineer (civil engineer) and performed by a different subcontractor (site utility subcontractor). This includes the water service to the building, a fire loop and hydrants around the building, storm system including catch basins, and sanitary sewer system. Typically, the utility subcontractor figures the piping systems to 5 feet outside of the building, and the plumbing subcontractor makes the connection there. The gas service is often provided by the local utility company, but the site utility subcontractor may be responsible for excavation and backfill. Site work OMs are described in the next chapter. Separate third-tier subcontractors that will be contracted direct through the HVAC and plumbing subcontractors include pipe insulation and pipe labeling.

Fire protection systems

The fire sprinkler system is often bidder-designed, meaning that the subcontractor bidding the work will design the system and then furnish a price based on their design. Fire sprinklers are typically bidder-designed for private negotiated work, but this is not always the case for public projects. The simplest OM estimate to prepare is again based on square footage of floor area. If the sprinkler system is designed, the GC estimator should count the heads shown, which often average one sprinkler head per 100 SFF and utilize a unit price per head, such as $300/head. But every room requires at least one head, so an 8-foot square room that measures only 64 SFF still requires one head. Assembly unit prices per head include all associated piping, pipe hangers, heads, escutcheons, labeling, and testing.

The fire sprinkler specifications for our case study state that the system is to be bidder designed, and only general information has been provided. The main concern for the estimator is the hazard classification, as this is the major cause of cost variations. Notes on the drawings indicate the National Fire Protection Agency specification for the system, and there are no extraordinary hazards noted within the structure. Both published and historical data on fire sprinkler systems usually are quite representative of actual costs. Unit price data for a wet pipe ordinary hazard system for an office building will range at about $3.40 per square foot. The garage is unheated; therefore, a dry system is used there which requires an air compressor. The GC should group all of the fire protection scope under one subcontractor for consistency.

$$265{,}000 \text{ square feet} \times \$3.40/\text{SFF} = \$890{,}000$$

A mistake that beginning estimators may make is to assume that the fire protection subcontractor will also provide the fire extinguishers and associated cabinets. This work is specified in CSI division 10 and will be purchased by the GC from a separate supplier and installed by the GC's carpenters. As mentioned earlier, the site fire protection loop and hydrants will also be separate and installed by the site utilities subcontractor. But instead of stopping five feet outside of the building, as is the case with water and sewer systems, the utility subcontractor brings the fire service within the building and stubs it just above the slab-on-grade. It is flanged and tested to that location. The building fire protection subcontractor picks up the service from that point.

Electrical

Electrical work generally is divided into two primary categories: line voltage and low voltage. Each of these also has several subcategories. All of electrical was formerly in CSI division 16 but is now split into divisions 26 (service and distribution), 27 (communications), and 28 (security). *Line voltage* includes service and distribution of primary electrical, including lights and power for receptacles, and equipment, including mechanical equipment. *Service* is the power from the utility connection to and through the main service panel. *Distribution* is the wiring, switches, receptacles, and connections within the building. *Low voltage* electrical includes several subcategories including fire alarm, controls for mechanical systems, low-voltage lighting control, audio-visual systems, televisions, telephones, data, and security. Security includes alarm systems, card readers, and cameras.

Order of magnitude estimates for divisions 26–28 are developed similar to other OM estimates we have discussed. Since these rough budgets are prepared only for the pre-bid day summary estimate, this type of OM estimate is adequate. Unit prices for each of these categories can be found in estimating references. The range for a mixed-use project such as our case study might be as follows:

Service including a 4,000-amp main panel	$1.00/SFF
Distribution: Power and lighting	$10.50/SFF
Low voltage and security	$0.50/SFF
Total:	$12.00 per square foot

This total is then multiplied by the overall building floor area to calculate an OM estimated cost. A GC estimator preferably would develop a more detailed plug estimate for electrical based on quantifying what they can and applying assembly unit prices such as:

- Site lighting poles: $2,500/EA, includes trenching, conduit, wire, and fixture;
- Building light fixtures: $300/EA, includes the fixture, conduit, wire, and termination;

- Convenience outlets and light switches: $200/EA, includes the box, outlet or switch, conduit, wire, and termination;
- Emergency generator: $/EA, based on size in kilowatts;
- Mechanical or service equipment utilizing horsepower of the equipment; and
- Low voltage devices: $200/EA, includes the device and wire, but not always in conduit. Low voltage cabling may be grouped in an open cable tray and priced as $50/LF. Security cameras do not fit within this rule-of-thumb, as they may cost $4,000 per each just for purchase, depending on performance.

Electrical outside of the building is contracted differently than plumbing work, which transfers to a site utility contractor once the pipe is five feet outside of the building. In the case of electrical, the building electrician provides trenching and backfill and conduit for the power company to pull its wire. The electrician also is responsible for all site lighting, including parking lot pole fixtures. Because the electrician does not normally operate backhoes to perform excavation and backfill, they will hire a subcontractor, which may be the same one the GC hired to dig the footings. Power for site lighting is typically in the electrical subcontractor's scope. If there is an emergency generator, the electrician will also supply and install that equipment, including conduit for power and controls. The electrical subcontractor will, however, typically exclude concrete for the site lighting and a pad for the emergency generator, and the GC will need to pick that up. The electrical subcontractor may attempt to exclude excavation and backfill; therefore, the GC needs to be diligent when analyzing subcontractor quotes on bid day.

Each of the operating systems described in this chapter would have separate QTO and pricing recapitulation (recap) sheets. Total subcontractor cost from each of these would be forwarded to the estimate summary sheet described in Chapter 19. For convenience of space, we have combined all of the systems onto one pricing recap sheet, Figure 15.1.

Because all of these systems 'operate', the building owner or architect may require their project to be commissioned. *Building commissioning* involves testing and retesting

---

City Construction Company
Cost Recapitulation Sheet

Project: Dunn Lumber   Date: 6/10/21
Location: Seattle, WA   Estimator: P. Jacobs
Arch: Flad   Estimate #: 1
Elevator & MEP Divisions 14–27

Code	Description	Qty	Unit	Subcontract Unit Cost	Subcontract OM
14.1	Elevators	23	Stops	50,000	$1,150,000
14.2	Cab Upgrades	6	EA	10,000	$60,000
21	Fire Protection	265,000	SFF	3.40	$901,000
22	Plumbing	265,000	SFF	4.00	$1,060,000
23	HVAC and controls	265,000	SFF	9.00	$2,385,000
26	Line Voltage Electrical	265,000	SFF	11.50	$3,047,500
27	Low Voltage Electrical	265,000	SFF	0.50	$132,500

Figure 15.1 Elevator and MEP subcontractor cost recapitulation sheet

all equipment to simulate all four seasons of the year. The goal with commissioning is to shake out any potential warranty problems before the contractors have demobilized. Buildings that are typically commissioned are those with intense MEP systems such as hospitals, biotechnology laboratories, and electronic clean rooms. Commissioning requires support from the major subcontractors discussed in this chapter, but subcontractors often exclude these costs from their scopes; therefore, the GC must verify inclusion on bid day. The commissioning agent is typically a third-party consultant employed by the owner direct, and their cost is substantial and may run 1% of the total construction cost.

Summary

General contractor estimators typically spend the most time on the work their company performs direct, which is to be expected, as this is the riskiest portion of their estimate. But the operating systems subcontractors, including elevators, HVAC, plumbing, fire protection, and electrical, are typically very expensive, and attention needs to be paid to their scopes and costs not only on bid day but in preparation of OM estimates in support of a pre-bid day budget total. The most accurate pre-bid day subcontractor estimates are provided by the subcontractors themselves after they have had a chance to review the bid documents. General contractor estimates based on a cost per square foot of floor area are fairly easy to develop but are not very accurate, due to the variety of building types and breadth of MEP scopes. Breaking down these scopes into assemblies such as stops for elevator, diffusers or tons of cooling for mechanical, plumbing fixtures, fire protection heads, and electrical light fixtures and outlets and applying unit prices that include all associated labor and material is better than $/SFF. By preparing these assembly estimates, the GC estimator or project manager will 'learn' the work as well and be better prepared on bid day and during buyout to make the best-value subcontractor choices. In our next and last chapter dedicated to subcontractor pricing, we will review many aspects of the civil work or site work that the GC must also consider.

Review questions

1   Which subcontractor connects to the other's piping work, the site utility subcontractor or the plumbing subcontractor?
2   If a building has two stories of underground parking garage, five occupied floors, and a green roof with outdoor spaces for tenants, how many 'stops' will an elevator have?
3   Why are small bore pipes and conduits not typically drawn on design drawings?
4   How many sprinkler heads should be estimated for a large assembly room that measures 24 feet by 42 feet?
5   Why should a GC estimator prepare an early ROM estimate? Why should they later refine that estimate with subcontractor OM estimates?
6   Why did we total all of the floor covering OMs in Figure 14.3 together but not the scopes included in Figure 15.1?

Exercises

1   Determine an OM estimate for a 2,500-pound capacity hydraulic passenger elevator for a two-story office building. Use commonly available estimating references.
2   Which costs more per stop, a hydraulic or traction elevator with the same capacity?

3 Have you ever worked on a project with an elevator? Was a down payment made before delivery? How long was the lead time? Was the cab finished per specification, or were upgrades made? Who built the cab upgrades – the elevator subcontractor, or the building finishes subcontractors?

4 Determine a reasonable unit price for HVAC of a two-story fire station. What kind of system is this?

5 How does the cost of the HVAC for a fire station compare with a two-story medical office building, both in unit price and percentage of total building cost?

6 Envision a 10-foot square office. Look up at the ceiling. Of the MEP systems described in this chapter, what will be located in the center?

7 Prepare HVAC, plumbing, and electrical subcontractor OM estimates for the Lee Street case study, available on the book's eResource.

8 Which is typically one of the first subcontractors to mobilize on a project site and one of the last to leave?

9 What building labor trades install the work described in this chapter: Elevator, HVAC, plumbing, fire protection, mechanical insulation, mechanical labeling, and electrical?

10 Why would a GC buy out all of the mechanical systems from one subcontractor or, conversely, separate the work out among several subcontractors? The same question applies to electrical.

11 Prepare an argument why a GC should partner with MEP and elevator subcontractors in preparation of a competitive bid or, conversely, why they should not partner with these companies.

12 Mechanical subcontractors may offer 'balancing' and 'commissioning' as extra services to be included with their scopes. Why should project owners provide these services?

13 The Lee Street executive townhomes did not include elevators. Assume there was space to add one and a potential buyer wanted to negotiate this high-end addition. Prepare an OM estimate for a small cab hydraulic elevator. Include $5,000 each for an elevator pit and an elevator penthouse.

14 What CSI division is 'commissioning' specified?

15 Other than the systems described in this chapter, what building features or equipment 'operate'?

16 Elevator equipment is often located in a rooftop penthouse. What contractors will typically build the penthouse?

17 Arrange these electrical activities in the order that the work will be performed: Test, install conduit, pull wire, inspections, install boxes, terminate wire at device and panel, install panel, install device, drill holes in studs and joist (if a wood-framed building).

18 In addition to elevators, what other systems or equipment may require a substantial down payment from the GC and, in turn, from the project owner?

19 Significant detail was included with the Vehicle project's MEP design. Quantify what is available and prepare a pre-bid day GC plug estimate for one of those systems.

16 Civil work

Introduction

Commercial general contractors (GCs) typically perform concrete, steel, and carpentry work with their own direct workforces. Other scopes of work are then performed by subcontractors. Subcontractors supply and operate equipment for scopes of work that require use of equipment. General contractors that specialize in heavy civil, utility, or industrial work usually own and operate their own equipment. Most of the civil site work for commercial or residential projects included in old Construction Specification Institute (CSI) division 02 and new divisions 31–33 is performed by specialty contractors that have access to specialized equipment.

Civil site work includes earthwork, shoring, underground utilities, paving, sidewalks and curbs, landscaping, and deep foundation systems. An estimator must determine which of these apply to a particular project from the work breakdown structure (WBS). A heavy civil contractor or a site work subcontractor develops a more detailed cost estimate than we cover in this chapter. Our goal here is to describe the process a GC estimator will go through to prepare plug estimates so that he or she has a realistic complete estimate before bid day. Of course, as stated elsewhere, the best source for subcontractor prices is from subcontractors, even if provided as a pre-bid day budget or order of magnitude (OM) estimate.

Since site work is now not only at the end of the CSI order, it is typically the last area a GC estimator will address. In addition to site work subcontract OM estimates, we discuss in the last section of this chapter the process the estimator will use to gather the entire subcontractor OM estimates in completing the pre-bid day summary estimate.

Mass excavation and backfill

Earthwork probably is the most difficult of all site work items to estimate. It is the mass excavation and backfill that are required to level the site and prepare it for the building. Structural excavation and backfill of foundations are often subcontracted, and the estimator may choose to quantify that work here rather than with the substructure. Earthwork subcontractors commonly quote unit prices for the cut, fill, and hauling. Thus, they transfer the risk of estimating the quantities of work to the general contractor. The general contractor should accept only bids for the total cost of the work from subcontractors.

Acceptable and firm earthwork conditions are not always present, and this major scope can present many problems for the GC's estimator, because subcontractors typically disclaim responsibility for unforeseen conditions. Even though a soils report is part of the project manual, it cannot predict everything that might be encountered. For example, a shallow rock outcropping or submerged spring may be discovered, and special work will be required to make the area suitable for the building. The subcontractor has a right to file a claim for more money for rock removal, to stabilize the ground of an underground spring, or for any other hidden condition. They cannot be responsible for unanticipated or undefined subsurface conditions. Most construction contracts contain a clause holding the project owner responsible for the cost of unforeseen subsurface conditions.

To forecast earthwork costs, an estimator can use a basic grid system to estimate the cut and fill. Identical rectangular grids are set up on the existing and proposed contour drawings, and then, by determining the difference of the average elevations for a given grid, the cut and fill can be estimated. Grids used for estimating cut and fill can be any size that the estimator finds convenient; however, a little forethought can simplify the work. A simple grid system to use is one in which area equals 270 square feet (SF). This may be squares of 16.43 feet or rectangles, such as 10 feet by 27 feet, or any other convenient size that equals 270 square feet. The key, however, in selecting the grid size is that with a 270 square foot grid, each foot of cut or fill equals 10 bank cubic yards (BCYs). This same concept applies to a grid of 2,700 SF and 100 BCY.

Once a grid has been established, the estimator finds the elevation change of a given grid on the existing and proposed contour drawings. The average elevation of a grid is determined by averaging the elevations at the four corners. The average elevation on the existing grid is subtracted from the corresponding elevation on the proposed grid. A negative answer represents a cut, and a positive answer represents a fill. Earthwork measurements and conversions from bank cubic feet (BCF) to bank cubic yards and truck cubic yards (TCYs) are the same as for structural excavation and backfill, discussed in Chapter 6. A GC estimator will use $/BCY unit pricing for earthwork subcontractor plug estimates, as shown in the next example for our case study.

Mass excavation: 33,500 BCY @ $44/BCY = $1,474,000
Under-slab drainage system: 2,600 LF @ $50/LF = $130,000
Capillary break: 60,700 SF of 6 inches thickness @ $2.25/SF = $136,575
Other mass excavation scopes: $170,525
Total mass excavation plug estimate: $1,911,100

Shoring and foundation systems

Major excavations are often accompanied with temporary *shoring systems*, especially in deep downtown skyscraper excavations and highway projects with steep embankments. For a commercial GC, shoring is typically quantified in an assembly of SF of wall, including all elements, as shown in our next case study example. Some of the typical shoring elements or systems include:

- Soldier piles, often wide flange steel columns with horizontal timber lagging;
- Tie-back cables or rods;
- Diagonal raker systems, often pipes, where tie-backs are not feasible;
- Soil nailing systems that do not use soldier piles or lagging;
- Shotcrete and other materials and systems.

The estimator typically estimates the SF of shoring needed and uses unit price plugs, as shown in the following example.

Shoring installation: 26,920 SF of wall @ $62/SF = $1,669,000
Tie-back removal in right-of-way: 125 EA @ $1,600/EA = <u>$200,000</u>
Total design-build shoring system plug estimate: $1,869,000

Figure 16.1 is a photograph taken from a drone which shows a recently erected horizontal jib tower crane. Drones are becoming valuable construction management tools to perform a variety of functions. The construction occurring on this project at this time was the slab-on-grade (SOG) and shoring wall scopes. This shoring system was similar to that of the book's primary case study: Soldier piles, wood lagging, tiebacks, and shotcrete.

Ideally, the geotechnical report allows the use of conventional spread footing foundations to support building columns. This was the situation with our case study project. But often the condition of the soil, even after a deep excavation, requires use of *piling*. These piles function as deep columns and are often drilled or driven to a depth where solid soil conditions are found. When piles are used, the footings that sit on top of them are typically referred to as pile caps and grade beams and are estimated in the same manner as conventional foundations, using the techniques discussed in Chapters 6 and 10. The difference is that they are often deeper and include more concrete

Figure 16.1 Photograph of tower crane and shoring operations

reinforcement (rebar) than do spread and continuous footings. Some common types of piling that are found on commercial, residential, and civil construction projects include:

- Auger cast: Drilled with rebar cages and cast-in-place concrete;
- Pre-cast concrete: Driven in;
- Steel pipes: Driven in and can be filled with concrete;
- Caissons: Large piles filled with rebar and concrete;
- Pin piles: Large groupings of smaller piles or 'nails', and others.

Piling is estimated by the GC estimator by noting the type and size, counting the quantity of each, measuring the length of each, and estimating the total LF of pile. A piling or shoring subcontractor will get into more detail than will a GC estimator who is developing pre-bid day plug estimates. The exact length of piling required for a project is often difficult to determine because of unknown underground soil conditions, so unit price bid items may be used by the project owner for pile construction.

Site utility systems

Site utility work includes all of the piping systems outside of the building. The building plumbing subcontractors install the piping from inside the building to five feet outside. The site utility subcontractor installs all of the exterior underground systems. Site utility systems include:

- Water service,
- Fire line,
- Sanitary sewer,
- Storm drains and sewer,
- Natural gas, and
- Electrical systems.

The service piping is a very involved estimate for a civil contractor, but for the general contractor's estimator, preparing a plug estimate is fairly straightforward. It is the piping from the nearest utility connections off site to approximately five feet outside of the building perimeter. The estimator should note pipe size, length, and depth of burial and then consult published cost data to obtain an anticipated unit cost. Some published pricing data will not include excavation and backfill; thus, the estimator may need to add these as well as any special valves and fittings. For most installations, the estimator should be able to estimate an installed unit cost per 100-foot section with minimal fittings and extend it to the length of line shown on the drawings. A reasonable allowance for connection and metering fees should also be included. Relying on the advice of vendors will greatly aid in developing a reasonable OM estimate for the service piping.

The foundation or footing drain is typically hard-piped to the storm system but is not typically installed by either the utility subcontractor or the plumber. The GC does this work with its laborers. The gas system is often provided by the public or private utility supplier. Electrical conduit will be installed by the building electrician, along with all associated work for the site lighting. The electrical subcontractor will excavate, install the conduit, and backfill, but the electrical utility company typically supplies and pulls the cabling to the building's main switchgear room. Usually, site utility systems OM estimates include everything necessary, such as:

- Excavation,
- Pipe bedding,
- Pipe and fittings,
- Testing and inspections, and
- Backfill.

Different pipe sizes and materials would be quantified separately. Permits and connection fees may be added or may be included by the project owner. A more accurate plug estimate can be developed by also adding costs of significant utility structures on a $/each basis. This includes expensive items such as hydrants for the fire system, manholes for the sanitary sewer system, and catch basins for the storm drain system. A GC estimator can then apply historical unit prices to these quantified scopes. A subcontractor OM pricing recap for all site utilities for Lee Street Lofts is shown in Figure 16.2.

City Construction Company
Cost Recapitulation Sheet

Project: Lee Street Lofts   Date: 1/13/21
Location: 301 Lee Street, Seattle, WA   Estimator: TS
Architect: Johnson Architects   Estimate #: 1
    Site Utilities

Code/ Description Det	Qty	Unit	UMH	Man Hours	Wage Rate	Labor Cost	Matl/Equip/Subs Unit Cost	Matl/Equip/Subs Cost	Total Cost
4" Storm Drain	238	LF				Sub	26	6,188	6,188
6" Storm Drain	7	LF				Sub	30	210	210
Storm Catch Basins	3	EA				Sub	2,500	7,500	7,500
Area Drains	7	EA				Sub	500	3,500	3,500
Downspout Connections	8	EA				Sub	500	4,000	4,000
4" Footing Drain	420	LF	0.1	42	35	1,470	1	420	1,890
Drain Gravel	60	TCY				Sub	38	2,280	2,280
Trench Drain	20	LF	2	40	35	1,400	100	2,000	3,400
Bioretention Vaults	4	EA	20	80	40	3,200	500	2,000	5,200
Storm Sewer Cleanouts	8	EA				Sub	550	4,400	4,400
6" Sanitary Sewer	200	LF				Sub	35	7,000	7,000
1" Waterline	214	LF				Sub	35	7,490	7,490
Water Meters	4	EA				Sub	6,000	24,000	24,000
Sewer Connection to City	2	EA				Sub	5,000	10,000	10,000
Page Totals (OM Estimate)				162		$6,070		$80,988	$87,058
Add labor burden @ 55% of labor								$3,339	$3,339
Total Cost									$90,397

Figure 16.2 Site utilities subcontractor cost recapitulation sheet

Site improvements

Pavement including concrete pavement and hot-mix asphalt

This scope of work involves placement of both concrete and asphalt pavements. Concrete pavements may be formed and placed by the GC's direct workforce, may be constructed by the site work subcontractor, or may be constructed by a specialized concrete paving subcontractor. Asphalt pavements require specialized equipment and typically are constructed by an asphalt paving subcontractor. Concrete is typically purchased by the cubic yard, and asphalt is typically purchased by the ton. Based on the specified pavement thickness, the GC estimator can develop a $/SF plug unit price for the OM estimate, as shown in the following example for our case study project.

Concrete pavement: 8,627 SF @ $26/SF = $224,302
Asphalt pavement: 5,273 SF @ $12/SF = $63,276
Subtotal pavements: $287,578

Walks, curbs, and walls

This scope of work, and associated OM pricing units, involves construction of the following:

- Walks and patios: $/SF,
- Cast-in-place curbs: $/LF,
- Curb and gutter: $/LF,
- Extruded concrete curbs: $/LF,
- Extruded asphalt curbs: $/LF, and
- Site retaining walls: $/CY or $/LF (may be cast-in-place concrete or masonry units). (Work may be quantified with other concrete scopes of work as discussed earlier in this book.)

Cast-in-place concrete work may be self-performed by the GC with their direct workforce or subcontracted to specialty subcontractors that have equipment and formwork unique to these work items. A GC may undertake some of this work if it is in small quantities as fill-in work for their crews. Some quantities and assembly unit pricing for our case study plug estimates are shown in the following example.

Sidewalks: 8,660 SF @ $11/SF = $95,260
Curb and gutter: 1,144 LF @ $43/LF = $49,192
Other hardscape work: $307,570
Subtotal walks, curbs, and walls: $452,022

Striping and signage

This scope of work involves painting striping on streets and parking areas and installation of traffic signage, wayfinding signage, and building signage. Excluded would be any electronic signage that is provided by the owner. To develop an OM estimate, the GC estimator determines the length of the striping and the number of signs required. Then, using plug estimates of $/LF for striping and $/EA for signs, the estimator can develop an OM estimate similar to the following example for our case study project.

Pavement and parking striping: 4,000 LF @ $4.35/LF = $17,400
Site signage: 6 EA @ $5,000/EA = $30,000
Subtotal striping and signage: $47,400

Landscape and irrigation

This scope of work typically involves placement of topsoil, installation of an underground irrigation system, and placement of sod and plants such as shrubs and trees. All of this work is performed by a landscaping subcontractor who, in addition, may be required to provide landscape maintenance for a specified period and an extended warranty on the plant material. The landscape scope may be awarded as a design-build subcontract, or a landscape architect may produce drawings and specifications that will be used by the subcontractor in developing a cost estimate. In developing an OM estimate, the GC estimator typically uses a plug estimate of $/SF, as shown in the following example for our case study project.

Green roof: 4,968 SF @ $30.10/SF = $149,646
Street level landscaping: 17,437 SF @ $26/SF = $453,354
Trellis landscaping: 4,097 SF @ $45/SF = $184,000
Subtotal landscaping: $793,000

Site specialties

This scope of work includes the miscellaneous site improvement tasks not included in earlier parts of this section. Plug estimates for some of the additional site improvement materials and systems for our case study project are shown in the following example.

Bike racks: 7 EA @ $800/EA = $5,600
Water feature: Allowance $38,800
Seating: 148 LF @ $250/LF = $37,000
Stainless railings: 15 LF @ $400/LF = $6,000
Subtotal site specialties: $87,400

Civil work as a general contractor

Heavy civil contractors typically self-perform a higher percentage of the total project scope of work than do commercial GCs. Heavy civil contracts tend to contain both unit price and lump sum bid items, so heavy civil GC estimators develop work packages for each bid item, whether it is priced as lump sum or unit price. An example work package for construction of 1,600 LF of sanitary sewer is shown in Figure 16.3. The work package may include both GC-performed work and subcontractor-performed work. It also includes labor burden for the GC's workforce. The objective is to estimate the direct cost for a specific bid item. A critical element in developing a cost estimate for the work package is estimating the productivity of the crew. In this case, the sewer line is to be placed 10 feet below the surface, so the productivity was estimated to be 200 LF per 8-hour shift.

High Country Construction

Tri-County Conveyance Pipeline - Estimate
Bid Item: 4 - Install 18" PVC Sanitary Sewer

Scope: Install 1,600 LF of 18" PVC Sanitary Sewer

CREW AND EQUIPMENT

Quantity	Description	Hourly Rate	Labor Cost per Shift	Equip Cost per Shift	Shift Total
1	Foreman	55	440		440
2	Flaggers	34	544		544
3	Operators	50	1,200		1,200
2	Laborers	35	560		560
1	Pipelayer	42	336		336
2	Truck Drivers	49	784		784
	Subtotal Labor Cost/Shift				3,864
	Add Labor Burden @ 55% of Labor				2,125
	Total Labor Cost/Shift				$5,989
1	Excavator			800	800
1	Wheeled Loader			500	500
2	Dump Trucks			400	400
1	70 KW Gen w/Pumps			200	200
1	Compactor			150	150
1	Tool Truck			120	120
1	Pickup			90	90
	Total Equipment/Shift				2,260
	Total Labor and Equipment/Shift				8,249
	Number of Shifts				9
	Activity Labor and Equipment				$74,241

CONSTRUCTION AND PERMANENT MATERIALS

Material	Quantity	Unit	Unit Price	Total Price
18" PVC Pipe	1,600	LF	28	44,800
Pipe Bedding	625	TN	8	5,000
Backfill (native)	0	TCY	400	0
Shoring	2	WKS	400	800
			Total Materials	$50,600

SUBCONTRACTORS

Subcontractor	Quantity	Unit	Unit Price	Total Price
Closed-Circuit Television Test	1,600	LF	3	4,800
			Total Subcontractor	$4,800

ACTIVITY DURATION

Activity	Shifts
Install 18" Sanitary Sewer Pipe	8
1,600 LF @ 200 LF/Shift	
Existing Utilities	1
Total	9

TOTAL ACTIVITY DIRECT COST

Labor and Equipment Cost	74,241
Materials Cost	50,600
Subcontractor Cost	4,800
Total Activity Direct Cost	$129,641

Figure 16.3 Sanitary sewer work package estimate

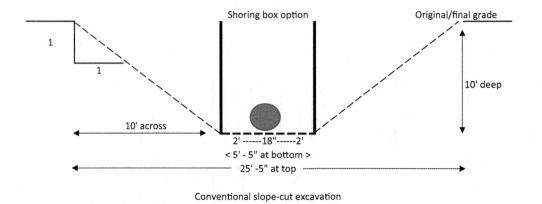

Conventional slope-cut excavation

Figure 16.4 Utility excavation sketch

Similar work packages are developed for each bid item on the bid form. Adding mark ups to finalize a civil bid will be discussed in Chapter 23.

As mentioned in other parts of this book, estimators often create sketches, such as Figure 16.4, of anticipated construction means and methods to assist with their QTO process and communicate their intentions for the builder's later use. In this case, the estimator chose a trench box shoring system in lieu of a 1:1 slope cut. This approach was faster, safer, and saved considerable costs. An exercise at the end of this chapter evaluates those two pricing options.

Completing the subcontractor list

As the OM estimates are completed, they are entered onto the subcontract list that is marked "OM Estimate". Once all line items have been filled in, the OM estimating work is complete and ready for use in determining the first run summary estimate total, as will be discussed in Chapter 19. Figure 16.5 shows a completed OM subcontractor list for the case study. The left-hand side of this same list will be used on bid day. The OM plugs will gradually be replaced with market competitive subcontractor quotes as they are received, as will be discussed in Chapter 20.

Summary

Civil site work includes earthwork, deep foundation systems, shoring, underground utilities, paving, sidewalks and curbs, and landscaping. A commercial GC estimator typically develops OM estimates for these scopes of work, while a civil GC estimator usually develops more detailed cost estimates. The reason is that the commercial GC typically will perform these scopes of work by subcontract, while the civil GC may self-perform many of them.

Mass excavation may be required to level a site and prepare it for building construction. Shoring systems may be required to support open foundations until the permanent foundation is constructed. Site utility systems are needed to connect the completed building

City Construction Company
Subcontractor OM Plug Estimates

Project:		Dunn Lumber	Estimator: Paul Jacobs Est. #: OM	
Bin	Spec	Description	GC OM Plug	Comments
1	24100	Demolition	486,000	Buildings & Site
2	31113	Form Elevated Slabs	0	w/Directs
	32100	Reinforcement Steel	0	w/Directs
	33000	Pump Concrete	0	w/Directs
	33500	Finish Slabs	0	w/Directs
3	39000	Shotcrete	803,600	
4	42200	Interior Garage CMU	32,200	
	51000	Misc. and Structural Steel	0	w/Directs
16.1	54100	Exterior Metal Stud Framing	906,200	Sub w/GWB
5	64100	Casework	35,000	
6.1	71400	Waterproofing	615,392	Sub w/Traffic
6.2	71800	Traffic Coating	400,000	Sub w/Wproof
7	72100	Insulation	266,000	
8	74619	Metal Siding	900,000	
9	74623	Wood Siding and 'Rain Shield'	462,500	
10	75300	Roofing and Flashing	546,000	
11	78100	Fireproofing	440,640	
12	79200	Caulking and Firestopping	37,000	City Allowances
13.1	81100	Buy Doors and Frames	223,976	Install w/Directs
14	83300	Garage & Retail Roll-up Doors	379,000	
15	84100	Windows and Storefront	3,700,000	Punch Install w/DL
13.2	87100	Buy Door Hardware	242,000	Install w/Directs
16.2	92100	GWB and Interior Framing	545,700	Sub w/Ext Studs
17	96100	Tenant Area Finishes/Floors	655,000	T.I. Allowance
18	99113	Exterior Painting	40,000	
	101400	Signage	0	w/Directs
	102800	Toilet Accessories	0	w/Directs
	104400	Fire Extinguishers	0	w/Directs
	108213	Louvers	0	w/Directs
19	111300	Loading Dock Equipment	75,000	
20	118123	Window Washing Equipment	200,000	Owner Allowance
21	129313	Bike Racks & Specialties	87,400	
22	142400	Elevators	1,210,000	
23.1	210000	Fire Protection (Design-build)	901,000	sub w/HVAC
23.2	220000	Plumbing	1,060,000	sub w/HVAC
23.3	230000	HVAC	2,385,000	
24	260000	Electrical	3,180,000	
25	312300	Mass Excavation	1,911,100	Includes Flaggers
26	314000	Shoring	1,869,000	Design-Build Sys.
27.1	311300	Pavement & Walks	739,600	sub w/Stripe
28	321400	Pavers	246,800	sub w/Lscape
27.2	321713	Pavement Striping	47,400	sub w/Pave
29	323113	Temporary Site Fence	62,400	Rental
28	329000	Landscape and Irrigation	793,000	sub w/Pavers
30	330000	Site Utilities	309,500	
31	334119	Temporary Dewatering	188,000	Owner allowance
		Total Subcontract OMs	**$26,981,408**	

Figure 16.5 Subcontractor OM plug estimates

to existing utility systems. Site improvements are required site features such as pavements, sidewalks, and landscaping.

Developing an OM estimate for these items ranges from relatively straightforward for paving and curbs to difficult for earthwork and landscaping. For much of the site work, the estimator usually has to perform a semi-detailed estimate and use published references for costing.

Because civil contracts typically contain both unit price and lump sum bid items, a civil GC estimator creates work packages for each bid item. These work packages may include both self-performed and subcontractor-performed work. The work packages are used to create a civil bid.

In this chapter, we have included only a few of the major civil work systems – there are many others. An estimator developing plug estimates for new scopes of work should look to how pricing is applied before developing quantity estimates and should seek the counsel of an experienced estimator. This work can be difficult to estimate and is often very risky, and the GC estimator should rely on a subcontractor specialist to assist with budget estimates wherever possible.

This concludes this section of the book covering preparation of GC in-house subcontractor plug estimates. In the next section, we begin the process of completing the GC's estimate, including development of a detailed jobsite general conditions estimate, estimating markups, preparation of the estimate summary form, and participation in bid day activities.

Review questions

1   Why would a commercial GC choose not to self-perform many of the scopes of work discussed in this chapter with their own craftsmen?
2   Why might a heavy civil GC estimator develop a detailed cost estimate for mass excavation, while a commercial GC estimator only prepares an OM estimate?
3   Why are unforeseen ground conditions an issue in preparing estimates for deep foundation excavation?
4   Why are shoring systems often needed when constructing deep excavations? How would you develop an OM estimate for soldier piles and timber lagging?
5   How is a grid system used to estimate the volume of cut and fill when preparing an earthwork cost estimate?
6   Why might a commercial GC decide to self-perform construction of concrete sidewalks instead of hiring a subcontractor for the work?
7   What are the major site utility systems that typically are required to support the construction of a new apartment building?
8   Which scope of work discussed in this chapter lends to it being performed by a design-build-operate subcontractor?
9   Why do civil GC estimators develop work packages for each bid item listed on the contract bid form?

Exercises

1   Which systems discussed in this chapter might be packaged by a GC all under one subcontractor?
2   Prepare an alternate complete stand-alone direct work estimate, including equipment and labor burden, for the concrete pavement, sidewalks, and curb and gutter OM

plugs presented in this chapter. Utilize the quantity take-off techniques and pricing included with direct work concrete examples from earlier in this book. Make whatever assumptions necessary.

3   Prepare an argument why a GC estimator may choose to utilize the subcontractor site concrete plug estimates as developed in this chapter, or later bids received on bid day, compared to performing all of this work direct.

4   Prepare a stand-alone site utility estimate for the vehicle maintenance facility based upon the drawings included on the book's eResource. Assume you are a civil specialty contractor providing a bid to a GC. Include the water, sanitary sewer, and storm sewer systems. Include all labor, material, equipment, labor burden, and all necessary markups. Make whatever assumptions as are necessary.

5   The Figure 16.1 tower crane photograph was taken during the shoring stage of another project. The shoring was bidder-designed based upon soldier piles, wood lagging, tiebacks, and shotcrete. The site measured 200′ × 200′, and the SOG was 20′ below original grade. The mixed-use project included several seven-story apartment towers over one floor of retail at ground level over two floors of underground parking garage. Prepare a GC-generated subcontractor OM plug estimate based upon (A) the unit price from the example included earlier in this chapter and/or (B) a unit price from *The Guide* or another source, such as a shoring subcontractor.

6   How many TCY of earth were generated from the excavation described in Exercise 5?

7   The contractor installing the sewer pipe in Figures 16.3 and 16.4 chose a moveable trench shoring system over a 1:1 slope cut excavation. Calculate the contractor's overall savings for this approach utilizing the excavation conversion process and pricing discussed with structural excavation in Chapters 6 and 10. Assume the native soils are sufficient for backfill.

8   Other than the tradeoff for less excavation and backfill versus the shoring box rental for Exercise 7, what are some additional complications and potential costs an estimator might consider for a slope cut of this magnitude?

Part V

Estimate completion

17 Jobsite general conditions

Introduction

After the estimating team has completed all of the direct work quantity take-off (QTO) and pricing and has plugged in order of magnitude estimates for subcontractors and major suppliers, it is time to begin the estimate summary process, which is the primary topic of Part V of this book. Up until this point, the team has been relying on the early rough order of magnitude estimate that was developed when the project first came into the contractor's office. Now a lot more is known about the project, and a much more accurate total estimate can be developed. The estimate summary sheet (Chapter 19) collects the direct work and subcontractor estimates and combines that with jobsite general conditions (this chapter) and several anticipated percentage markups or add-ons (Chapter 18). We have used the abbreviation GC for general contractor throughout this book, as is customary in the industry. But GCs is also a common abbreviation for general conditions, and we will utilize that definition in this chapter. Part V completes the estimating process all the way up to and through bid day activities (Chapter 20).

The general conditions portion of the lump sum estimate is represented by step 8 of the lump sum estimating process shown in Figure 1.1. Prior to determining the GC's requirements, it is usually necessary to prepare a summary schedule for the project (step 7 in Figure 1.1) so that the duration of the project can be estimated. General conditions costs are the jobsite overhead and project indirect costs that are not attributable to specific work activities but occur throughout construction. Estimating references commonly refer to these costs as Construction Specifications Institute (CSI) divisions 00 and 01 or general requirements costs. The general conditions cost estimate often connects to the project manual (or contract) general conditions and/or special conditions sections, which define many contractual and managerial aspects of the project.

Indirect labor and materials are those not utilized directly in the performance of the construction activities on the project. For example, the labor of the general contractor's superintendent and the project management staff are not attributable to any of the specific work activities but are necessary for the execution of the entire project. Similarly, a jobsite pickup truck or forklift is used for many tasks, but to prorate these equipment costs to each construction activity is difficult and meaningless. Some other general conditions items include trash hauling, safety, jobsite office rental, and temporary utilities. Conversely, direct work scopes include earthwork, structural steel, carpentry, plumbing, and others.

The general conditions should be thought of as the jobsite or project overhead. It includes costs associated with administration of the project, indirect equipment usage,

temporary construction, and general requirements costs. This is different from the contractor's company overhead and profit. The jobsite overhead pertains strictly to the
project, while the contractor's company overhead and profit relate to the home office
operation of the construction firm. Home office overhead and profit, or fee, will be discussed in the next chapter.

Project summary schedule

In order to develop a general conditions estimate, an estimator must know the project
duration. This may be determined in more than one way. The duration may be defined
in the contract, or a project owner may request that bidders submit a schedule along
with the price. In either case, the estimator or project manager (PM) should develop a
project summary schedule prior to completing the general conditions estimate. A summary schedule differs from a project schedule in its level of detail. It is a summary of the
work required to complete the project and shows only gross activities, such as major
concrete elements, structural steel, interior finishes, elevator, mechanical, electrical,
and site work. It does not necessarily show a detailed breakdown of these activities. The
80–20 rule applies to scheduling as it does to estimating. A project summary schedule
should exhibit 20 to 30 significant activities for the project, those being the 20% that
account for 80% of the time or effort. Several formats are available for creating a project
summary schedule. The format that is easiest to create and understand is the horizontal
time-scaled bar chart. Figure 17.1 is an example of a project summary schedule for our
commercial case study project. The detailed schedule for this project is included on
the book's eResource along with another summary schedule for the Vehicle case study
project.

Man-hours can be used as a guide to determine general activity durations. For example, suppose that the summary recap for foundation and SOG concrete shows that it
will take 1,897 man-hours to complete. The estimator determines that the average
crew will consist of six carpenters, three laborers, two ironworkers, and two concrete
finishers. The crew makeup will vary depending on the specific task being done at a
given time, but in general there will be a 13-person crew for this work. The duration
calculations are:

(13 craftsmen) × (8 hours per day per person) = 104 man-hours per day
1,897 man-hours ÷ 104 man-hours per day = 18 workdays duration
Just short of four weeks: Checks? Yes

This then becomes the length of the bar on the summary schedule for concrete, and
the procedure is repeated for the remainder of the direct work. Duration of subcontract
activities is more difficult to determine. The best method to determine subcontractor
durations is to call major subcontractors the general contractor has a relation with. This is
similar to obtaining early plug budgets, as was discussed in Chapters 14–16. Without that
input, the estimator may use either published estimating references or past similar projects
to determine typical durations for subcontracted activities. The estimator must also be
aware of any unusual or difficult situations that may affect crew productivity and factor
that into the summary schedule. As indicated, many of the GC's estimate line items are
time dependent. After completion of the summary schedule, the estimator can proceed
with populating the GCs estimate template.

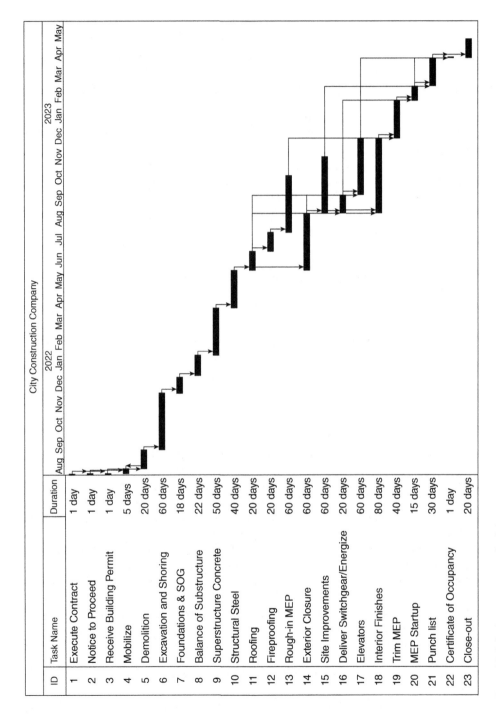

Figure 17.1 Project summary schedule

Alternative techniques

There are three methods for developing a general conditions estimate. A first quick and easy method is to calculate jobsite overhead based on a historical percentage of the total direct costs prior to the addition of the final fee (overhead and profit), for example 10%. While this method is used in developing budget estimates, as discussed in Chapter 2, it is not recommended for lump sum bidding because it may not account for some project-specific items that will affect general conditions costs. For example, estimates for industrial projects tend to have high general conditions costs because of more field supervision, additional jobsite equipment, and the use of precision small tools.

Another shortcut approach is to base the jobsite GCs estimate on a historical monthly cost, such as $50,000 or $75,000 or $100,000 per month. This also does not consider project nuances such as location, construction cost, equipment requirements, etc.

The third and preferred method is to do a detailed estimate for all items that will be required. This is the method that is discussed in this chapter. A QTO is not required, because the quantity being used is generally the project duration, or in some cases a portion of the duration. Some items, such as the building permit, business taxes, or insurance, may be prorated as a percentage of the overall cost of the contract and included as markups on the summary estimate sheet and not within the jobsite GC's cost.

Preparing a detailed estimate allows the estimator to analyze the needs of a specific project and estimate costs accordingly. A good use of historical percentages in lump sum estimating is to check the magnitude of the detailed general conditions estimate to the anticipated bid amount. If the project is such that the estimator would expect the general conditions to be approximately 8% of the subtotal prior to the fee and it actually turned out to be 11%, the team should determine why such a variation exists and modify if necessary.

Elements of the general conditions estimate

The general conditions estimate template featured in this book is divided into four parts: *administrative expense, equipment, temporary construction,* and *general operations.* The general conditions template that we use is a four-part form that lists many items that might be included as indirect costs to a project. A live blank version of this complete template is included on the book's eResource. For most projects, only a portion of the line items are used, but other different items might need to be added for some special requirements or to satisfy company policies or contract requirements. The GCs form also serves as a checklist to ensure that scopes have not been overlooked.

Before proceeding with the general conditions estimate, an estimator should thoroughly review CSI divisions 00 and 01 of the specifications and any supplemental or special conditions of the contract to look for specific requirements or limits that will affect how the costs are determined. As previously discussed, the estimator also needs to determine the project duration. The project owner of the case study specified a 20-month duration. This defines the duration of several line items within the general conditions estimate, such as the field supervision and staffing, job offices, and temporary utilities. It also establishes the start and finish of the project summary schedule shown in Figure 17.1.

When developing the general conditions estimate, an estimator should keep in mind that jobsite overhead must be kept reasonably low so that the final bid is competitive. Each of the pages of the general conditions estimating form should be carefully examined to

Description	QTY	Units	Direct Labor U. Price	Cost	Material/Equip/Subs U. Price	Cost	Total Cost

City Construction Company
Jobsite General Conditions Estimate
(Abbreviated)

Project: Dunn Lumber
Owner: Dunn Lumber Enterprises

Estimator: Paul Jacobs
Estimate # 1
Date: 6/18/2021

Description	QTY	Units	Direct Labor U. Price	Cost	Material/Equip/Subs U. Price	Cost	Total Cost
Administrative Expense:							
Project Manager	86	wks	2,500	215,000		0	215,000
Project Superintendent	86	wks	2,750	236,500		0	236,500
Project Engineer (3 PEs)	258	wks	1,550	399,900		0	399,900
Assistant Superintendents (2)	172	wks	1,850	318,200		0	318,200
Safety Inspector	50	wks	2,000	100,000	0	0	100,000
HO BIM Engineer	50	wks	1,750	87,500	200	10,000	97,500
ST Administrtive Expense:				$1,722,700		$259,700	$1,982,400
Equipment Expense:							
Tower Crane (See separate recap)	1	LS		407,940		542,710	950,650
Pickup Trucks (4 EA)	80	mos			1,000	80,000	80,000
Compressor	w/cost of the work					0	0
Welder	w/cost of the work					0	0
Forklift	20	mos	8,800	176,000	3,800	76,000	252,000
Small Tools	3,000,000	DL $			2.0%	60,000	60,000
ST Equipment Expense:				$613,940		$882,060	$1,496,000
Temporary Construction:							
Job Office	20	mo			2,500	50,000	50,000
Dry Shacks/Tool Vans (2 EA)	40	un/mo			500	20,000	20,000
Temporary Power	265,000	SFF			1.25	331,250	331,250
Temporary Lighting	20	mo	500	10,000	1,000	20,000	30,000
Radios, 10 ea	200	un/mos			75	15,000	15,000
Handrails	5,500	un/mos			15	82,500	82,500
Fences	1,400	LF			15.00	21,000	21,000
ST Temporary Construction:				$35,300		$786,800	$822,100
General Operations:							
Street use Permit	1	allow			300,000	300,000	300,000
Flaggers (2)	172	wks	1,500	258,000	0	0	258,000
Periodic Cleanup	86	wks	2,200	189,200	0	0	189,200
ST General Operations:				$472,200		$585,250	$1,057,450
Continued....							
Grand total jobsite general conditions:				$2,844,140		$2,513,810	$5,357,950

Labor burden on salaried labor:		Carried on estimate summary page		$0
Burden on craft labor:		Carried on estimate summary page		$0
Total jobsite general conditions:				**$5,357,950**
% of total anticipated bid @	$46,500,000	CHECKS? **High, but includes tower crane, OK**		11.5%
Monthly rate @	20 months	CHECKS? **Yes**		$267,898

Figure 17.2 Jobsite general conditions estimate

select a winning strategy for this part of the bid. Figure 17.2 is a condensed form of the detailed four-page general conditions estimate for the case study that is included on the book's eResource. The unused line items in the generic template have been deleted to save space. Another detailed general conditions estimate for the industrial case study is included on the eResource.

Because some members of the built environment confuse a general contractor's jobsite general conditions cost with their home office overhead and consider them both part of the fee, contractors are often under pressure to keep the jobsite GCs as low as possible. This is also part of *activity-based costing*, which involves estimating and charging costs to specific work activities wherever possible. Because of this, general contractors will attempt to shift the jobsite general conditions exposure to direct work items (such as compressors and welders to concrete and steel) or subcontractors (such as mechanical hoisting to the mechanical subcontractor) wherever possible to keep the overall GCs percentage on the lower range.

Administrative expense

The administrative expense page contains the field supervision and will reflect most of the indirect labor. An insufficient amount of supervision will be detrimental to the project, and too much could cause inefficiency. The estimator should consult with the PM to determine the most efficient staffing that will ensure continuity of the project. Quite often the PM slated for a proposed project will prepare the GCs estimate with input from the project superintendent and therefore have buy-in. The administrative expense sheet prices the field management and supervision, office and field support personnel, and material that are required to run a project field office efficiently. The first section in "Management and Supervision" lists the supervisory staff required for a project. Principal line items include:

- Project manager,
- Project superintendent,
- Assistant superintendents,
- Project engineers, and
- Field surveyor and/or field engineer.

Projects the size of the case study typically will have the full-time services of a project manager, three full-time project engineers (PEs), and a full-time superintendent. Assistant superintendents usually are necessary only on very large projects; two have been estimated for this project. The field surveyor is an in-house company person who lays out the building corners and grid lines. The project owner's surveyor typically will have located the property corners.

The PM's effort will be needed full-time at the beginning of the project when planning and controls are being developed and the project management procedures are being established. Monitoring the controls, administering the change order process, and processing pay requests occupy the PM throughout the middle of the project. If project engineers are on the staff, they will perform the day-to-day administrative duties such as processing requests for information and submittals, handling document control, and supporting the field supervision team. At the end of the project, the project manager and all of the PEs will be busy again with close-out activities.

The superintendent is a full-time field supervisor and often the chief safety officer for the project. This project also uses home office safety and quality specialists to support the project superintendent. Detailed discussions of a project superintendent's responsibilities are included in *Construction Superintendents, Essential Skills for the Next Generation*. The interested reader should look to a resource such as that for additional roles of field superintendents. The superintendent will be on the project for its entire duration. For the case study, this means that the time for the superintendent is 20 months. Field surveying is the work of laying out the building. The duration of the field surveying will be less than the total project duration, and assistance will be provided by other personnel already working on the jobsite such as the carpenter foreman.

The second section in the administration portion of the template lists items primarily for site engineering staffing, scheduling, quality control, and safety. A PE is commonly used to perform the day-to-day functions of contract administration. These include estimating change order proposals, monitoring cost controls, preparing status reports, conducting meetings with the owner, design team, and subcontractors, processing requests for information and submittals, and detailed duties relating to closing out the project.

Three other line items that are used on most projects include planning and scheduling, drawing reproduction, and safety/medical. Planning and schedule development can be done either in-house or by a consultant, which would result in both labor and material costs. A project owner usually furnishes one hard copy set of drawings, and the general contractor purchases additional copies as needed or prints their own from electronic sources. Subcontractors purchase the sets that they need or rely on electronic drawings. Safety and medical supplies are required for every job. The company safety department should be able to provide guidelines about what is needed, and historical data will help to determine typical costs for similar types of projects. Job size and type will dictate whether or not items such as additional engineers, concrete form-detailing, and/or professional surveying will be required. Safety training for the case study will be in the form of a weekly toolbox safety meeting, subcontractor orientation, and morning stretch. Some projects have special safety requirements, and the cost will be included in this section. Larger projects may require a full-time safety and/or quality control inspector, but most projects typically rely on one day a week visits from home office specialists.

The final section of the administrative page deals with the field office. Typically, on lump sum projects, the project owner allows the contractor to determine what is necessary to administer a project and include the costs in the bid. The estimator should plan for the smallest staff that can do the work effectively. Excessive GCs increase the total price and may result in a noncompetitive bid or proposal. Many items are required for running the field office, and these may include utilities, telephones, printer and copier, and office supplies. Home office personnel may perform some of the duties listed in this section, but costs are not typically estimated for them, because they are included in the company overhead and are therefore part of the fee. Some general contractor estimators include a small amount for public relations, which is used at the discretion of the field supervision to enhance relationships with the owner, architect, or neighbors or have a crew-appreciation luncheon. This is especially relevant for private negotiated projects.

Several line items on the administrative page and throughout the GCs template are not used for every project. Lump sum projects need to be staffed lean to be price competitive, and open-book negotiated projects will utilize more of the GCs line items to comply with owner and contract requirements. In our example, most of the unused items have been removed for convenience.

Quantities used on the administrative page are in units relative to the overall project schedule rather than specific to construction activities. Productivity factors are not an issue for project administration labor. Units are listed in weeks or months, and the labor unit price is the weekly or monthly pay of the position or line item. Many of the material items are one-time costs, and the units are shown as lump sum (l/s or LS). Some items, such as cellular telephones, are quantified as unit-months, which equals the number of telephones on site for the number of months. If there will be eight cellphone lines used for 20 months, the quantity is 160 unit-months, and the material unit price will be the anticipated monthly bill per telephone line per month.

Equipment

The equipment page is used to list any equipment that is not specific to a given direct work activity but will be needed on the project. This includes a jobsite pickup truck and often a forklift that is used for general purposes throughout the work. Equipment owned by the general contractor is rented to the job just as if it were acquired from a rental company. It was discussed in previous chapters that equipment used for specific activities, such as hoisting concrete tilt-up panels or structural steel erection or wood framing, are priced as part of those activities and not shown on this equipment page. This is a principle of activity-based costing.

Indirect equipment includes equipment that is used for multiple activities on a project, such as a jobsite forklift. The estimator determines the number of indirect units required and their duration. Extension is then a simple multiplication of the quantity and the rental rate. Some items, such as a welding torch, may be purchased, and the units are lump sum.

General contractors have found that for certain sites, a tower crane is needed. While these cranes do not have the capacity to lift heavy items, they will perform most of the lighter hoisting and can prove to be more economical than other equipment that might be considered. The estimator should include all costs associated with the tower crane on the equipment sheet and price it for the duration of the general contractor's direct work. Due to its use for multiple work activities, the tower crane is considered a piece of indirect equipment to be used throughout a major portion of the project.

The lines in the center of the equipment page pertain to both tower cranes and labor and material hoists. The estimator should make a special QTO and recap page for these items, as they can be quite expensive. There are costs associated with constructing a concrete base or footing and the erection and dismantling of the equipment. The estimator needs to take the time to do a detailed estimate on this type of equipment so all costs are included. The process to develop detailed estimates for tower cranes and personnel and material hoists was included in Chapter 13, along with sample estimates of each. A question arises many times regarding who furnishes the heavy hoisting, such as lifting mechanical air conditioning units to the top of a roof. On a lump sum bid, subcontractors are typically responsible for their own hoisting as well as any other equipment and tools needed to perform their work. It may be advantageous for the GC to perform hoisting for subcontractors, but this is an item to be negotiated after the project is won.

Estimated costs for the last four line items on the equipment page are relatively consistent for most projects. Saw sharpening is a necessity for most carpentry work, including concrete form construction, rough framing, and finish carpentry. Saw sharpening is therefore often considered an indirect job cost.

Small tools are those tools not furnished by the craftsmen but are required for them to complete the construction work. As a general rule, these tools are ones that cost less than a prescribed amount, commonly $500 apiece. An example would be a portable power circular saw, which is used to build forms as well as for rough and finish carpentry. Because these are minor costs and used throughout the project, it is not practical to assign the cost of hand tools to specific work activities. For this reason, small tool costs are estimated as a proportion (commonly 2%) of the direct labor cost for the project. Smaller projects may have a higher percentage. Most companies have good historical guidelines for determining this cost. On open-book negotiated projects, the estimator should review the special conditions of the specifications regarding small tools. If an estimator calculates the small tool percentage assuming a tool cost of $500 or less but the project owner specifies the maximum cost at $250 or $1,000, a different percentage should be used.

Consumables involve miscellaneous items like welding supplies or continuous cleanup materials that are used for general purposes. The type of project will influence this cost, and the estimator must be aware of what they are. Discussions with superintendents will help in determining an allowance to be used for consumables. Equipment fuel and maintenance is another GCs item that is difficult to quantify. A generally accepted method is to use 15% of the equipment rental cost as the amount needed for fuel and maintenance.

Temporary construction

Temporary construction is the office and site facilities that will be used during the construction process and then removed once the project is complete. These items include field office trailers, tool trailers, utilities installation, temporary roads, rubbish chutes, fences, and signs. The general contractor's estimator needs to find out what the rates are or anticipate monthly bills, in the case of utilities, and the durations they will be needed, usually the full schedule duration. Most line items will have only material costs, but some labor is necessary for items such as setting up trailers. The costs of some of the temporary construction items for our case study project are based on:

- Job office: It is typical to include rental of a single trailer or other structure for the duration of the project. It should have adequate office space for the superintendent, project manager, project engineer(s), and a small conference room. Because the excavation for this project takes up the entire city block, City Construction Company has chosen to rent adjacent warehouse space for an office.
- Owner/architect's jobsite office: This cost is not included unless specifically required and defined in the front end of the contract specifications.
- Dry shacks: These are very simple trailers or rooms where the workers can hang up wet clothes overnight to dry and have their lunch out of the weather.
- Warehouse: Required only for those projects with limited on-site storage area such as our case study.
- Tool shed: May be combined with the dry shack or a separate trailer for larger projects with a lot of self-performed labor.
- Trailer transportation: The cost to move the field office and dry shack to and from the jobsite.
- Trailer setup: The labor necessary to set the trailer on blocks and any work necessary to make the inside usable for the field staff. This might include building plan tables and structures on which to place office equipment such as printers.
- Saw line: Associated with predominantly wood-framed buildings. These usually are employed on large projects where considerable production sawing is required. The cost is the labor and materials to set up a large saw and build tables for it, and sometimes a full-time operator.

Not all line items on this page or any of the GC's template pages will be used for a specific project estimate. The estimator needs to determine which ones will be required and develop typical costs for them according to the job parameters. For example, large projects will require a sizeable telephone system and jobsite radios. Renovation of a major high-rise building may require a rubbish chute, and certain concrete pre-cast or tilt-up projects may require a separate pre-cast yard other than the building floor slab. Some typical items that will be required for the case study project are:

- Temporary power hookup and distribution;
- Temporary lighting: In winter, to support second shift or allow construction on cloudy days;
- Project sign;
- Layout and batter-boards: The markers that show the outside corners and grid lines of the building;
- Signs and barricades: For open trenches and pits; and
- Fire protection: Fire extinguishers placed throughout the jobsite.

The estimator also needs to include monthly allowances for temporary utilities. Projects that are built on a new site usually do not have services available (water, electricity, sewer or communications), and the general contractor must provide them. Temporary power is an expensive item. Electrical subcontractors usually do not include this in their bid, and the estimator should consult with one to determine what a typical temporary power installation will cost. Temporary lighting and heating systems are weather-related, and the estimator needs to anticipate their use. Job communications, such as radios, are determined by the size of the project and company policy.

Telephone installation is an item that requires input from the jobsite team. The cost will depend on how extensive a system is needed and where the lines are coming from. If the telephone lines will be used for data transmission, the systems will be more extensive and expensive. Cellular telephones have helped reduce this cost, but high-speed data landlines may still be needed.

As is the case with many line items on the generic GCs estimate template for any specific project, some of the site facilities line items are not applicable to this case study project. The estimator, however, must understand the conditions that may require them on other projects. For example, if a project is being built on a site that retains a lot of surface water during a rainy season, some temporary roads may be needed. Most sites will have a project sign. The layout and batter-board line item contains the labor and minor materials that the carpenters will use in helping the surveyor lay out the building. Erosion control materials will be needed to prevent surface water from leaving the project site and should be included in the general conditions estimate. The jobsite may need to be fenced to keep the public from entering the construction site. The cost of installing and removing the fence should also be included in the general conditions estimate.

One item in the protection category required on all project sites is fire protection. This covers the material cost of supplying fire extinguishers at various locations on the jobsite. Usually, state safety requirements will dictate how many and what sizes are needed. The cost is relatively modest, but necessary for a complete estimate. All other protection items are dependent on the jobsite conditions and the project owner's contract requirements.

General operations

The general operations section includes line items that are specified by the project owner to be furnished by the general contractor and those that are required to maintain the temporary facilities and other parts of the site. The first section pertains to permits, licenses, and taxes. An estimator needs to determine whether the building permit is to be paid for by the general contractor and the cost included in the bid or whether the permit is to be furnished by the owner. It is common for the owner to submit for the building permit and pay for it. If it is the general contractor's responsibility, a call to the local building department or researching its web page will provide the estimator with parameters for determining what the permit will cost. Some cities have a published graduated rate schedule, and others simply use a straight-line percentage calculation. If required by the GC, a rule of thumb (ROT) for budgeting purposes is 1%. Most other permits, such as plumbing and electrical, are furnished by the respective subcontractors, and the cost will be included in their bids.

There can be other permits, such as demolition, jobsite environmental and storm water control, excavation and grading, and foundation permits. Some are part of the building permit requirements, and others may be in addition to them. The estimator should become familiar with all of the permitting requirements and make sure that he or she knows who will pay for them and whether or not the owner requires these costs to be included in the contractor's bid or proposal and at what cost. The city requires the contractor to pay to take portions of sidewalks and streets out of use. Street use permit costs are the general contractor's responsibility, as they are in control of schedule and means and methods. Anytime a street is blocked off for construction purposes, a street use permit must be obtained. The estimator needs to find out what the unit cost is, extend it by the anticipated duration, and include it in the estimate.

The line items below permitting on the blank template pertain to various taxes. Similar to permits, an estimator must understand which taxes are applicable to a project and how they are calculated. Usually a contractor's accounting department will have this information or be able to determine it. State sales taxes, local sales taxes, and excise taxes will vary from one location to another. Some states require that sales tax be included in the bid price, while others exclude the sales tax from the price. There is no universal procedure that can be applied to sales taxes, and the estimator must know which ones apply to the specific location of the work. The case study is located in a state that does not include sales tax in the bid price, so the owner must determine and add this cost to contractor bids, but excise tax is required. Information regarding specific building permit rates as well as those for taxes, insurance, and bonds should be kept in the estimator's personal notebook. Taxes are often not included with the GCs but rather considered as markups on the estimate summary sheet, as discussed in the next chapter. This allows for automatic adjustment if bid day prices come in higher or lower than expected. This information should be updated annually. Because taxes are a percent of construction cost, many contractors include them on the estimate summary page and not in the GCs estimate.

Performance and payment bonds are a requirement that will be stated in the bid documents. Project owners may request separate quotes for bonds identified as an alternate on the bid form and then decide at the time of award whether or not they wish to purchase them. Bonds are obtained through contractors' bonding agents or surety, and estimators can obtain rates from them. Bonds, if required, are also typically considered a 'below-the-line' markup in that they are listed below the subtotal of direct and indirect work on the estimate summary sheet.

Insurance is a requirement on most construction projects. Many owners are specific about coverage, and the contractor's insurance provider will furnish either a quote or a method for calculating the cost. Two similar line items are shown in this subsection, broad form and liability insurance. These cover personal liability and property damage. Earthquake and flood insurance is a specific choice of the project owner and in most cases is not included in the general contractor's bid. Similar to taxes and bonds, insurance is often a 'below-the-line' markup on the estimate summary sheet.

The estimator can determine most general contractor material pricing by getting quotes from suppliers and/or using historical information. A jobsite should have a good source of clean, cool drinking water. Commonly one or two insulated containers filled with ice water each day will be sufficient. While this is minor, it is a cost and should be included in the bid. The estimator must establish the average number of persons who will be on the job to determine the number of chemical toilets required. Many state laws are very specific about this. On most projects, the general contractor furnishes chemical toilets for the subcontractors' use as well. A good ROT is to provide one chemical toilet per 10 on-site craftsmen, serviced twice weekly. Separate restrooms for women, Americans with Disabilities Act (ADA) accessible restrooms, and running hot water are considerations for many jobsites today.

Specifications will dictate what inspections and testing are required. A general contractor may be contractually responsible for all inspections and tests required by a building department. If the owner's requirements are more than those of the building department, the contractor must satisfy the contract requirements. Inspections usually are not a cost item if performed by the building department but must be accounted for if done by an independent agency and not paid by the project owner. An estimator must review the general and special conditions of the contract to determine who furnishes the testing for earthwork, concrete, roofing, and structural steel. Some project owners assign the testing to the general contractor, while many municipalities consider it a conflict of interest to do so. All other inspections are part of subcontract work, and costs are included in their respective contract prices. The estimator needs to research what these requirements are and obtain unit prices or quotes from a testing agency for any that are the contractor's responsibility. A reasonable ROT rate for testing and inspection if required of the GC is also 1% of construction cost. Company historical data also may be helpful for this.

Ongoing jobsite cleanup, dumpster rental, and final cleanup are significant jobsite GCs expenses on every project and finish out this last section of the GCs template. An estimator who assumes subcontractors will take care of all clean-up is shortsighted and should consult with the superintendent on this and many other portions of the jobsite GCs estimate.

Summary

The general conditions estimate defines and prices the jobsite overhead (indirect work) for a specific project. It lists all of the indirect labor, material, and equipment that are required to manage, supervise, and complete work activities. When developing the jobsite overhead estimate, the estimator must be aware of what the contract requires. For example, a full-time superintendent is usually required to be on the job for its entire duration. Other items, such as the jobsite pickup, will not be defined but are left up to the bidder to determine whether or not they are needed. The GCs template we relied on lists numerous potential indirect work items. For a typical project, only some of the line items will be necessary. This generic listing can be used as a checklist as well as a pricing template, similar to a grocery list. An estimator is encouraged to use these forms and modify them as needed. For example, a contractor may be more inclined to use an extending-boom forklift in lieu of a boom truck or light crane. The rental rates are different, and the estimator may want to add a specific line item to the template for this equipment to make sure it is not forgotten on future projects. After all of the direct work labor, material and equipment, subcontractor plugs, and jobsite GCs are calculated, they will be posted to the estimate summary in preparation of adding markups, which is discussed in the next chapter.

Review questions

1 What does the general conditions section of the estimate represent?
2 What are indirect work items?
3 What are the titles of each of the four pages of the general conditions estimate?
4 How does the general conditions section of the estimate relate to the general conditions of the specifications?
5 What is another name for the general conditions part of the estimate? There are multiple uses of the term.
6 Why should the general conditions be estimated in detail rather than using a percentage of the project costs?
7 How are taxes and insurance commonly estimated?
8 What is the basis of costing of the building permit?
9 What is the basis of costing performance and payment bonds?
10 What is the advantage of placing percentage add-ons such as permits, taxes, insurances, and bonds on the estimate summary sheet rather than in the general conditions estimate? There are a couple of good answers here.
11 Why might a negotiated project have a higher GCs percentage than a bid project?

Exercises

1 Develop an administrative estimate for a project involving two buildings with an estimated cost of $7 million each. Both buildings will be constructed at the same time but on opposite ends of a large site; 2,500 feet apart. Project durations for both buildings are 28 weeks, and they will be done concurrently but have a one-month stagger-start.
2 Prepare an equipment estimate of an eight-story high-rise building over two stores of underground garage that uses a tower crane and personnel and material hoist. Assume the project will be completed in one year. Each of the eight floors will cost approximately $2 million to build. Expand this for a complete jobsite GCs estimate. Make whatever assumptions necessary.
3 Obtain the building permit rates from your nearest major city and determine the permit cost for a building with an estimated construction contract amount of (A) $3.3 million (B) $10 million, and/or (C) $60 million.
4 City Construction Company prepared their general conditions estimate based on a 20-month preliminary schedule, from mobilization through receipt of a certificate of

occupancy, which is a common approach. Explain why this approach is flawed – in that they might come up short in this portion of their estimate.

5   What is an expected percentage range of GCs for a lump sum commercial project?

6   Compare the industrial project's general conditions estimate with its site logistics plan and summary estimate, all located on the book's eResource. Do the three documents match? What changes would you recommend to make them more coordinated?

18 Estimating markups

Introduction

Estimate markups, also known as *below-the-line* markups, pertain to the overall total estimated cost of direct and indirect work. The addition of the markups to the cost of the work will be the contract amount if the project is won or negotiated. These include the general contractor's (GC's) contingency, liability insurance, taxes, fee, and others. Many of these were introduced along with general conditions in the last chapter. The fee represents the project's portion of the home office overhead and profit (OH&P). Home office overhead covers the basic services provided to each project by the company's home office. Home office OH&P is also known as fee. This includes accounting services, sales, estimating, the home office facilities, cost of project managers (PMs) and superintendents between jobs, and the salaries of the home office personnel including corporate officers.

The fee is only one markup that is placed below the line in the estimate summary; there are several others. All of the estimated costs *above the line* are considered costs of the work, which have been the focus of the last several chapters of this book. Similar to fee, other items below the line are percentage add-ons to the estimate and are volume dependent and therefore are variable costs. Labor burden is applied to direct and indirect labor only, not to the entire direct cost subtotal. Labor burden is not applied to material costs, and subcontractors are expected to have covered their own labor burden in their bid prices. Different burden rates should be applied to direct and indirect labor. Labor burden was introduced in Chapter 10. Typical percentage add-ons that will show up below the line on a construction summary estimate include contingency, liability insurance, excise tax, fee, labor burden, sales tax, and bonds. The 'line' we refer to in estimating is a horizontal line on the estimate summary page that separates the cost of the work from markups. The estimate summary page will be discussed more thoroughly in the next chapter.

Risk assessment

There are numerous potential risks that can be influenced by the project design, the client, subcontractors, design team, and other internal and external sources that can influence a contractor's estimated and final fee. The greater the risk, the higher is the estimated fee. If the risks are too great, there might not be any fee that would make the job attractive. Risks are also resolved for the contractor by purchasing insurance, bonding subcontractors, increasing estimating contingencies, or taking on a joint venture partner for a specific project. A project that has too great a risk or combination of risks may not be pursued as a business decision of the contractor. Contractors may choose to *raise their estimated fee* on one particular project for a variety of reasons, including:

- To provide a practice bid, although this is expensive to do and potentially dangerous if the contractor accidently becomes low bidder;
- Market conditions allow a higher fee;
- The type of work is not the contractor's specialty;
- Disproportional requirement for direct craft labor;
- The client may be a high-risk client that is prone to lawsuits;
- The construction documents are not complete, although some contractors would see this as a change order opportunity;
- Tight schedule mandated by the client, possibly with liquidated damages;
- Other risky contract clauses, such as slow payment terms or high retention withheld;
- To provide a courtesy bid to a client they do not want to offend, if they have a sufficient backlog; and/or
- If the project is difficult or complicated, it may require a higher fee due to associated risks.

Contingencies

The estimator may want to include a contingency below the direct cost total line or added to the fee. Contingencies are appropriate for negotiated projects where design information may be limited; contingencies generally are not used on competitive lump sum bids. If GC 'A' includes a contingency with its bid but GC 'B' does not, B has an improved chance of becoming the successful low bidder. The bidder is being asked to furnish a price for the work as defined in the contract documents. It is the project team's duty prior to executing the contract to make sure that the project scope of work has not been increased. This then minimizes the need for a contingency add-on. An explanation of some of these different contingencies follows, and their relative ranges with respect to design completion are reflected in Table 18.1.

- *Design contingency*: This is usually held by the owner and protects them from contractor estimate increases due to design discrepancies, especially early in the design process, such as design development documents. This is also known as lack of design definition or 'design creep'. An example would be that the owner always intended to include window coverings and had told their architect to include them, but the architect forgot or just hasn't gotten to that level of detail yet. During development of multiple budget estimates by a preconstruction GC, the estimator may include this contingency line item in the budget, but at contract time, it would likely roll over to the project owner.
- *Document contingency*: Similar to design contingency, this would also be held by the project owner but is intended more for errors discovered in the documents after the project has been bid. This category could also be known as document inconsistencies. For example, anchor bolts were shown for all of the columns but missed in the foundation wall or a requirement to acoustically insulate all interior office walls was not called out. For a negotiated guaranteed maximum price project, the GC may

Table 18.1 Contingency values

Contingency types	Estimate type		
	Budget	GMP	Lump sum
Owner contingencies:			
Design contingency	10–20%	5–10%	NA
Document contingency (1)	NA	10%	5–10%
Permit contingency	NA	1%	1–3%
Scope contingency	5–10%	5%	0–2%
Unforeseen conditions (1)	NA	10%	5–10%
Contractor contingencies:			
Escalation contingency	NA	2–3%	included in cost
Estimating contingency	5%	1/2–1%	included in profit
Construction contingency	NA	2%	included in profit
Totals:	(2)	(2)	(2)

Notes: (1) Document and unforeseen conditions are a combined 10%.
(2) Most of these contingencies are not cumulative, it is generally
either one or another, or they function independent of each other.

want control of this contingency to cover 'minor' changes, but 'major' changes likely still would result in change orders to the owner.

- *Permit contingency*: During permit review by the authority having jurisdiction, often the city, the permitting agency may require changes to be made to the documents that the architect had not anticipated. This fund would also be under control of the owner unless the project was design-build, and then the contractor would hold this contingency line item. Examples would include requirements to revise city sidewalks which are adjacent but not directly impacted by the new project or protection of significant trees or not allowing work during evenings or on the weekend.
- *Scope contingency*: The owner wants to add additional scope, such as one floor of the building had been shelled but now finishes have been added, or the owner's program changes, and instead of a dry laboratory a new scientist has been hired who requires a wet laboratory. No one is at fault for these occurrences, and there would be no way for the contractor to have estimated for them. This is also known as 'scope creep'. This can be one of the largest values of change order dollars. The owner would again be in control of this contingency fund.
- *Unforeseen conditions contingency*: During earthwork activities, unsuitable soil or buried debris may be discovered that the GC could not have anticipated. This is also the case with remodeling projects, when rot or asbestos might be discovered. The design documents attempt to discover these situations, and language is often included in the contract to attempt to shift responsibility to the contractors, but this is difficult to uphold. This is another owner-controlled fund.
- *Escalation contingency*: Similar to design contingency, a negotiated preconstruction GC will include this line item in its budget. Economists can only make educated guesses as to what inflation may occur in the following years, so the amount to assign to this line item is very subjective. It is also very difficult to define what is covered – for

example, increases in prices of copper pipe and electrical wire or steel tariffs – and when the contingency would either be realized or removed. Natural disasters have a way of substantially increasing the cost of equipment such as an emergency generator. There is no way a contractor could have anticipated this in its bid. It is this author's recommendation to project owners that they control this fund as well and require the contractors to prove that escalation occurred.

- *Estimating contingency*: Estimating is a blend of art and science, and the contractor will not estimate every item exactly correct. This contingency will show up as a below-the-line add-on in a GC's estimate on negotiated projects more than bid projects. Even for negotiated projects, a case can be made that the contractor/estimator is the expert, and if they cannot estimate accurately, they are not earning their fee. An example would be missed scope such as concrete shown on the mechanical drawings but not on the structural drawings. Additional examples would be if a subcontractor pulls its bid or a subcontractor suffers financial default.
- *Construction contingency*: Even with very good construction documents and a thorough and complete estimate, things happen in the field that fall under the contractor's control or are part of their means and methods and are not subject to an owner change order. This may be the case with poor subcontractor performance, schedule delay, or a rained out concrete slab that requires repair. Because negotiated private project owners tend to remove contingencies from the contractor's side of the ledger, estimators may find ways to build in estimating and construction contingencies within their budgets or guaranteed maximum prices (GMPs) but not necessarily label them as such.

Insurance

The contract terms will dictate what types of insurance a contractor must provide and the insurance limits the policy must cover. Similar to our discussion of performance and payment bonds later, the estimator will send the bid documents to the contractor's insurance company when they first arrive in the contractor's office, especially Construction Specification Institute divisions 00 and 01 concerning the special conditions of the contract. Insurance types and areas of coverage are covered more thoroughly in a construction project management book such as *Management of Construction, A Constructor's Perspective*. The reader should look to a resource such as that for expanded coverage. Some of the more common types of insurance and their approximate markup rates include:

- Liability insurance: The most common insurance type carried by all contractors, which protects from claims made by third parties. An average rate is 1%, less for larger GCs and higher for smaller GCs or specialty contractors.
- Equipment floater insurance: Protects the contractor against financial loss due to physical damage to equipment from named perils or theft. Coverage is for owned, leased, and rented equipment not operated on streets and highways.
- Automobile insurance: Protects the contractor against claims from another party for bodily injury or property damage caused by contractor-owned, leased, or rented automobiles and equipment operated over the highway. Coverage may include damages to the automobiles and equipment.
- Umbrella liability insurance: Provides coverage against liability claims exceeding that covered by standard general liability or automobile insurance. For example, a

contractor may have a general liability insurance policy covering up to $2 million per occurrence and an umbrella policy covering up to $50 million per occurrence.

- Workers' compensation insurance: Protects the contractor from a claim due to injury or death of an employee on the project site. Workers' compensation insurance is no-fault insurance in which the employer cannot deny a claim by an insured employee, and the employee cannot sue the employer for injuries sustained on the job.
- Errors and omissions insurance: This is typically provided by design professionals and protects the project owner due to potential building failures.

In addition, on certain projects, the owner will require *builder's risk insurance* or *property insurance*. This is in addition to typical liability insurance, and the estimator must price it accordingly. Builder's risk insurance is a policy that protects the owner against financial loss due to damage to the uncompleted project. The project owner may provide this insurance, or the general contractor may be required to furnish it. Regardless of which entity provides the insurance, the other party will want its interests protected, will customarily be named as an 'additional insured', and should have a chance to read the policy. It is important to realize that this is an additional cost to the contract and treat it accordingly. If the GC is required to provide builder's risk insurance, the percentage add-on markup may be in the range of 0.1% to 0.2% of the estimated construction cost.

Taxes

Business and occupation (B&O) tax or excise tax is a markup based on the total contract amount and covers the taxes for doing business in a state, city, and county where the project is located. Most states and many cities have some sort of B&O, but their rates and applications vary considerably. Construction company accounting departments can usually furnish a percentage markup rate for the estimator to use for business taxes. The case study in this book is being constructed in the area where City Construction Company normally works, and a combined state and city 1% markup for the B&O taxes has been factored in. If the project is in a state and city where the contractor does not normally work, the estimator must also determine the costs for a *business license* and add it to the markup or include it in the general conditions estimate, as well. *Sales tax* becomes an issue in an estimate in three different places:

- Items that do not become a permanent part of the project, such as concrete form material, also known as temporary structures;
- Equipment rental; and
- Sales tax on the full contract value.

The estimator needs to understand the procedures for sales tax in the state in which the project is being constructed and apply those procedures accordingly. When purchasing form material, for example, the unit prices used may need to include sales tax. If a construction company's historical information is used as the basis for pricing, it typically includes appropriate sales tax. However, if vendor pricing or estimating references are used, sales tax may not be included. Equipment rental also is subject to differing state rules regarding sales tax. At the location of the book's case study, equipment that is rented without an operator is subject to sales tax, while equipment that is rented with an operator is considered a subcontract and is not subject to sales tax. Both of these sales tax examples (formwork and equipment) are part of the cost of the work and are factored above the line.

When tendering a final bid, the inclusion or exclusion of sales tax below the line on the total cost may be a matter of state law. Some states require that sales tax be added to the contract amount to calculate a bid total, and others leave it out. Some states tax labor only, and others tax materials only. This becomes an important issue when a contractor is estimating a project in a state other than its home location. In addition to state sales tax, the estimator needs to determine if any local sales taxes apply. Washington State, where the case study was constructed, does not require sales tax to be included in the bid or contract, but the GC still invoices the client monthly for state sales tax and pays the state on behalf of the project owner. This qualification and exclusion is an important item to note in the qualifications and assumptions document discussed in Chapter 2. The bottom line with taxes is that it is complicated and inconsistent, and the estimator must do his or her research before finalizing the estimate summary sheet.

Home office overhead and profit

Home office overhead for general contractors varies, depending mostly on the contractor's size and volume. For small to medium-size contractors, it ranges from about 3% to 7% of the total contract cost. Larger general contractors have a home office overhead as low as 1% to 2%. The contractor for our case study project has an annual volume of $400 million and anticipates $6 million in home office overhead; therefore, a 1.5% home office overhead percentage needs to be covered by each project undertaken. Home office overhead is only part of OH&P. A fee of just 1.5% covers this project's proportional share of home office overhead but does not return any additional profit. This is known as the *breakeven fee*. A higher fee is needed beyond the breakeven fee if a profit is to be returned to the company's ownership. The breakeven fee for our case study would therefore be:

$$\$43.4 \text{ million} \times 1.5\% = \$651,000$$

There are several considerations for determining the profit to include in the project bid. The first priorities for the general contractor are to cover home office overhead and to limit the risk of losing money. The biggest single risk for the general contractor is whether or not its direct labor will achieve the productivity that was used in the estimate. Some GCs limit this risk by subcontracting all or most of the work and then managing the subcontractors. The risk then shifts to the contractor's jobsite management capabilities and the technical capabilities and financial condition of the subcontractors. The downside is that a contractor which is innovative in performing direct work benefits by generating savings, which are added to the profit. When all of the work is subcontracted, the contractor will not typically realize savings beyond the profit figured in its original estimate. There are several factors to consider in calculating an appropriate fee. Some of these include:

1   The fee may be influenced by the amount of estimated direct labor. The fee must first include the home office breakeven point, as discussed earlier, and then add a desired amount of profit. A profit goal that is about 50% of the direct labor cost is considered adequate protection against possible labor estimating errors or labor cost overruns. This is saying that if the estimator missed the labor estimate by half, they will still come out even. Using the total direct labor for our case study project, the labor risk the profit needs to cover would be as calculated here.

$3 million × 50% = $1.5 million profit
$1.5 million profit/$43.4 million cost = 3.5% profit goal

Adding this to the 1.5% breakeven home office overhead requirement results in a fee of 5% of the subtotal estimated cost prior to adding B&O taxes and insurance. Although this feels low, it might be appropriate in a competitive market.

2   Another approach to fee determination is to factor in the earning power of the project superintendent and PM. This is also known as their opportunity cost. The capability of the team to organize and run a project successfully, including keeping costs under control and maximizing profit potential, are factors that should be considered. The PM and superintendent are soft assets with earning power. They are expected to earn a profit for their company. In many cases, the superintendent and PM may not be known at bid time, but the type of project usually dictates the capabilities required. For example, a very large job that has a high labor risk will require a very experienced superintendent and project manager. If they cannot realize a profit of $20,000 per month each on project 'A', then the contractor should pass on it and pursue project 'B'. An experienced project engineer (PE) can be added to this equation as well. A highly skilled superintendent and PM for a job with a 20-month duration should therefore generate a profit of:

2 supervisors × $20,000/month × 20 months = $800,000 profit
$800,000/$43,400,000 cost = 1.8% profit goal

Note that this is considerably less than the required labor protection calculated in the first method. Combined with home office costs, this represents only 3.3% of the estimated cost of the work before business taxes and insurance are added. This would be considered a reasonable fee on a very large project but may be a little low for the commercial case study and current market conditions.

Other fee determination methods should also be considered. A job that is straightforward and is largely done by subcontractors has a lower risk, and the work of the superintendent is primarily that of ensuring that the subcontractors perform in accordance with the contract documents. A PM or project PE will do much of the subcontract documentation work, and the superintendent can be one who has less experience but is working to increase his or her knowledge. Smaller projects that employ mostly subcontractors can be built successfully with a team of lesser experience. In these instances, the estimator may decide that an opportunity cost fee of only $15,000 a month for each is warranted.

3   A company's backlog of work also enters into the fee consideration. The estimator and the officer-in-charge (OIC) usually discuss the company's backlog and decide how important the project is. If the backlog is low, they may decide they need the job and propose a lower fee. If the backlog is high, they might decide to use a higher fee. If they do not win the bid, at least they have responded reasonably, which usually keeps them on the owner's preferred bid list for future projects.

4   A survey of market conditions is also used to determine the fee. When the company's backlog decreases, it usually is low for an entire class of competitors. Consequently, all of them probably will be bidding to get work and use lower fees. The contractor has to be careful not to bid the job so low that there will not be any profit at completion.

There is a fine balance between keeping personnel occupied on a project at almost any cost versus losing money. The final decision rests with the OIC.

5   The bid team also must assess the availability of personnel to supervise the construction activities. If all of the contractor's key personnel are committed to other projects, a higher fee may be proposed to cover the cost of hiring a new superintendent and possibly some other key field people and promoting PEs to become PMs, all of which can be risky if not performed diligently.

6   The contract type can significantly influence the fee decision. Lump sum projects have increased risk for the contractor and deserve more fee than a cost plus project. A project with a guaranteed maximum price is somewhere in between. There are many contract issues and contract clauses that can raise or lower the fee, such as liquidated damages and definition of reimbursable costs in the AIA A102 contract, Articles 7 and 8. If there are additional opportunities for home office costs to be considered job cost and therefore reimbursable, this allows money to be moved from the home office general conditions to jobsite GCs and allows a lower fee, all the while retaining the same profit goal.

7   The construction company owners have invested their personal money, also known as 'equity', in the company and expect a high rate of return, likely greater than 15% return on equity, for what everyone considers a risky business.

The estimator should review the project with the OIC in detail before bid day. This will help the OIC to assess risk in relation to determining a fee. If the project is straightforward and is the type that the contractor normally undertakes, the risk is relatively low and a modest fee can be used. If the work is a new type for the company or there is a very large direct labor force, the risk is high, and the fee should reflect this. At this stage, the estimator prepares the bid summary for bid day. For the case study the team has decided to use a fee of 4.5% based on the following:

- The home office overhead (HOOH) is currently running about 1.5% of total work under contract. This is the breakeven fee but with 0% profit, which is unacceptable to the company's equity partners.
- The high direct labor risk coverage combined with HOOH produces a fee of 5%.
- The subcontract work is significant, which lowers risk.
- An experienced project team's earning power for 20 months, combined with HOOH, reflects a fee of 3.3%, which feels too low.
- The company's backlog is reasonably strong but could use more work for the upcoming year.
- Market conditions seem to indicate that projects are being bid with fees in the range of 4% to 6%.
- The project's direct cost is $43.4 million and, at 20 months, is larger and longer than most in this marketplace; therefore, a fee on the lower end is anticipated.
- The work is the type that the company normally performs, and a project manager and superintendent with this experience are available to build the project.

In this case, the CEO decides a 3% profit keeps the fee at the lower end of the market range and competitive. It also is within the earning power of an experienced superintendent and project manager. A fee calculation of 4.5% (1.5% overhead + 3% profit) is therefore entered on the pre-bid day summary estimate.

Contractors set many goals, including building quality projects, meeting schedules, keeping everyone safe, and developing a good reputation with clients and subcontractors. But contractors are also in the business of making a profit. Construction is risky and is not a 'not-for-profit' industry. Project managers will be reminded of their fee goals throughout the course of construction on their projects by the home office. One of the unique aspects of construction is that the true fee will not be known until the project is complete. In the home office, corporate executives will not know what the total yearly profit will be until the year is complete and all job costs and revenues have been factored, along with the actual home office general conditions expenditures.

Additional markup considerations

When preparing the estimate summary form for bid day activities, the estimating team should verify markup requirements from the owner-provided *bid documents*. Frequently, the project owner will ask for markups that the general contractor intends to use for change orders. This can become an issue because the owner may assume that the stated markup includes those of subcontractors, while the general contractor assumes that they do not. If there is a question, an estimator should submit a request for information to get a clarification. Sometimes an owner also will require markups the general contractor plans to use for subcontractors. The estimator must ensure that he or she understands the contractual markup structure prior to finalizing the bid, as it can have an effect on how the final fee is determined. Procedures for sales tax should also be reviewed.

In general, a *jobsite general conditions* estimate that is about 5% to 10% of the total expected bid is considered adequate for a project. Is the amount calculated by the estimator within a reasonable range? Straightforward projects can be built with a little less field management, and 8% might be considered high. On bid day, this should be discussed with the OIC, as it may affect what, if any, adjustments are made. The pre-bid day summary estimate for the case study shows a jobsite overhead exceeding 11%, which feels high but is appropriate for this type of job. The project is longer than many and includes the cost of a tower crane, which increases its general conditions estimate significantly. If a project is difficult, a higher jobsite overhead will be required. On some heavy civil and industrial projects, the general conditions can be as high as 15% due to additional management and supervision and expensive specialized construction equipment. In early budget estimates, general conditions may be added to the estimate summary as a percentage add-on or markup. But any project that proceeds into or beyond the schematic design phase will warrant a detailed line-item general conditions estimate, as was prepared in Chapter 17.

Not all contractors approach how they estimate for *labor burden* the same. As presented in Chapter 10, labor burden is a combination of labor taxes imposed by the government and labor benefits, which may be associated with labor unions or a particular contractor's choice. Some contractors utilize loaded wage rates in their estimates, especially early budget estimates, where labor burden is combined with the actual wage the craftsman receives. In this case it is difficult to modify the burden percentage rate for an entire project, which may be required in a competitive bid environment on bid day. Other contractors split out the labor burden and use a combined or blended rate, which may combine all crafts together into one percentage markup, such as 55% for union crafts, or split the burdens out between different crafts, such as 50% for cement masons and over 70% for ironworkers. Merit shop craftsmen are typically lower, such as 30% to 40% for a labor burden markup.

Labor burden for indirect labor, such as the PM and superintendent, is at a considerably lower rate than crafts, approximately 35%. Some contractors will split the markup out for indirect labor separate from craft labor, and other contractors will combine the two into a blended rate. In the case of a closed-book lump sum bid project, the owner usually cannot audit the contractor's accounting books and will not know or have influence on how labor burden is applied. But a project owner in an open-book negotiated project will likely have audit capabilities, and the rate for labor burden should be stipulated in the contract and verification of how it is accounted for will need to be addressed.

Performance and payment bonds are typically a below-the-line markup. In fact, bonds are below the bottom line in that they are usually excluded from the base bid and contract amount and are an additive alternate if chosen by the project owner. The estimator should have sent the bid documents to the contractor's surety when they first arrived in the office. The surety will gauge the risk level of the project with the contractor's backlog and available bonding capacity. A graduated bond scale similar to the example here will be sent to the estimator to be posted in the bid room and factored into the computer dedicated to calculate the final bid amount. A rule-of-thumb performance and bond rate is 1% of the contract amount. But because of the graduated bond schedule, smaller projects (and smaller contractors) have higher bond rates and larger projects (and larger contractors) have lower bond rates. Also, an individual contractor's bond rate is indicative of their success, reputation, and financial standing. Those with more have lower rates, and those with less have higher rates. Bond rates are not flat percentages, such as 0.75%, but rather are determined from a graduated scale and are dependent on project size, project complexity and risk, and the strength of the contractor providing the bid. Lower bond rates can result in a second low-bidding contractor being awarded a project if the project owner elects to include the bond alternate with the combined base bid.

Sample general contractor's graduated bond schedule	
Contract price:	Cost per $1,000 of contract price:
$1 to $100,000	$30
Next $400,000	$20
Next $1,000,000	$15
Next $2,000,000	$10
Next $2,000,000	$7
Next $2,000,000	$6
Over $7,500,000	$5

A sample bond calculation for a $5 million project using this graduated schedule is as follows and, coincidently, results in a total bond cost of just over the rule of thumb rate of 1%. A smaller project would have had a higher percent, and a larger project would have had a smaller percent total.

Bond cost example for $5,000,000 project			
Contract price:	Cost per $1,000:	Extension:	Total:
$1 to $100,000	$30	$30 × 100 =	$3,000
Next $400,000	$20	$20 × 400 =	$8,000
Next $1,000,000	$15	$15 × 1,000 =	$15,000
Next $2,000,000	$10	$20 × 2,000 =	$20,000
Next $2,000,000	$7	$7 × 1,500 =	$10,500
Total:		$5,000,000 =	**$56,500 = 1.1% bond cost**

Most GC sureties will require the GC to also obtain performance and payment bonds from their subcontractors, especially the major subcontractors or those above certain bid amounts, such as $50,000. But different than the GC's bond being an alternate add-on, the subcontractor bonds will need to be factored into their bid prices. If the owner chooses to have the GC purchase a bond and agrees to the additive bond amount, the owner will not also agree to purchase bonds on the subcontractors for benefit of the GC. General contractors on negotiated projects may be able to have the owner pay these added costs, but only if discussed before the contract is executed.

Summary

In this chapter we introduced many different markups or percentage add-ons that a contractor may put 'below the line' on an estimate summary form. There are many factors a contractor will consider regarding which markups to include and at what percentage and where in the estimate to include them. The general conditions estimate template discussed in the last chapter allowed for some of these to be included there, but most contractors apply these percentages on the total cost of construction – including direct, indirect, and subcontract costs – and therefore inclusion on the estimate summary form may be the most effective application. Evaluation of project risks plays an important role in markup decisions. Contractors can choose to accept risks, mitigate them, share them with project owners, insurance companies, and subcontractors, or increase markups such as fee and contingencies. There are a variety of contingencies applicable to a construction estimate; some of them are managed by the project owner and some by the contractor. Contingencies generally decrease as design progresses, and negotiated projects have higher contingency markups than do lump sum bids – if they have any contingency at all.

There are also several different types of insurances that contractors may either choose to purchase or may be required by the contract. Some of the most common ones include liability, builder's risk, and worker's compensation, which is typically part of labor burden. Different states and cities will also require sales and business or excise tax, also known as B&O tax. The estimator must research not only the contract requirements with respect to insurance and taxes but also local jurisdictions and include applicable rates.

Profits are not what contractors add to the bottom of construction estimates; they add a proposed fee. But for some participants in the built environment, they see the fee as all profit. The fee is also known as the 'margin', or generically the 'markup'. As discussed in this chapter, the fee first needs to cover home office overhead, and any money remaining after all costs and overhead is accounted for may be considered gross profit. Net profit for the company is determined after deducting taxes from gross profit. Net profit is also known as pure profit or after-tax profit.

Additional potential markups include jobsite general conditions costs (for early budget estimates only), labor burden, and performance and payment bonds. General conditions and labor burden are not fees but are costs of the work. Bonds often are stated as bid alternates and are on the bid form below the bottom line and not added to the base bid. Public bid projects tend to require bonds more than do private projects. All of the markups discussed in this chapter have been carried forward to the estimate summary sheet, which will be discussed in the next chapter.

Review questions

1   Placing labor burden below the line on the estimate summary versus incorporating it above the line in loaded wages provides the contractor with what advantage?

2   What is the advantage of estimating general conditions on a detailed line-item basis versus plugging it in as a percentage markup?
3   What effect on the bottom line does reordering the liability insurance, fee, excise, and contingency markups have? Assume a $20 million construction cost and 1%, 5%, 1%, and 2% markups, respectively.
4   Where is the 'line' drawn in our 'above-the-line' and 'below-the-line' discussions?
5   List three items included in home office overhead and three items included with jobsite general conditions.
6   What are the components of the fee?
7   What are four factors to consider in determining the fee?
8   Who on the contractor's estimate team typically makes the final decision on markups?
9   Of all of the markups discussed in this chapter, which one would not be considered a cost of the work?

Exercises

1   Many reasons why a contractor might choose to increase its bid fee were discussed in this chapter. Why might a contractor choose to lower a lump sum bid fee or negotiated proposal fee?
2   If a contractor suspects a project might be risky to pursue, they have several options available to them. What might some of those be?
3   What is the minimum fee, in both dollars and percentage of construction contract, a general contractor would need on an upcoming $10 million bid project? Assume the following parameters:

   • $100 million expected yearly corporate volume
   • 3% home office overhead budget
   • Owner's equity is $4 million
   • This project is estimated to have $1 million in direct craft labor
   • This project is expected to last 12 months
   • There is a full time project manager and superintendent without a project engineer
   • Cost accounting is performed out of the home office
   • Construction company equity owners expect a 15% return on equity

4   How would your answer change in Exercise 3, if:

   A   Market fees are at 7%,
   B   This was a negotiated project,
   C   The GC will perform $1.5 million in direct labor, and/or
   D   The GC has sufficient backlog and is submitting a complementary bid to a past client?

5   Where would an estimator choose to order the fee markup on the estimate summary form on an open-book negotiated project proposal?
6   Determine what business taxes are required for a general contractor in your local area.
7   Calculate the bond cost and percentage of contract for the following three projects utilizing the graduated bond schedule included in this chapter or another one you may have access to: (A) $2,000,000 project, (B) $7,300,000 project, (C) $25,000,000 project, and/or (D) $100,000,000 project.

8 In Table 18.1 we stated that not all of the contingencies were additive. Why not? Add together the high-end range of each type of contingency. What is the total percentage? What would happen to a project's pro forma (look ahead to Chapter 23) if an owner includes this cumulative amount?

9 Provide another example for each of our contingency categories.

10 Prepare a spreadsheet comparing the various markups included on the book's three case study projects. What similarities and differences exist? Is one more accurate than another?

19 Estimate summary

Introduction

The estimating work has progressed from initial document review to quantity take-off (QTO) and pricing recapitulation (recap) of the self-performed work. Order of magnitude (OM) estimates have been developed for subcontractor work, and the project summary schedule and general conditions estimate are complete. Pre-bid day markups have been factored on this detailed work. This concludes the first eight steps of the lump sum estimating process shown in Figure 1.1. The two remaining steps represent work that is done on bid day. It is now time to complete the estimate and determine a pre-bid day total based on the information developed thus far. The completion process is an excellent time to review the work, not only for its completeness but to check transferred numbers. Even though an estimator has diligently performed all reviews and error checks during the development of the estimate, a final review and check is warranted. We introduce the estimate summary page in this chapter and populate it with all of the estimating work accomplished thus far, including:

- Direct work pricing recaps,
- Subcontractor OM plug estimates,
- Detailed jobsite general conditions estimate, and
- Pre-bid day markups.

In addition to filling out the pre-bid day summary estimate, there are several other activities to prepare the bid room and estimate team for the busy activity of successfully completing a lump sum bid. This chapter completes everything right up to bid day and our next Chapter 20 captures all of the actual activities that occur on bid day. All of these pre-bid day activities are reflected in the top half of Estimate Elements Figure 19.1.

Final document review

The first part of completing the estimate is to conduct a thorough review of the bid documents including all addenda. This should be done in the same order that they would have been read before beginning the estimating work; that is, looking at the invitation for bid, instructions to bidders, the bid form, and any supplemental and special conditions. The bid form is especially important because it defines what pricing is required and how the bid is to be tendered. This is also a time to verify that the markups proposed on the estimate summary form and with the bid conform to contract requirements. Markups were discussed in the last chapter.

Estimate Elements

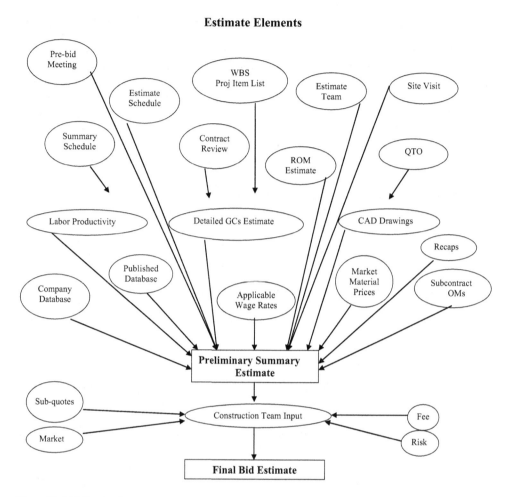

Figure 19.1 Estimate elements

Another item that may be requested is unit pricing for a given list of work activities. These are more common for industrial and heavy civil construction, but occasionally a project owner of a commercial project will request them. Unit price determination will be discussed in Chapter 23, but the estimator should know that if requested, their inclusion is required in order to tender a qualified bid.

The next part of the document review is the drawings that were used for estimating. These are the ones that were marked up as items were quantified. The estimator should look at each drawing to verify which work items are self-performed and which ones are to be subcontracted. It is very important to look at the mechanical, electrical, and civil drawings because they may have items, such as concrete pads for transformers or air conditioning units, that the general contractor (GC) is to construct and may not be shown on the architectural and structural drawings. During the drawing review, the estimator should ensure that all items of direct work have been marked. It is not necessary to perform another dimensional check of the quantities, but rather to make sure that all items of work have been quantified. Those items not marked should be double-checked as required and verified as not having been estimated, and a miscellaneous QTO and recap sheet should be prepared for them. All of the sections and details should be reviewed thoroughly. Sometimes an element on a plan was marked as taken-off, but the section or detail was not. While the information in the estimate may be correct, the detail may show things that are different than the estimator thought. When finished, all sections and details that pertain to self-performed work should have been marked. This is an especially important check.

Some additional items in the specifications should be verified. Typically, specification sections are basic standards that are modified for a specific project. Additional specifications are included on some drawings, and the estimator must ensure all are accounted for. Much of the specifications concern standards to which the work is to be performed and the methods to accomplish it. For most projects, the labor productivity used in the estimate is adequate to cover minor variations in the specifications. A quick scan will verify this. The estimator should also pay attention to the material specifications. For example, 3,000 pounds per square inch (psi) concrete may be required for footings and slabs-on-grade, while 4,000 psi may be required for elevated decks, and the estimator must ensure that the proper quantities have been priced.

Once the document review has been completed and any miscellaneous items have been identified, it is time to review the estimate sheets. Every sheet should be reviewed to ensure the totals that are to be used are double underlined. The estimator should ensure that each double underlined total has been circled, signifying that it has been transferred to the next applicable sheet in the estimate, and check that sheet to ensure the number was transferred correctly. A spot check should be done to make sure the spreadsheet formulas are working properly. Finally, the assembly unit costs (such as $/SF, $/CY, $/SFF, $/ton, and/or $/MH) at the bottom of the recap sheets should be checked to determine if they are within a reasonable range for the work to be performed.

At this point, the estimate should be ready for the final summary. All totals on the recaps and summary recaps should be double underlined, and those that have been transferred from a recap to a summary recap should be circled and all pricing forwarded to the estimate summary page. The estimator then submits the estimate to the project manager (PM) or chief estimator for review. A properly prepared estimate file makes it easier to spot errors or anomalies that may need to be corrected. The reviewer can spot other items that the estimator may have overlooked or work procedures that may not have been considered.

Completing the estimate summary

The bid summary template is used for summarizing the work and calculating the total estimated cost. There are a variety of different layouts for this document used by contractors, but they all include the basic concepts we describe here. Our bid summary template is laid out in five sections. Starting at the top, the first section is for entering the totals from the general conditions estimate. The lines are already labeled to conform to each of the four pages of the general conditions estimate. The second section is for listing the amounts from the various self-performed work packages. The line descriptions are blank, and the estimator enters appropriate descriptions as numbers are ready to be transferred. Different projects may have different descriptions, but standard Construction Specification Institute categories or assemblies should be used whenever possible. The total bid shown on the bottom line, once determined, is the one that will be tendered to the project owner.

The estimator may fill in the bid summary in a variety of fashions. Some will start the sheet at the beginning of the estimating process and fill it in as various sections are completed, while others will wait until all self-performed work, subcontractor OMs, and general conditions have been estimated and fill it in all at once. Whatever system the estimator chooses, consistency should be used for all estimates.

When transferring numbers from pricing recap or summary recap pages to the bid summary, only the total man-hours, labor, and material amounts are transferred. A summation formula in the far-right column will calculate the total cost. This total is then checked against the total on the recap sheet to make sure they both match. This serves as an additional guard against transposition errors of numbers. The transferred numbers on the recap sheets are circled, and the total is also circled when it has been verified that the total of the transferred numbers is correct.

The third section of the bid summary serves to add necessary adjustments to the direct labor. Labor burden is a cost that is in addition to the base wage rate being earned by employees and was discussed in Chapters 10 and 18. Contractors have different ways of determining labor burden, and it is the estimator's responsibility to know the basis of his or her company's calculations. A common way is to cost all labor within the estimate, both direct and jobsite indirect, using unburdened rates, that is, those without fringes and payroll taxes. Labor burden can range from 30% for merit shop craftsmen to as high as 75% for certain union trades, such as ironworkers. Some companies calculate a blended percentage markup to be used for all work, which is based on an average crew mix used throughout the company. This percentage is used for the jobsite administration labor as well as for the trades, and the entire labor burden is considered a material item.

Understanding how labor burden is calculated is crucial for the estimator in that it affects what wage rates are used throughout the entire estimate. Our case study GC, City Construction Company, has chosen the method of treating fringes and payroll taxes as a material item; they determined a company-wide average that is 55% of direct labor and 35% of indirect labor and extended these on the summary sheet. Their calculation of labor burden is:

Labor burden on direct labor: $3,093,199 × 55% = $1,701,259
Labor burden on indirect labor: $2,844,140 × 35% = $995,449

Other lines in the labor section apply to special situations. If a jobsite is at a remote location, travel and subsistence costs may be incurred; then the estimator would need to calculate these costs for any affected direct labor. If the project is in another city where labor is readily available, there are no additional costs. Wage rates for some labor categories may increase during construction of the project, often after the first of June each year. The estimator needs to determine the increase and add it on the appropriate line if necessary. The project summary schedule should be reviewed to determine how much of the work will be done prior to the increase and how much will be done after award. Figure 19.2 shows the completed pre-bid day bid summary with estimates forwarded for jobsite general conditions and direct work. The labor burden on direct and indirect labor has also been calculated and input, but no adjustments have been allowed for labor increases or travel expenses because our case study project is not affected. A live blank version of this estimate template is included on the book's eResource.

City Construction Company
Pre-bid Day Summary Estimate

Project:	Dunn Lumber		Bid Date & Time: June 22, 2021, 2:00 p.m.		
Owner:	Dunn Lumber Enterprises		Estimator:	Paul Jacobs	Est #1
			Square Foot:	265,000	

Code	Page	Description	Man-hours	Labor	Material	Total
		Administrative (See attached detailed estimate)		$1,722,700	$259,700	$1,982,400
		Indirect Equipment		$613,940	$882,060	$1,496,000
		Temporary Construction		$35,300	$786,800	$822,100
		General Conditions		$472,200	$585,250	$1,057,450
		Total Jobsite General Conditions	11.6%	$2,844,140	$2,513,810	$5,357,950
		Demolition/Support	200	$78,000	$15,000	$93,000
		CIP Concrete (See summary recap	56,281	$2,216,687	$2,803,082	$5,019,769
		Structural & Misc. Steel	7,531	$322,852	$2,264,660	$2,587,512
		Rough Carpentry/Backing	436	$17,876	$3,575	$21,451
		Finish Carpentry	1,222	$50,102	$25,545	$75,647
		Doors, Frames, and Hardware	2,297	$135,373	w/subs	$135,373
		Punch Windows	1,200	$49,200	w/subs	$49,200
		Specialties	3,232	$132,512	$158,880	$291,392
		Sitework/Support	2,323	$90,597	$22,242	$112,839
		Subtotal	74,722	$5,937,339	$7,806,794	$13,744,133
		Direct Craft Labor Burden @ 55%		$3,093,199	$1,701,259	$1,701,259
		General Conditions Labor Burden @ 35%		$2,844,140	$995,449	$995,449
		Trade Travel & Subsistance				$0
		Labor Rate Increase				included
		Subtotal: Checks' Yes v/		$5,937,339	$10,503,502	$16,440,841

Subcontracts (See attached list)	$26,981,408
Subcontractor Bonds (Allow)	$10,000
Adjustments	$0

Wage Check: $/Hour
   $41.40 Checks ~ Carp's wages
SF Check: $/SF
   $174 Low, but garage and shell spaces reduce $/SF

Subtotal	$43,432,249
Fee @ 4.50%	$1,954,451
Subtotal	$45,386,701
Liability Insurance @ 0.95%	$431,174
Business Taxes @ 0.72%	$326,784
Contingency @ 0%	$0
Subtotal	$46,144,659
Adjustments	$0
Performance & Payment Bond	on bid form
Total bid	**$46,144,659**

Figure 19.2 Pre–bid day summary estimate

Summary estimate

An important part of estimating is to compare the pre–bid day summary estimate with the rough order of magnitude (ROM) estimate that was developed when the project documents first arrived in the contractor's office. If there is a significant difference, the estimator needs to compare the components of the ROM against like components of this summary estimate to determine where differences exist. Totals that are within a reasonable variance (±10%) validate that the ROM and detailed estimate match. When creating a pre-bid day summary estimate, the estimator should prominently mark the top of the summary as "Pre-bid". This is extremely important so that it does not get used for the final bid. Figure 19.2 shows a summary estimate marked as a "Pre-bid" estimate.

The summary estimate is created from the information that has been developed to this point. The self-performed work and the jobsite general conditions already have been transferred from their corresponding estimate pages, and labor burden calculations and adjustments have been made. These parts of the estimate will remain the same on subsequent bid day summary estimates. Competitive subcontractor and major supplier pricing usually have not been received when the summary estimate was generated. The estimator should use the total from the subcontractor list on the OM plug estimates developed in Chapters 14–16 (see Figure 16.3). The total from that list is transferred to the appropriate line in the fourth section of the summary sheet, as shown in Figure 19.2. This completes all the self-performed work, job site overhead, and subcontracts, and the estimate is now ready for the pre-bid day markups.

Two other lines in this section are not used for this summary page but may be required in the final bid. On bid day, a very low bid may be received from a subcontractor with whom the general contractor has not worked. At first, there may be a tendency to disregard this bid as being unrealistically low, but it must be considered that the GC's competitors have also received this bid and may use it. The greatest risk to the contractor is that the bid was made in error, and the subcontractor may either decline to do the work or may suffer bankruptcy while work is in progress. The GC may decide to require the subcontractor to furnish performance and payment bonds as one potential means of financial protection. An anticipated cost of these bonds would then be entered on the subcontractor bond line. The subcontracts section is subtotaled and, as before, includes the subtotal of the previous section.

Preparing the estimate for bid day

As the time for tendering the bid approaches, things can get hectic in the bid room. This makes the process especially vulnerable to errors. Much can be done ahead of time by setting up a universal bid summary template and modifying it for the particular project being bid. In most instances, modification will not be necessary, and the first run bid estimate can be used to check all calculations.

All calculations on the summary sheet are straightforward, and the formulas should be entered in the appropriate cells. All lines should sum across to the total column, and columns should sum down. In addition, the sum of the labor and material column totals should equal the sum of the total column. A physical check mark or 'OK' is a responsible 'checks and balances' effort from the estimator. Although the computer is an essential estimating tool, checking these totals with another calculator is a worthwhile estimating quality control check. The final preparation of the estimate summary page, including anticipated costs for all direct work, subcontractor plugs, general conditions, and markups, will lead into an efficient bid day process discussed in Chapter 20.

Pre-bid day setup

Setting up the bid room

Most general contractors have a designated room that is used on bid day specifically for processing subcontractor bids, determining the final price, making last minute adjustments and forwarding information to a person who will deliver the bid to the designated location. This is commonly called the *bid room*. It should be organized so that it will be functional for the work that needs to be done on bid day. The bulk of the setup is done the evening before bid day.

First, the lead estimator will make sure that all proper documents are in the bid room. This includes the complete set of drawings that were used to create the estimate and the owner's project manual, including the specifications and all addenda. Many times, subcontractors or major suppliers will call in with last minute questions, and someone on the bid team must be able to provide answers immediately. A complete set of bid documents in the room will reduce the time needed to research an answer and respond. Also, some subcontractor bids may list exclusions, and the bid team will need to review the documents to evaluate their impact.

A properly labeled bid package containing a signed original bid form and all documents to be tendered should be prepared. The documents are put into an envelope that is left unsealed. The bid form should be filled in completely with all requested information except the final price(s). Failure to do this can be a basis for disqualification. A copy of the signed bid form will be filled in with the final price on bid day, then copies of all submitted documents attached to it. This then becomes the contractor's record copy of the bid.

A very important item on the bid form is acknowledgement of all addenda issued during the bidding process. It is the GC's responsibility to know how many addenda have been issued and to make sure all information has been incorporated into the bid. On bid day, the bid takers must verify that subcontractors and major suppliers have seen all addenda, or if they have not, that they are willing to do the work for their bid prices as submitted. Failure to acknowledge all addenda – or any other inconsistency in the bid documents – may disqualify the contractor, especially for public projects.

Other documents that need to be in the bid room are all the bids and bid forms from the subcontractors that have been received prior to bid day. Many subcontractors will submit a form to the GC listing conditions and qualifications of their bid and leave blank the cost amount, which is provided on bid day. The estimator must be familiar with each of these bid forms to ensure that they conform to the bid documents. If they do not, a choice must be made whether to disqualify subcontractors or suppliers or to get them to bid according to the documents.

A relatively large stack of bid proposal forms should be prepared for recording the subcontractor telephone bids. See Figure 20.1 in the next chapter; a live version is also included on the book's eResource. Most of the bid proposal forms will be given to the bid takers on bid day, and some will remain in the bid room for early receipt of bids. It is a good idea to fill in the title block of the bid form with the project name so the bid taker can easily recognize which forms are to be used for your project.

Written instructions should be prepared for the bid takers. They should indicate the number of addenda issued and emphasize the need to ensure the subcontractor's bid is correct and that it fully conforms to the drawings and specifications. In some cases, alternate pricing or certain price breakdowns will be asked for. These instructions should be brief but complete. Many subcontractor bids may be received via electronic mail on bid day. These bids must be managed like those received by telephone. One or more computers may be needed to receive vendor bids by e-mail. The estimator should assign the receipt of e-mail bids to certain bid takers.

It is common that the final bid summary is completed on a computer. Two computers are recommended, so that if one crashes the other is ready to take its place. The previously prepared summary estimate with all the general conditions, self-performed work, and related information should be loaded onto both computers. All markup percentages should also be entered. Operations on both bid summaries should be thoroughly checked using fictitious numbers. Most estimate summary forms began as Excel spreadsheets. It is easy to move rows and columns, and occasionally formulas get crossed. Once it is deemed that they are operating correctly, the words "Final Bid" should be typed at the top.

Other computers may be used for subcontractor bid evaluation. The proper forms should be loaded and their operations checked. All forms on the computers should be properly titled. In addition to computers, one or two adding machines, preferably with tapes, and a hand calculator should be available in case they are needed for evaluation purposes. This old-school system still has a place in construction. The work pace will become more intense as the time for bid submission approaches, and the potential for errors to occur increases. At some point, all access into and out of the room should be limited, so the bid room must be supplied with everything needed to complete the work.

Subcontractor bid evaluation spreadsheets should be made up for use in the bid room. All OM subcontract estimates are entered onto the worksheet and noted as a 'plug'. This signifies that the number is a GC-generated OM estimate that should only be used if no other bids are received in a specification section. Subcontractor quotes received early may also be posted to the spreadsheet at this time. The complete detailed estimate should be readied for the bid room. There should be an OM estimate for every section of the subcontractor or for major supply items on the spreadsheet. Some subcontract prices may include work that is in the direct work estimate, and the bid team must be able to evaluate quickly how to handle this situation. The total from the first run summary estimate should be on hand for comparing to the early runs of the final bid that will be done on bid day.

Bid forms

The owner's bid proposal form received with the initial request for quotation is at the heart of the bid tendering process. This is the first page that the project owner sees upon reviewing the bids. In addition to the price, the owner will look for completeness and professionalism by the bidder. Any information that can be filled in ahead of time should be done. It is crucial that the bid form be completed in its entirety and done neatly so that it is easy to read. It is especially important that the final price be neat and readable. Many project owners tend to customize their contracts, and their bid forms may require a variety of information. A typical bid form may contain any or all of the following:

- Project name,
- Spaces for the total bid amount,
- Signature line,

- Spaces to acknowledge addenda,
- Cost of a performance and payment bond add alternate,
- Spaces for any desired breakdown costs of the bid,
- Spaces for any desired alternate prices and/or unit prices,
- Proposed change order markups by the GC and subcontractors,
- Treatment of sales tax,
- Time of completion or duration of the project, and others.

The signature page should have all of the information typed in ready for the officer-in-charge (OIC) to sign. It is a good idea to have the OIC sign the bid form the day before bid day. Sometimes he or she will be called out of the office on an urgent matter and may not be available to sign the document until after the bid runner has left for the tendering site.

Some project owners are very specific about the envelope in which the bid is to be tendered. It may need to be a certain size and have a very specific label. On occasion, an owner will include the envelope label with the bid documents. All documents requested must be in the bid envelope. These might include a project summary schedule, bid bond, and company financial statement, and on occasion, a request will be made for value engineering (Chapter 5) suggestions.

A bid bond is a type of bond that assures the owner that if the low bidder withdraws a bid after bid opening, the cost difference between the low bid and that of the next lowest bidder will be paid by the bonding company. This difference is usually limited to 5% of the total bid amount. The bid bond is often therefore known as the 5% bid bond. The bonding company usually furnishes it at no cost. If the GC cannot furnish a bid bond, they may be required to provide other bid security, such as a certified check or cashier's check, with the bid. It is important that either the bond certificate or check, if required, be in the bid envelope at the beginning of bid day. Bid security is required for most public projects but rarely required by private owners.

Some project owners, in lieu of a bid proposal form, include a contract within the request for quotation that includes instructions for the general contractor to fill in the price(s) and sign it. If it is accepted, the owner executes the contract and returns a copy to the bidder. This style of bidding usually limits further negotiation, and the contractor must ensure that the bid is complete, contractual terms are acceptable, and the work can be accomplished successfully and profitably. In some cases, private owners will interview the contractor after the bid but prior to executing the contract to make sure all work has been included and that they feel comfortable with the relationship.

The bid day team

An important part of the bid day organization is the makeup of the bid day team. Each person or position has specific responsibilities, and when the team is well coordinated, even the most difficult bids are easier to complete. Some positions require more than one person, and multiple positions may be handled by the same person on smaller projects. The quantity of team members that are needed will depend on the size and complexity of the project. Personnel requirements and duties of each position are as follows:

- Officer-in-charge: An executive officer of the company who is a financial decision-maker. He or she will make the final review of the estimate, oversee the bid day activities, and make any last-minute adjustments to the final bid, including all markups and the fee percentage.

- Project manager: The person who will eventually manage the project if it is won. The PM may or may not have been involved in the estimating process. He or she reviews the estimate and the project with the estimator and monitors the subcontractor bidding process. Specific areas of focus include construction schedule, jobsite general conditions estimate, and best-value subcontractor selections.
- Chief estimator: The person who is ultimately responsible for the quality of the company's estimating procedures. He or she may or may not take an active role on bid day but is always available for problem solving and assessing special conditions.
- Estimator: The person who is responsible for preparing the estimate. He or she reviews the estimate and project with the PM and the OIC and manages the subcontractor bid receiving and evaluation process. The estimator determines the low bidder for each subcontractor and supplier category and establishes the cutoff point for posting bids. He or she reviews the selection of low bidders with the chief estimator or PM prior to final posting. The estimator may also be the project manager who will be assigned to the project if the contractor is successful.
- Bid posters: One or more individuals who receive the subcontractor bid information from the bid takers and post it on the bid evaluation work sheets(s). These positions are usually accomplished by the aforementioned team members.
- Bid takers: Personnel who receive subcontractor or major supplier telephone bids and record them on the bid proposal forms. They immediately forward the information to the bid room. These are often project engineers brought in from other jobsites to assist on bid day.
- Bid runner: The person who will deliver the final bid. He or she will take the bid documents in their envelope to the location for tendering the bid, coordinate the bid room time with the owner's clock, fill in the final price, seal the bid and deliver it. If the owner specifies a public bid opening, the bid runner usually stays and records the results.

Only critical personnel should be in the bid room as the pace of the work becomes more intense. As the time for bid submission approaches, subcontractor bids will be received at an increasing pace, and the last few minutes can be extremely hectic. The personnel within the bid room need to be able to work without distraction. At a certain point, the door to the bid room is closed, and all others, such as the bid takers and clerical personnel, should not be allowed in. Bids should be received through a slot in a door or window. This procedure allows the estimator, project manager, and OIC to maintain control and avoid making a catastrophic last-minute error.

Subcontractor bid evaluation procedures

The primary work of bid day is receiving and evaluating bids from subcontractors and major suppliers. Bids are posted as they are received, and the volume dictates the need for good procedures so the lowest bids and combinations can be determined quickly. As the bid time approaches, the frequency of bids increases, and fast decisions must be made. Certain bid posting formats aid in carrying out the evaluation process so that the best bids can be determined during this period. Two methods have proven reliable for accomplishing this. One method is to use spreadsheets for positing and evaluating bids. The second

Subcontractor bid posting sheet

Bin 10	Roofing and Flashing 075300		
Sub	Base Bid	Insulation	Roof Hatch
CCC Plug	$546,000	w/insulation	w/specialties
Sub 1			
Sub 2			
Sub 3			

Bin 11	Fireproofing	
Sub	078100	Comments
CCC Plug	$440,640	Verify cleanup is included
Sub 1		
Sub 2		
Sub 3		

Bin 12	Spec Section 079000		
Sub	Caulking	Firestop	Combination
CCC Plug	$25,000	$12,000	$37,000
Sub 1			
Sub 2			
Sub 3			

Figure 19.3 Subcontractor pre-bid day posting sheet

method is to use an individual bid analysis sheet also known as a *bid tab* for each work item or specification.

Spreadsheets are usually used for manual operations where personnel are posting bids by hand. The spreadsheet is made using one or more large accounting sheets. Boxes that represent each work specification section are laid out on the sheet and are sized to include the pertinent information expected to be received. The boxes are also known as *bins*. Figure 19.3 is a partial spreadsheet prepared pre-bid day by City Construction Company ready for posting and evaluating bids for the case study project. Two optional formats are presented in Chapter 20.

The boxes laid out on the spreadsheet may not be of equal size, because the amount of information will vary depending on the specification section. Also, more bids will be received for some work sections than for others. For example, more bids can be expected for the drywall than for skylights, and drywall contractors are more likely to submit combination bids or to bid on more than one specification section. Thus, more columns will be needed in the drywall box than for the skylights. Each box is titled with the specification section number and work description. A sequential bin number is written into the top left corner. Boxes are sized to show the name of the bidders, their price(s), and any short pertinent notes. No other information is necessary on the spreadsheet. It is preferable for the boxes to be laid out in the general order of the specifications, starting from the upper left corner of the sheet, with each succeeding work section below the last one to the bottom of the page. This is the arrangement shown in partial Figure 19.3.

It becomes easier to find particular work sections this way than if they were placed randomly. Modest-size projects will require only one spreadsheet, while more complex ones may need two or three. When multiple spreadsheets are required, it is a good idea to have a separate bid poster for each one.

The advantage of the spreadsheet is that it is easy to review many subcontract categories at same time. The team can also diagnose key bins that contain combination bids. Empty space or an extra spreadsheet should be available for quickly calculating the best bid combination. Typical combination bids on commercial projects include site work (excavation, utilities, pavement), finishes (drywall, paint, insulation, ceiling tile), and mechanical (HVAC, plumbing, fire protection). The bid evaluation spreadsheet is set up the day before bid day. Once prepared, all OM estimates are entered into their respective boxes, and the vendor name *plug* is used. This tells the bid team that if no bids are received in a particular category, then the OM number may be used so a bid total can be calculated. Utilizing in-house GC plugs on bid day is also a risk factor for the contractor. Any subcontractor bids that were received early are also entered into their respective boxes.

Some bid teams prefer to use an individual sheet for each work section rather than one large spreadsheet, and they are worked similarly. Individual bid evaluation sheets can be worked manually or electronically. Evaluation sheets for each specification section are made up and titled with the section number and description, and a bin number assigned to them as well. As on the spreadsheet, OM numbers are entered as plugs, and any pre-received bids are also written in. The sheets are then put into the bid room ready for bid day. The estimator needs to decide which method to use and define the procedures accordingly. It must also be decided how many bid posters are needed and how many sheets each one can manage. It is important that these sheets are readily available for review by the estimator and others, who will constantly be monitoring the process, and safeguards are in place so that they will not be misplaced. The evaluation sheets must not be removed from the bid room during bid day.

Bins also correspond to physical places where all posted subcontractor and supplier bid forms are kept during bid day. They may be anything from accordion files to built-in cubbyholes. They are numbered sequentially and correspond to the bin numbers on the spreadsheets. As the bids are posted and reviewed by the estimator, they are then put into the bins, where they can be quickly accessed during the bidding process. When reviewing the bid evaluation sheets, the estimator may need more information, especially regarding combination bids or exclusions. He or she can pull the bid forms from the bins, find the ones for specific vendors, and review their proposals quickly. This system helps the bid team to make informed decisions very fast. It is important that the estimator sets these systems and procedures in place the day before bid day.

Summary

An estimate is near completion when a final check has been made of the work and cost figures have been entered onto the estimate summary sheet in preparation for bid day. A thorough document review is performed to ensure that all components of self-performed work have been properly quantified and priced. Estimate pages are scanned to make sure all usable totals are double underlined, signifying that they have been checked, and all of these totals are circled, indicating that they have been transferred to the next summary level.

A key point in this process is validating the results. This is done by making a first run summary estimate and comparing the results against the ROM estimate developed at the beginning of the process. The pre-bid day summary is prepared prior to the receipt of market subcontractor bids and utilizes the total of the OM plug estimates developed in Chapter 16. Anticipated markup percentages are entered, which results in the total estimate. This summary is validated if it compares within a reasonable range to the ROM estimate. It is very important that the summary estimate be prominently marked as "Pre-bid" and that it is not confused with subsequent totals before the final summary page is produced in the bid room on bid day.

During the development of the summary estimate, all calculations on the summary sheet are checked for proper ranges and operations. Cells containing formulas are then locked to avoid inadvertent entry of other numbers on bid day. A final summary sheet containing above-the-line totals for general conditions, direct work, and labor burden is then made up in preparation for use on bid day. The subcontractor OMs and markup extensions will not be carried forward on the final bid day estimate summary sheet in preparation for bid day activities.

A key to tendering a complete and accurate bid is organizing the bid day process. Much of this work is done on the evening before bid day. The estimator is responsible for ensuring that the proper documents are in the bid room, the computers are set up and spreadsheets are operating properly, and other materials are ready for use. Spreadsheets for the subcontract bid evaluation and the bid proposal forms are prepared and sufficient copies made for bid day. Instructions for the bid takers and the bid runner are written, and personnel are recruited for both jobs. The bid package is made up and readied for delivery by the bid runner. The original bid form is signed by the OIC, and copies are made for the bid package and for use in the bid room. A properly labeled envelope is prepared, and all documents to be submitted are readied, with the bid form original on top. The estimator verifies that all personnel needed to complete and tender the bid will be available on bid day. In addition, office personnel are alerted so they know how to route telephone calls and keep visitors away from the bid room. A well-organized bid day process will keep last-minute stress to a minimum and ensure that a complete and accurate bid is tendered. In our next chapter, we discuss the exciting bid day process, including error-prevention techniques.

Review questions

1   The estimate summary form is organized into five general sections. What does each section represent?
2   How are the general conditions and direct work subtotals checked?
3   What percentage range is used as a guide when comparing the total of the general conditions estimate against the total bid amount?
4   What is labor burden?
5   When would an amount be entered for a subcontractor bond?
6   What is a pre-bid day estimate, and how is it used?
7   What is the function of the pre-bid day estimate on bid day?
8   List five documents that should be put into the bid room in preparation for bid day.
9   What types of documents typically go into the bid package to be delivered to the owner?

Exercises

1   Determine the percentages of the pre-bid day total that are attributable to the direct work versus that from subcontractors from our case study. Use the direct work after the addition of the labor burden. How might these percentages be useful as a guide on similar estimates?

2   It is customary for the OIC to take the pre-bid day estimate home the evening before bid day. What type of information would he or she be checking for? What other documents related to the estimate might they also be interested in reviewing?

3   How might a GC be 'sure' they will receive subcontractor bids on bid day?

4   Prepare a wage check from the Vehicle project's summary estimate. Is it within a reasonable range? What labor craft does it most closely match?

20 Bid day process

Introduction

After the first run summary estimate has been prepared, the remaining work toward tendering a bid to the project owner is represented by steps 9 and 10 of the lump sum estimating process shown in Figure 1.1. Step 9 is receipt, evaluation, and totaling of the competitive quotations or bids from subcontractors and major suppliers. Step 10 is entering the subcontractor total onto the bid summary, making any necessary adjustments to markups, and then completing and tendering the bid form. All this work is done on bid day. Bid day can be chaotic, especially as the time for bid submission approaches. A well-organized bid room will retain order and reduce a tendency to panic in the last few minutes. Setting up the bid room a day ahead of time helps to impart order to the process such that everyone involved can perform his or her assigned duties efficiently, as was discussed in the last chapter.

Most project owners prescribe a specific bid tendering process. Many contracts have been lost because the general contractor (GC) did not explicitly follow procedures. This may be anything from submitting the bid late to improperly filling out the proposal form or not supplying other requested documents with the bid. Some owners even define the envelope and label for submitting the bid.

Requests for quotation may require submission of other information after the initial bid has been tendered. Frequently, the owner will ask for a list of major subcontractors and the value of their work, either with the bid or within a short time after the initial bid was tendered. Public owners may require a two-envelope bid in which the GC submits first a statement of compliance with minority subcontracting requirements and, if acceptable, can then submit a bid price. The important thing is that regardless of what an owner's requirements are, a well-organized process on bid day will ensure that they are met.

Preceding bid day, all the direct work estimating (Chapters 6–13) is required to have been completed. Order of magnitude estimates have been created for subcontract work (Chapters 14–16), and a project summary schedule and general conditions estimate are also prepared (Chapter 17). The first run summary estimate has been completed with anticipated markups and reconciled with the rough order of magnitude estimate, which was developed in Chapter 2. The estimator has prepared the bid room and assembled the owner's bid package with all required documentation and a signed copy of the bid form. The bid room is ready to receive and evaluate subcontractor bids, fill in the subcontractor bid totals on the bid summary, and complete and deliver the owner's bid form to the tendering site. All of this was reflected in Figure 19.1.

Bid day activities occur in three places. The bid takers, who are people throughout the contractor's office, will be receiving and recording subcontractor and supplier telephone prices on bid proposal forms. The bid room team posts the bids on the evaluation sheets, determines the low prices, completes the bid summary, and forwards the total(s) to the bid runner. The bid runner, who is located near the tendering site, receives the final numbers from the bid room and completes the bid form. He or she then delivers the bid package to the project owner's specified location and learns the results, if it is a public bid opening.

The bid team leader, either the estimator or the project manager (PM), must control the tempo of bid day. As the tendering deadline approaches, the atmosphere becomes stressful, and emotions can destroy the entire process and jeopardize the final bid. A smooth bid day operation minimizes emotional effects and results in the best possible bid being tendered on time. At the beginning of bid day, the team captain should make sure that the organization is in place and that all team members understand their duties. Doing this will help make the work of tendering the bid go smoothly and provide the bid team some time for final strategizing. The team leader should verify that the officer-in-charge (OIC) will be present in the bid room during the hour prior to the time the bid must be submitted.

A few things remain to be done after the bid has been tendered, regardless of whether the contract was won. These are issues involving the bid opening, vendor queries, preparations for post-bid interviews, and record keeping. In addition, whether or not the bid was won, an effort should be made to glean information from the estimate that will be useful in preparing future estimates. Another important issue is post-bid negotiations with private owners. Understanding the project owner's position helps the contractor develop a plan for the post-bid interview. There are many ethical estimating issues raised before, during, and after the bid process. These topics are the focus of our next chapter.

Subcontractor and supplier bids

As the bidding deadline approaches, some subcontractors and suppliers will quickly give their prices and minimal information so they can proceed with calling other bidding general contractors. The information can be too incomplete for a comprehensive evaluation. When this happens, someone in the bid room may be required to call the subcontractor or supplier back and get the remaining information – a time-consuming and sometimes difficult task. If the bid room activity level is high, there may not be time to do this, and a quick decision must be made whether to use the subcontractor or supplier's bid.

When receiving a call, the bid taker should initially indicate that he or she will ask the questions, thus pacing the call. The bid taker must be aware that subcontractors and suppliers have other calls to make and are motivated to get off the call quickly. Most subcontractors and suppliers know what information GCs need and will provide it. Conversely, some subcontractors may want to be over-detailed, which takes too much of the bid taker's time and restrains him or her from receiving other bids.

Subcontractors and suppliers may elect to submit their bids either by mail, e-mail, or hand-delivered hard copy. When using mail, a bid form or scope letter is received that lists the terms and conditions of the bid, including exclusions and inclusions, and has blanks for pricing. The estimator should review the form to ensure that it conforms to the bidding requirements and to the GC's subcontracting procedures. If there are discrepancies, the estimator or PM must attempt to resolve them before bid day. If a subcontractor is firm on his or her conditions, a decision must be made as to whether to use that particular bid. The subcontractor or supplier then calls in prices on bid day, and the bid taker either enters the price on a copy of the submitted form or fills in a regular bid proposal form and attaches it to the subcontractor or supplier's form.

Subcontractor and supplier bids are recorded on bid proposal forms like the one shown in Figure 20.1. The bid team leader must stress the importance of filling out the form

	City Construction Company		Date:	6/24/2021
	Subcontractor Bid Proposal		Time:	1:35 PM

Bidding Firm:	Klose Door Company				Job:	Dunn Lumber
Phone:	360.441.2020		Union:	Yes	Merit Shop:	NA
City:	Everett, WA		WBE:	NA	MBE:	NA
Estimator:	Jim Stevenson	Quoted by:	Jim Stevenson		Bondability:	$15 mil
Prices Valid Until:	30 Days				Bond Premium Included:	NA

Spec Section	Bid Item/Inclusions	Amount
81100	Supply hollow metal door frames	$91,000
81100	Supply hollow metal door leafs	$62,290
	Subtotal	$153,290
Alternate 2, see below	Supply wood doors	$80,000
	Subtotal	$233,290
87100	Supply door hardware	$244,440

Addenda Nos.:	1 Thru:	3	**Basic Bid:**	$477,730

Per Plans and Specs?:   Yes   Supply Only?: Yes   Installed: _____   Erected Only?: _____

Furnish Only:: _____   FOB: Jobsite: Yes   Trucks: _____   Other: _____

Weight: _____   Delivery Time: _____

Exclusions and Clarifications:	Alternates or Unit Prices **	$ Amount
Wood doors	1 Install D/F/HW	$500/leaf
Access doors	2 Supply wood doors	$80,000
Door relight kits	3 Supply relight kits	$5,000
Installation	4 Supply access doors (not MEP)	$3,000
Grout HM door frames		
Electric door strikes or card readers	** City must accept 100% base	
	bid before any alternates are	
	considered	
	Received By: Paul J. Bin:	13

Figure 20.1 Subcontractor phone bid form

completely and correctly. The following describes the individual items on the bid proposal form and their importance:

- Firm: The name of the subcontractor or supplier should be filled in completely. Sometimes there are different companies with similar names, and filling in a partial name may not define adequately which company submitted the bid.
- Job: Identifies the project for which the bid was received. This is especially important on days when more than one job is being bid by a GC.
- Vendor's telephone number and quoted by: This provides a contact if the bid team has any questions.
- Bid item/inclusions: It is very important to list all of the specification sections and inclusions. This will also be a factor in writing the subcontract or purchase order. The bid taker should also repeat the price(s) back to the bidder to verify that it has been recorded correctly.
- Addenda numbers: This verifies that the specialty contractor has submitted a complete bid. If this is not done, it provides the vendor with an opportunity for negotiating for more money if he or she is the low bidder.

- Per plans and specifications: A "yes" answer by vendors along with verification of all addenda reduces their opportunity to try to raise their price when executing a contract. So long as the vendors have had the opportunity to review the project owner's bid package, they are responsible for all items of work affecting them.
- Exclusions and clarifications: These should be brief, and while many items may be obvious, they should always be written down.
- Alternate or unit prices: Not all bidders will be required to submit alternates or unit prices complying with the GC's bid to the project owner, but these are helpful for buyout if successful.
- Received by: The bid taker on the GC's team who received the bid.

All construction materials should be quoted free on board or freight on board (FOB) jobsite, which means that the cost includes the shipping. If the general contractor is requested to purchase owner's equipment as part of the project, the suppliers of the equipment rarely include shipping. The GC should accept prices only with a freight allowance added; otherwise, separate shipping bids will have to be solicited, which sometimes can be difficult to obtain.

After receipt of the call and completion of the form, the bid taker forwards subcontractor and supplier bids to the bid room. Early on bid day when the activity level is low, bid takers can deliver the bid proposal forms when convenient. As the deadline approaches, the frequency of calls may make it more difficult to send the bids to the bid room. It is always important for the bid takers to be available full time to receive bids during the last two hours before the deadline, as this is when the activity is most intense. The bid team leader should monitor the bidding activity and, as it gets more intense, should designate a separate person to pick up bid proposal forms from the bid takers. This allows the bid takers to receive bids uninterrupted and ensures that all prices are delivered to the bid room expeditiously.

Posting subcontractor and supplier bids

The bid posters receive the bids in the bid room from the bid takers and post the relevant information on the bid evaluation sheet. Notes should be just a few words such as "excludes doors" or "includes section 099000". More detail tends to clutter the evaluation spreadsheets. If the bid team needs more information, the bid forms are in the bins, available for review. Figure 20.2 shows three bids received for masonry specification section 042200 that appear close to each other. Single bid tab number 4, or bin 4, was utilized in this case, as the GC did not anticipate any other complicated bids with this scope. Although Adobe had the low initial quotation, their bid was incomplete, and City Construction Company went with S&J Masonry on bid day. They specialize in landscape pavers, which is a much larger scope, and were successful in that area in this case. Adobe's bid will not be discarded but may be used later for buyout opportunities.

Some subcontractors or suppliers may submit pricing for more than one specification section of work. Their price may be a single amount that includes several sections or may list the individual sections and a cost for each. A typical example might be wall insulation, drywall, and acoustical ceilings. Some bidders will supply an individual price for only one of these sections, while a drywall contractor may include two or three all in a single bid. The team must determine whether the combination bid is lower than the sum of other

City Construction Company
Subcontractor and Supplier Bid Evaluation Sheet

Project: Dunn Lumber   Estimator: Paul Jacobs
Location: Seattle, Washington   Estimate No.: 1
Architect/Engineer: Flad   Date: June 24, 2021
Bin 4: Masonry, Spec section 042200

Subcontractor:	City Plug	S&J	Burns	Adobe
Base Bid	$32,200	$36,000	$40,000	$35,000
Include rebar installation?	Assume	Incl	Incl	Excl
Include rebar supply?	w/direct $	Excl	Excl	Excl
Scaffold	Assume	Incl	Incl	Incl
Addenda	3	3	3	2
Graffiti protection	Assume	Incl	Incl	Excl
Landscape pavers	w/landscape	Excl	Excl	$250,000
Adjustments:	0	0	0	NA
Adjusted bid:	$32,200	$36,000	$40,000	DQ
Low Bid		**$36,000**		

Figure 20.2 Individual subcontractor and supplier bid evaluation sheet

individual bids. These can be easily seen, evaluated, and noted on the bid evaluation spreadsheet. Figure 20.3 illustrates how a combination bid may look. Other scopes that may be bid separately or in combination would be several areas of site work or mechanical combined with plumbing and fire protection.

There are some cases where three or more specification sections are involved. Some subcontractors or suppliers will bid on a single section, while others may bid on two or more of them. Prices may be individual or combined. When this occurs, the estimator usually evaluates them by making separate boxes on the spreadsheet of the various combinations and listing the prices in them. Estimators 'know' what scopes may be bid as combinations from experience or from talking with potential bidders prior to bid day. Figure 20.3 shows how combination bids are entered onto the individual bid evaluation sheet. In this case, Dunham Drywall bid a combination package that was slightly less than if the GC had utilized the low bidder in each of these four categories. This detailed analysis facilitates the bid day process, but if the GC becomes low bidder, there are many different opportunities for further analysis, which may result in a lower and more complete combination. Bid shopping, as discussed in the next chapter, is unethical and may be illegal in the case of publicly financed construction projects. But buyout, or buying-smart, is not unethical, especially in the private construction industry.

If multiple computers are being used for bid evaluation, the bid captain should consider putting the mechanical bids on one and the electrical bids on another. Typically, during the last minutes before the deadline, many electrical and mechanical bids are received. With a few minutes to go, the bid captain may decide to select all other bids except the mechanical and electrical and have someone monitor the other incoming

Subcontractor Bid Day Spreadsheet

Bin 7 — Insulation, Spec Section 072100

Subcontractor:	City Plug	Div 7	Dunham	Thermal Co.	McKinville
Acoustical	$100,000	$295,555	w/bin 16.2	$275,200	Excl
Thermal	$166,000	in	w/bin 16.2	in	$152,000
Roof insulation	w/roofer	excl	excl	+$125,000	Excl
Foundation insul	w/directs	excl	excl	+$45,000	Excl
Addenda	3	3	3	2	3
Per plans & specs	Assume	Yes	Yes	No response	Yes
Adjustments	0	0		Assume OK	$100,000
Adjusted bid:	$266,000	$295,555	w/bin 16.2	$275,200	$252,000

Bins 8 & 9 — Siding

Spec	Subcontractor: / Scope	City Plug	Robinson Company	General Constr.	McKinville Siding
72100	Metal Siding	$900,000	No bid	$1,000,000	$1,466,000
74619	Wood Siding	$462,500	$599,000	No bid	Included
72100	Thermal insul	w/bin 7	Excl	Excl	w/bin 7
	Addenda	3	3	3	3
	Scaffold	Assume	Included	Included	Included
	Adjustments	0	$1,000,000	$599,000	0
	Adjusted Bid:	$1,362,500	$1,599,000	$1,599,000	$1,466,000
	w/o thermal insulation				

Bin 18 — Painting, Spec Section 99113

Subcontractor:	City Plug	Division 9	Joe's Paint	Dunham
Base bid:	$40,000	$41,000	$48,000	w/bin 16
Addenda	3	3	NA	16
Per documents	Assume	Yes	Assume	Yes
Adjustments	0	0	Assume OK	0
Adjusted bid:	$40,000	$41,000	$48,000	w/bin 16.2

Note: Italics indicates an adjusted bid and not verified

Bins 16.1 & 16.2 — GWB including Exterior Metal Studs

Spec	Subcontractor: / Scope	City Plug	Dunham	Mike's GWB	Regency	Experts
54100	Exterior Studs	$906,200	$1,800,000	$1,035,555	$2,000,500	excl
92100	GWB	$545,700	$1,800,000	$622,240	included	$444,200
	Thermal insul	w/bin 7	excl	excl	excl	excl
	Acoustical insul	w/bin 7	included	excl	+$125,000	excl
	Paint	w/bin 18	included	excl	included	excl
	Addenda	3	3	2	3	1
	GWB Adjusts:	0	0	0	0	$1,035,555
	ST GWB Bid:	$1,451,900	$1,800,000	$1,657,795	$2,000,500	$1,479,755
	Insul & Paint	$306,000	included	excl	$125,000	excl
	Combo Adjusts:	0	0	$316,200	$152,000	$316,200
	Combination Bid:	$1,757,900	$1,800,000	$1,973,995	$2,277,500	$1,795,955

Figure 20.3 Combination subcontractor bid day spreadsheet

bids in case a significantly lower one is received. The bid team will then concentrate on evaluating the mechanical and electrical bids. These divisions can contain several combinations, and if subcontractors are lowering prices and providing alternate pricing in the last few minutes before the bid is due, the team needs to be able to make quick informed decisions.

Whichever bid evaluation method is used, the posting of bids is essentially the same. A designated team member receives bid proposals from a bid taker and posts the subcontractor's name, price, and any pertinent notes on the spreadsheet. Notes are kept to an absolute minimum and must be brief so as not to clutter the evaluation process. Once posted, the bid poster writes the designated bin number in the box at the lower right-hand corner of the bid proposal form and puts it into the appropriate file. The use of bins makes it easy for a team member to find and review all the bids of any specification section during the bidding process. Bins may also commonly be referred to as bid tabs. They are a tabulation of vendor bids for a given work category. Bid tabs are used by the project team after a project is won to reevaluate the bids for other combinations or situations which may result in an even lower price. This is part of buyout and will be discussed in Chapter 24.

It is important for the bid team to be aware of the pace of a typical bid day. The case study has a bid time of 2:00 p.m. Subcontractor and supplier bid activity will be relatively slow until midmorning. This time is used to post previously received bids and make sure the organizational work is complete. From midmorning to about noon, the frequency of bidding will increase but will still be at a manageable level, and a single bid poster can probably handle the work. The entire bid team, including the bid takers, typically works through lunch. From noon until about 1:30 p.m., the bid frequency will increase, and more bid posters may be necessary. Bids during this time typically will be from subcontractors and suppliers of all specification divisions, but usually there will still be only a few for the mechanical and electrical work. The bidding level will become very intense during the last half-hour, when most of the bids for mechanical and electrical are received.

Evaluating subcontractor and supplier bids

The bid team must be prepared to determine quickly which the best prices are and when to decide on a posting cutoff point on bid day. For most of the bidding, the subcontractors and suppliers submit prices for single categories of work, and determining the low price is straightforward. Besides combination bids, previously discussed, the bid team needs to be aware of the specific inclusions and exclusions of each bidder. While most of them bid in accordance with the plans and specifications and note all addenda, some will exclude pertinent items that may result in them incorrectly but apparently having the low bid. The bid team must identify these situations and decide whether to call the vendor and get them to include the item or to disqualify their bid. This then becomes an ethical issue, which we will discuss in the next chapter.

Special care must be exercised if a subcontractor bids the work that the general contractor typically performs. While both prices reflect the same scope of work, there are differences in how each treats variations. For example, a general contractor's cost for concrete reinforcing steel installation has some flexibility in it, and any minor discrepancies that are encountered are just part of the job. Usually the cost to correct them can be absorbed with minimal impact to productivity. A subcontractor, on the other hand, bids the job as shown, and any discrepancy will be an extra cost. If the delivery of any direct installed material is interrupted, the general contractor can usually shift manpower to another task or even another project temporarily. The subcontractor cannot do this and will charge extra for every interruption. The estimator should discuss the inclusions and exclusions with each subcontractor.

It is important for the bid team to keep on top of subcontractor and supplier bid evaluations so that the final minutes of the process remain under control. At about one hour before the bid is due, the team captain should determine the state of bid coverage and proceed with calculating an early bid total. The low bid in each bin is transferred to the bid day subcontractor list. General contractor-prepared plug numbers will be placed on the subcontractor list only for those sections for which no bids have been received. The subcontractor list is totaled and checked, and this amount is then transferred to the bid summary. Both pages are prominently marked "First Run Bid". This designation differs from the pre-bid day summary estimate discussed earlier. Each successive total is more accurate than the last. Just as the pre-bid day summary was more accurate than the rough order of magnitude early budget, this first run bid is more accurate than the previous day's preliminary total using all in-house plug values.

The earlier the first run bid total is generated, the more likely GC plug numbers will continue to be used. As each succeeding total is made, most of the plug numbers will be replaced by market-driven bids. The final bid run ideally should have no plug numbers in the subcontractor list, or if it does, they are for small work categories with low risk.

The first run bid is used as a baseline on bid day. First, it is compared with the previous day's summary estimate total to see if the two are reasonably close. If they are not, the bid team must quickly search for a transposition error. As the day progresses, the bid captain may make several more bid totals, numbering each one successively. They will then be compared with the previous total to see that they are still reasonably close and there are not any significant swings. Each progressive total should be printed so that key members of the bid team can review it for potential errors. As time progresses, there is less opportunity to analyze differences between in-house subcontractor plugs and market quotes. Further analysis can be done after the bid has been tendered.

At a predetermined time, usually about 15 minutes before the deadline, the bid captain declares a cutoff for receiving bids for most specification sections. Low bids in each bin are double underlined and entered onto the subcontractor list. A bid team member receiving bids for closed sections compares them to the selected low bid. Only significantly lower bids are now called to the attention of the bid captain. If a bid is received for a work section that has been closed which is lower than the one already declared, an adjustment is made on the 'cut and add' column of the subcontractor list. The bid poster determines the difference between the bid used and the new bid received. This amount is entered into the cut and add (+ or −) column on the proper line of the subcontractor list, and a revised total is calculated. Figure 20.4 shows the bid day subcontractor list for our case study project and illustrates how the cut/add column is used.

City Construction Company
Final Subcontractor Bid Day List

Project: Dunn Lumber   Estimator: Paul Jacobs Est. #:   1

Bin	Spec	Description	Bid	+ or −	Revised	Subcontractor
1	24100	Demolition	472,000			Nuprecon
2	31113	Form Elevated Slabs	0			w/Directs
	32100	Reinforcement Steel	0			w/Directs
	33000	Pump Concrete	0			w/Directs
	33500	Finish Slabs	0			w/Directs
3	39000	Shotcrete	813,300			BB Shot
4	42200	Interior Garage CMU	36,000			S&J Masonry
	51000	Misc. and Structural Steel	0			w/Directs
16.1	54100	Exterior Metal Stud Framing	w/bin 16.2			Dunham
5	64100	Casework	w/bin 17			Division Nine
6.1	71400	Waterproofing	1,200,000			Rain Gear
6.2	71800	Traffic Coating	w/bin 6.1			Rain Gear
7	72100	Insulation	w/bin 16.2			Dunham
8	74619	Metal Siding	1,466,000			McKinville
9	74623	Wood Siding and 'Rain Shield'	w/bin 8			McKinville
10	75300	Roofing and Flashing	523,000			ACB Roofers
11	78100	Fireproofing	450,000			Cementech
12	79200	Caulking and Firestopping	37,000			City Plug
13.1	81100	Buy Doors and Frames	233,290			Klose Door Co
14	83300	Garage & Retail Roll-up Doors	381,800			Stanley
15	84100	Windows and Storefront	3,627,740			Hurley Glass
13.2	87100	Buy Door Hardware	244,440			Klose Door Co
16.2	92100	GWB and Interior Framing	1,800,800			Dunham
17	96100	Tenant Area Finishes/Floors	655,000			T.I. Allowance
18	99113	Exterior Painting	w/bin 16.2			Dunham
	101400	Signage	0			w/Directs
	102800	Toilet Accessories	0			w/Directs
	104400	Fire Extinguishers	0			w/Directs
	108213	Louvers	0			w/Directs
19	111300	Loading Dock Equipment	75,000			EZ Lift
20	118123	Window Washing Equipment	200,000			Bid Allowance
21	129313	Bike Racks	88,088			Specialties, Inc.
22	142400	Elevators	1,222,950			High Rise, Inc.
23.1	210000	Fire Protection (Design-build)	1,210,000			Fire West
23.2	220000	Plumbing	1,350,000			Alabama Plumbing
23.3	230000	HVAC	2,788,000			Holaday Mechanical
24	260000	Electrical	3,380,900			Homes Electric
25	312300	Mass Excavation	1,950,000	(10,000)	1,940,000	Haus Construction
26	314000	Shoring (Design-build)	1,920,200			BDM Systems
27.1	311300	Pavement	728,600			Dean Company
28	321400	Pavers	250,000			Adobe
27.2	321713	Pavement Striping	47,400			City Plug
29	323113	Temporary Site Fence	62,400			City Plug
28	329000	Landscape and Irrigation	803,000			Adobe
30	330000	Site Utilities	319,990			Graystone
31	334119	Temporary Dewatering	188,000			Bid Allowance
		Total Subcontract Bids	**$28,524,898**	**($10,000)**	**NA**	

Figure 20.4 Final subcontractor bid day list

After reviewing the subcontract bids for the case study, the bid captain and the OIC have noticed two very low prices from subcontractors with whom the company has no prior experience. A quick discussion ensues about the validity of these bids and the risk of lack of performance, and the bid captain and the OIC decide to add an amount to bond these subcontractors. This provides some financial protection in case one or both default in the performance of their work. The gravity of this decision may cause the GC to not be the low bidder, but the bid captain and the OIC have decided to take that chance and added $9,000 to cover the cost of the bonds. This is entered in the subcontractor section of the bid summary.

During the last few minutes of subcontractor bidding, the frequency of mechanical and electrical bids typically increases. As bids are submitted, the subcontractors may ask how their prices look. Bid takers must be instructed not to answer these questions, as they may not be in the best interests of the GC. This is also a bid day ethical issue. A bid poster is assigned to each of the remaining open bins, and the bid captain monitors them closely. A few minutes before the bid must be forwarded to the bid runner, the remaining categories are closed out, the prices are entered in the subcontractor list, and a bid total is determined. The OIC then makes a decision regarding any adjustments including the final fee markup, and the final bid total is calculated. The bid captain forwards the total to the bid runner, and a final run is printed. The final bid summary for our case study project is shown in Figure 20.5.

Subcontractor bids may continue to be received after the bid room team has made their cutoff. Some bids will even be received after the owner's bid time. These can be valuable during buyout, and the bid takers should continue their diligent process of documenting their receipt. The bid captain will eventually post these to the spreadsheets and file the bids in their respective bins. The potential use of these late subcontractor bids will be discussed in the next chapter.

Delivery of the bid

About an hour before the time for bid submission, the bid captain gives final instructions to the bid runner. The project owner's bid form is reviewed so that the bid runner knows exactly what is to be filled in. Directions to the bid submission site are given, clocks are checked to ensure that the bid runner has the correct time, and communication procedures are established. The bid runner uses a cellular telephone or portable computer for receiving the final information that is to be entered onto the bid form.

The bid runner is sent to the bid submission site well in advance of the time when the bid must be submitted. An allowance for unusual traffic conditions should be made to ensure that the runner arrives on time. Once the runner arrives at the bid submission site, he or she should contact the bid captain and inform the bid room of the time required to submit the bid. The bid captain should then instruct the bid runner to call back at a designated time to receive the final bid price.

The final bid amount is to be read to the bid runner by the bid captain and written onto the original and all copies of the bid form. The number should be read back to the captain for verification. The bid package will then be placed into the prepared envelope, sealed, and delivered to the project owner at the designated bid submission site. The bid runner also will note the exact time the bid was turned in; often bids are time-stamped by the recipient. If the bids are to be opened publicly, the bid runner usually remains at the bid site and records the bids that are received. The results are then reported back to the bid captain.

		City Construction Company			
		Final Bid Day Summary Estimate			

Project:	Dunn Lumber			Bid Date & Time: June 24, 2021, 2:00 p.m.	
Owner:	Dunn Lumber Enterprises			Estimator: Paul Jacobs Est #1	
				Square Foot: 265,000	

Code	Page	Description	Man-hours	Labor	Material	Total
		Administrative (See attached detailed estimate)		$1,722,700	$259,700	$1,982,400
		Indirect Equipment		$613,940	$882,060	$1,496,000
		Temporary Construction		$35,300	$786,800	$822,100
		General Conditions		$472,200	$585,250	$1,057,450
		Total Jobsite General Conditions	11.2%	$2,844,140	$2,513,810	$5,357,950
		Demolition/Support	200	$78,000	$15,000	$93,000
		CIP Concrete (See summary recap	56,281	$2,216,687	$2,803,082	$5,019,769
		Structural & Misc. Steel	7,531	$322,852	$2,264,660	$2,587,512
		Rough Carpentry/Backing	436	$17,876	$3,575	$21,451
		Finish Carpentry	1,222	$50,102	$25,545	$75,647
		Doors, Frames, and Hardware	2,297	$135,373	w/subs	$135,373
		Punch Windows	1,200	$49,200	w/subs	$49,200
		Specialties	3,232	$132,512	$158,880	$291,392
		Sitework/Support	2,323	$90,597	$22,242	$112,839
		Subtotal	74,722	$5,937,339	$7,806,794	$13,744,133
		Direct Craft Labor Burden @ 55%		$3,093,199	$1,701,259	$1,701,259
		General Conditions Labor Burden @ 35%		$2,844,140	$995,449	$995,449
		Trade Travel & Subsistance				$0
		Labor Rate Increase				included
		Subtotal		$5,937,339	$10,503,502	$16,440,841

	Subcontracts (See attached list)	**$28,524,898**
	Subcontractor Bonds (Allow)	$9,000
Wage Check: $/Hour	Adjustments	$0
$41.40 Checks ~ Carp's wages		
SF Check: $/SF	Subtotal	$44,974,739
$180 Low, but garage and shell spaces reduce $/SF		
	Fee @ 4.50%	$2,023,863
	Subtotal	$46,998,603
	Liability Insurance @ 0.95%	$446,487
	Business Taxes @ 0.72%	$338,390
	Contingency @ 0%	$0
	Subtotal	$47,783,479
	Adjustments	-$10,000
	Performance & Payment Bond	on bid form
	Total bid	**$47,773,479**

Figure 20.5 Final bid day summary estimate

After the bid has been submitted, all of the bid room documents, including spread-sheets, interim bid totals, subcontractor bids, and other documentation is neatly gathered and organized. This completes the lump sum bidding process.

Validity of verbal prices

Verbal bids taken by telephone should constitute commitments by the bidders. The fact that these bids were in response to a request for quotation makes them a price for which the subcontractor should be willing to sign a contract. If a subcontractor backs out after a bid has been tendered, the GC who won the project is still required to perform the work for the bid price. Either the low subcontractor must do the work for the price they bid, or the GC will have to find another subcontractor who will. That may be difficult to do.

It is rare that a subcontractor backs out of a bid commitment. There are cases where a major error might have been made resulting in the bidder submitting an unusually low price. On bid day the bid team may choose to contact the subcontractor and ask for verification of his or her bid and review all inclusions and exclusions. This is also an ethical issue. The subcontractor might be told that the price looks to be out of line without indicating which way or by how much. Whether to use the low price and hope it is not in error or to query the subcontractor is a decision that must be made. If a subcontractor has made an error, a corrected price may be provided, and the bid team can hope that all other general contractors also have been notified. There is no guarantee the subcontractor will do this or the other GCs will acknowledge the change. A low-bidding subcontractor may also attempt to later raise its bid, qualify its bid, or pull its bid after a competing or unsuccessful GC has informed them how low they were. All these potential scenarios raise ethical questions.

Bid opening

The opening of the bid by a project owner may be done in one of two ways. If it is opened in the presence of the bidders and the results announced immediately, it is a public bid opening. A private bid opening is when the owner elects to open the bids without the bidders being present. Private bids are opened at the owner's convenience and may be done on bid day or later. Owners frequently state in the bid documents when they intend to enter into a contract regardless of whether or not the bid results are announced.

Public projects, such as highways or government buildings, usually are required by law to have a public bid opening. The timing of the opening is specified in the bid documents. It is also a requirement that the contract be awarded to the lowest *responsive, responsible* bidder. Responsive means that the bid conforms to ALL requirements specified in the *invitation for bid*, and responsible means that the construction company is qualified to construct the project. This may appear to be a subjective judgment and may result in a bid protest. Bid results are published in either a government document or a recognized trade journal.

Private owners have the option of selecting a public or private bid opening and usually will indicate which is to be used in the bid solicitation documents. They decide the timing of the opening and whether to make the results available to all bidders. In some cases, it is stated in the bid documents that the owner retains the right not to award to the lowest bidder. In some cases, the owner might reject all bids because they exceed the project budget. A new round of bids may be requested after some design modifications or the owner may select one of the bidders and contract for help in developing construction alternatives to try to reduce costs, similar to value engineering discussed in Chapter 5.

Post-bid day

If the general contractor is not the low bidder, an analysis should be performed to determine why the bid was too high. When doing this, it is assumed that everyone had essentially the same total for the subcontractor and supplier bids. It is risky for a subcontractor to provide a preferred quote to one GC; other contractors will find out. The self-performed work should be analyzed to see if the most efficient construction methods were planned. The fee should be examined in relationship to the difference between the GC's bid and that of the successful bidder. Many times, this will be the deciding factor. It may be that another contractor wanted to establish a relationship with this owner and used a lower fee. Another contractor may have determined that the project has a high potential for change orders and decided to bid a low fee, with the intent of making money on them. Or the low bidder may have made an error. Contractors often feel their own price was the correct one and any lower bids were in error. Whatever the reason, this is a chance for the estimator and PM to get a feel of the bidding methods of the competition (market conditions) so that appropriate adjustments can be made on the next bid effort.

Many private owners conduct a post-bid contractor interview. The two or three lowest bidders are asked to review the project with the owner and the architect. The intent is to make sure that they have included all the work that is required by the contract documents. The project owner also may want to determine the contractor with whom he or she feels most comfortable.

If the general contractor was second bidder or even a close third, the estimate should be retained until the owner executes a contract with the low bidder. On occasion, conditions may change such that the low bid is rejected, or the parties cannot agree to contract terms. The bid documents should be researched to see if there is a specified period that the bid is to remain valid. The estimate should be kept at least for this time-period, and if the contractor was second bidder, consideration might be given to keeping the files until the project is built. If the low bidder should default, the owner may want to hire the second bidder to finish it. Having the estimate available will save estimating time. Once it is deemed that a bid estimate is no longer needed, the estimator should review it for possible historical or reference use and input it into the company's database.

Summary

Bid day is the day the final bid is completed and submitted. Early in the day, the bidding activity usually is relatively quiet. As the day progresses, the activity generally increases and can be very intense just before the final bid is submitted. The bid captain must have procedures in place for controlling traffic into the bid room and making sure that each team member understands his or her duties. It is important for bid takers to understand what information is needed and to make sure that proposal forms reach the bid room in a timely manner. Bid posters are assigned to process bids for a group of specification sections. Clear and concise instructions are given to the bid runner, who will deliver the bid to the location designated by the bid instructions and remain at the site to retrieve results if it is to be a public opening.

As the bid time approaches, the bid captain determines when a first run bid total is to be made. Low subcontractor bids at the time are used on the subcontractor list, and those items for which no bids have been received retain GC plug numbers. The total is compared against the prior day's summary estimate total that was made using subcontractor order of magnitude plug estimates. The bid captain determines how many preliminary bid totals will be made and when they will be done. As the bid time approaches, the final numbers are reviewed with the OIC, who decides if any adjustment should be made to the bid. Once this is determined, the bid summary total is forwarded to the bid runner, who completes the bid form, seals the envelope, and delivers it to the owner.

There is some wrap-up work to do after the bid. Prices not posted should be reviewed and filed. Bid proposals are removed from the bins and filed in order by their specification sections with the bid evaluation sheets on top. The final bid summary and electronic files are backed up. This completes the bidding process.

Certain work needs to be completed after the bid has been tendered to the owner. Subcontractor queries must be handled with care, so that the contractor, if awarded the contract, is not put into a disadvantageous position. Upon notification of a post-bid interview, the contractor must consider the owner's methods of doing business and develop an interview plan accordingly. Lastly, whether or not the contract was won, the estimate should be gleaned for its use in future estimates. As in the actual bidding process, ethics remains an issue after the bid is submitted.

Record keeping also must be considered. Most bids have a time period for which bids are to remain valid. The estimate should be kept for at least this time period, and if the bid team believes that there is a situation where the low bidder may be rejected or other negotiations will ensue, then the records should be kept until it comes to conclusion. If the project is like other work that the company does on a regular basis, the records may be kept for future reference. Additional pre-bid, bid day, and post-bid day ethical issues are evaluated in the next chapter.

Review questions

1 Name three positions that are required on the bid team.
2 What is the responsibility of the officer-in-charge?
3 List three activities which need to be done for bid day to run smoothly.
4 On a bid day that is expected to be moderately busy, how should the estimator anticipate the activity level to change from the start of the day until time to tender the bid?
5 Who records bids coming in from the subcontractors and suppliers, and on what form?
6 What are the consequences of the bid form not being filled in completely?
7 What are combination bids, and how are they evaluated?
8 Who evaluates the bids and determines which ones will be used in the final bid?
9 When should a preliminary bid day total be made, and why?
10 What are two common methods for the owner of a project to open bids?
11 Is the owner of a private project required to award to the low bidder? Why might they choose not to?
12 Can the owner of a private project conduct a public bid opening? Why would they choose to do that?
13 How long should the bid files be kept for an unsuccessful bid?

Exercises

1 Have you participated in bid day at your place of employment? What was your role? Was your firm successful? Would you like to one day be bid captain?
2 Other than rebar installation, what are some work packages a GC may typically self-perform but choose to hire a subcontractor for if their proposal is complete and more competitive?
3 Is it ethical to use subcontractor bids received after bid time?

4   Assume there are ten minutes left before a bid is due and you are the lead estimator. The OIC is gathering information on whether to make a last-minute adjustment and has come to you for input. What factors do you believe you need to address in your response? List each one and state why it should be considered. With only ten minutes available, your answers need to be quick, clear, and concise. Each answer should be no more than two sentences long.

5   Congratulations, you were the lower bidder, but five minutes after the owner's bidding deadline, a key subcontractor whose price was used in your bid claims error and withdraws its bid. This is a $200,000 impact that will move you out of first place. As the estimator, what do you do now? Write a paragraph on considerations and a recommended course of action to take.

6   What are some of the issues you would consider when preparing for a post-bid interview with the owner?

7   Other than drywall, what are some areas for which several specification sections may be bid separate or combined and a bid day subcontractor spreadsheet should be prepared to facilitate this comparison?

8   Fill in the available information on a sub bid proposal form from the following call by a vendor. What other information is needed, and who should call him or her back? Four addenda have been issued. List five mistakes in this telephone proposal.
    "Hi, this is Joe at M & B, (255) 555–4430. My price for the insulation is $262,500. I have seen two addenda; Good-bye."

9   Prepare an argument why the estimator either should or should not on bid day take the boxed-in quote presented at the end of the cast-in-place concrete systems in Chapter 11 for forming the elevated concrete slabs.

10 Looking only at Figure 20.4, what are the potential costs or bid risks for the GC, and what are the risks for the project owner?

11 Prepare a list of the advantages or disadvantages of the GC using the direct work estimate on bid day for doors/frames/hardware from Chapter 12 compared to the alternate bid received from Klose for the same scope.

12 Why should the estimator stay with the concrete pump cost of $15/CY already included in the direct work estimate and not modify concrete costs with the concrete pump quote (Figure 13.2) received on bid day?

Part VI

Advanced topics

21 Estimating ethics

Introduction

We now enter into the last section of this book, Part VI, which focuses on more advanced estimating topics and applications of construction cost estimating leading into the next phase of construction, project management. In this chapter, we discuss ethical issues faced by subcontractors, general contractors (GCs), and project owners. Issues involving ethical questions are raised before bid, during bid day, and post-bid. In the next chapter, several estimating technology tools are introduced. Many of these connect to our previous topics as well as connecting estimating to project management. Chapter 23 introduces a variety of other types of estimates prepared by estimate and construction teams, and our final chapter bridges the gap from estimating to project management.

A contractor's ethical behavior affects their reputation among other GCs, project owners, subcontractors, the city, and potential employees – both positively and negatively. Ethical awareness is embedded in all aspects of construction, including safety and quality control. Ethics are moral standards used by people in making personal and business decisions. They involve determining what is right in a given situation and then having the courage to do what is right. Each decision has consequences, both to the contractor and to their subcontractors, suppliers, and customers. The focus of this book has been estimating expected construction costs, and this chapter addresses ethical issues during pre-bid, bid day, and post-bid processes.

Pre-bid day ethics

There are many instances before bid day that may raise ethical questions for all parties. We present some of these in the form of questions in this section and throughout this chapter rather than including them all at the end in the form of exercises. This allows the reader to ponder potential answers – often there is no one exact correct answer for all instances.

Ethics 1: Should a GC partner with a subcontractor and agree to only work together in pursuing a project? Will other subcontractors still bid to the GC, and if their prices are more cost-competitive, should the GC use them and ignore their informal partnership?

Ethics 2: Would it be ethical for a GC and architect to have pre-bid clarification questions, just between those two firms, and not have the results documented in bid addenda available to all contractors? Would your answer change if this were a bid or negotiated project?

Ethics 3: Is it ethical for a GC to market the project owner or architect pre-bid?

Ethics 4: General contractors and subcontractors discover many inconsistencies and errors in the bid documents when preparing a detailed estimate. Should they hold back their questions until post-bid or bring them up during the bid cycle?

Ethics 5: Some contractors ask many questions during the pre-bid meeting. Some contractors will not ask any questions until after the bid is awarded. Other contractors will pull the owner or architect aside during the pre-bid walk-through and ask questions one-on-one. What are the advantages and disadvantages of each of these scenarios, and are these strategies ethical?

Ethics 6: Is it fair for an owner and their design team to issue large addenda the day before or even the day of a bid and expect that the GC has "informed all bidding subcontractors and suppliers" (or similar language typically included in the addendum)?

Ethics 7: Most contractors will diligently schedule their estimating process, as was discussed in Chapter 1. Most contractors would prefer to get the bid turned in and either proceed with the project or move on to the next opportunity. But there is always one firm that cannot seem to get organized and will ask for a bid extension. Should project owners allow this?

Ethics 8: A concrete pipe supplier contacted a utility contractor who was preparing a bid for a major sewer project and offered discounted prices if the contractor purchased the needed pipe from the supplier. Was the supplier's action ethical?

Ethics 9: The site work for the construction of a medical clinic was unit priced. During review of the contract drawings, the estimator determined that the quantity shown on the unit price bid sheet for asphalt pavement was considerably less than what would be required for completing the project. The estimator decided not to notify the project owner and to inflate the unit price for the asphalt bid item because of the anticipated overrun. Was the estimator's action ethical?

Bid day ethics

General contractors may find themselves a victim of bid day *bid shopping*; this can occur without them instigating it or even knowing about it. Your firm may be the most ethical in town, but subcontractors will find out from other GCs where they stand, and your firm may unwillingly become involved in bid auctions. This happens when subcontractors determine from a bidding contractor how their prices were in relation to those of their competition. Some general contractors will divulge this information to bidders. This is a way of driving subcontractors' prices down and thus lowering the GC's own bid to the owner. This is known as bid shopping and is considered unethical. Subcontractors expect general contractors to treat their bids as confidential information until the subcontracts are awarded. Subcontractors will also ask a GC, apparently innocently, "how their price looks" and engage in a bid day auction on their own.

The intent of the lump sum bidding process is for bidders to submit their lowest and best price the first time for specific scopes of work. By bid shopping, general contractors are implying that subcontractors did not submit their best price and have the ability to lower their prices for the work. Some subcontractors will respond accordingly to win the work. Others will stand with their original bids. The question then arises as to whether the subcontractor who engaged in the bid shopping process really submitted the best price to start with or whether a sacrificial contingency had been included. The other condition is that the subcontractor may have submitted the best price initially and consequently by lowering it may not be able to complete the work for the reduced price. Yet a third condition is that the subcontractor is in financial trouble and is trying to increase cash flow to avoid insolvency. Any of these situations could result in the subcontractor defaulting during the course of the work, which could increase the GC's risk exposure for both short- and long-term potential cost increases.

Ethics 10: Should a GC communicate with a potential owner on bid day or, vice versa, should the owner communicate with the GC? What might some of these communications entail?

Ethics 11: A reverse bid action is where GCs post their bids electronically with the project owner for all competitors to see pricing simultaneously. Contractors then raise or lower their prices not necessarily on the expected cost of the work but to beat the competition. Is this fair?

Post-bid ethics

Many activities can and do continue after the bid has been turned in which may involve questionable ethical situations for the project owner, designer, general contractor, and subcontractors. In this section we have described just a few of these potential issues.

Subcontractor and supplier inquiries

After a bid has been tendered, subcontractors and suppliers will begin calling and inquiring about how their bids compared against their competition. Answering them too soon after the bid has been tendered can cause problems for the winning GC. If a subcontractor is told that its price was low and the owner awards to the general contractor, the subcontractor will expect to be awarded a contract also. There are times, however, when a reevaluation of subcontractor bids uncovers a combination that changes the results, and the subcontractor ends up not being the low bidder. However, if the subcontractor was told that it submitted the low bid on a public project, the subcontractor may have legal recourse if not awarded the subcontract from the general contractor. The estimating team must be very cautious about responding to subcontractor and supplier inquiries regarding their prices until final subcontractor and supplier selections have been made.

Ethics 12: Do subcontractors bid the same price to all of the GCs? Do they give some preferential bids? Why would they do this? What risks are the subcontractors taking? How does the GC respond if they discover that they did not receive the same bid from a subcontractor as did their competitor GCs?

Many public bid projects require the GC to list its chosen subcontractors on the bid form – at least the major categories such as mechanical and electrical. This is intended to eliminate the potential for a successful GC to shop subcontractor bids post-bid day. But by listing the subcontractors, the GC has seriously impacted their opportunity to verify that the best-value subcontractor will be awarded the bid.

A subcontractor who has canvassed other GCs, especially those that were not in contention for the project, may find out their exact bid placement. This is not an ethical violation, and the third or fourth placed GC will earn future favors by informing the subcontractors. They also do not want the subcontractors to be unfairly 'shopped' by the low-bidder. Not all GCs, however, receive bids from the same subcontractors or may not analyze the subcontract pricing in the same way. There is no obligation by the winning contractor to execute a subcontract with the specialty contractor who believes their bid is low but has not yet been verified by that GC, especially in private work.

Sometimes a bid team member will let subcontractors or suppliers know where their prices stood if they are totally out of contention. This is done without divulging the low bidder and by indicating an approximate percentage difference from the winning bid rather than stating the low price. The GC feels they are doing the subcontractor a favor. Release of any subcontractor bid results should wait until the owner has issued a notice of intent to award a contract.

When conducting a subcontractor and supplier bid review, the project manager or estimator must make sure that a vendor that bids more than one item is willing to contract for only one item at the price bid. Also, vendors that bid a single price for a combination should be queried for a breakdown by specification section and asked the same question. Working the bids by the estimator may result in a lower combination cost. The difference between this cost and the bid used is commonly known as *buyout* and is in effect a potential savings for the GC; this is different than bid shopping. A buyout analysis may result in a different vendor being the low bidder. Clear notes should be made regarding how the buyout was determined and attached to the subcontractor list in the bid files. This way, if the contract is awarded, the project team has a record of all price determinations. Buyout will be discussed in more detail in Chapter 24.

Some GCs, upon winning a bid, will attempt to issue subcontracts or purchase orders for a few percent less than the price submitted by the lowest bidders. This would be known as 'buying down', which is different than 'buying smart' or ethical buyout. When the subcontractors complain, they are told that if they want the work, they will sign the subcontract for the stated amount. This may work initially, but eventually this GC will have difficulty being low on future bids because the subcontractors and suppliers will add a percentage to the bids that they know the general contractor will force out of them. This is similar to bid shopping, which was discussed earlier. Although the construction industry is large, in each city or county it is a close community. Treating subcontractors unfairly will not bode well in the long run for a GC's reputation. Your author shares this mantra with young project engineers and project managers: "You will need a favor from each subcontractor at some point; if not on this project, then the next one. Don't burn any bridges."

Buying subcontractors fast when they are too low is also a strategy for some GCs. This is essentially an attempt to tie them to their bid day price before the subcontractor has realized its error and potentially withdraws its bid. Bid bonds (somewhat) protect against this, but they do not exist on all projects. There are differences between the terms 'buying down', 'buying fast', 'buying smart', and 'bid shopping' during the buyout process. Some are solid professional practices and others are unethical.

In Chapter 20 we shared that subcontractor bids will continue to trickle in after the chief estimator has come to his or her total. Some phone calls and electronic bids will even continue after the owner's bid time has arrived. Once a successful bid is announced by the owner, that GC will benefit from additional bids that may have only been sent to its competitors.

Ethics 13: Is it ethical to use late subcontractor bids (a) on a public project, (b) on a private project, and/or (c) if the GC did not receive a valid bid in that category and used its own plug number?

Ethics 14: Is it ethical for a private project owner to utilize a late GC bid? Would your answer change if it was a public or private bid opening? Would your answer change if you were the low bidder at (a) bid time or, conversely, (b) post bid time?

Post-bid negotiations with the project owner

The owner may issue revised bid documents that change or add only a minor amount of work and ask one or more contractors to submit a revised bid price. These may be in response to late receipt of requests for information or value engineering (VE) proposals. They may also be a strategy of the owner and its designer. Without saying so, the owner essentially is auctioning the project to the contending GCs. The owner is asking them to take a second look at the project and submit a new and presumably lower bid. Often these 'changes' in the documents are minor, yet because all GCs will now know each other's previous bid amount (in a public bid opening) they may price the 'change' as needed to win the project. We have seen prices reduced even though the 'changes' obviously add work. The contractor submitting the lowest price usually wins the project.

Ethics 15: Is it ethical on a private project for a GC to communicate with the project owner post-bid to find out how their price looks and/or inquire whether there is anything else the project owner needs?

Ethics 16: Some private owners request voluntary VE proposals to be submitted with GC bids. There is a strategy from the GC's perspective whether they respond with much detail on this open request. Some owners may then collect all of the VE proposals, share them with all of the GCs, and ask for new pricing. This is obviously an example of a reverse bid auction as well. How should GCs respond to this request?

Another approach used by some private owners is to request a 'best and final offer'. The owner often indicates that the project is over budget and requests the contending contractors to review their bids and submit best and final offers. The contractors each must decide how badly they want the project and whether to lower their prices. The best and final offer approach is legal in private work and often part of the contracting process. An owner may believe that the bidders did not submit their best bids, and this method is used to obtain the lowest possible price. During the initial bid, GCs generally do not increase their fees, because it may take them out of contention. Upon being asked for a new price, they usually examine combinations of vendor bids and the self-performed work to see how much they can reasonably reduce their bid and still make a fair profit. Fee reduction is considered the last choice, depending on the results of the other reviews. Lowering prices places risks on all of the parties. Any reduction in price by the GC and its subcontractors must be recovered at some point during the course of construction, often in the form of change orders or claims.

Some project owners bid the mechanical and electrical work direct as separate contracts. During post-bid negotiations, they may attempt to assign these contracts to the general contractor for no additional fee. Administering these contracts can cost money by extending the jobsite overhead, and the contractor deserves a profit for doing so. The risk from preselected subcontractor performance usually is lower, and a reduced fee may be requested by the owner.

During post-bid negotiations, the contractor needs to determine issues that they are willing to concede and those that are not negotiable and to gain an understanding of the owner's objectives. The contractor may want the work and yield to the owner's demands to the point of not being able to do the work profitably. An owner may try to obtain as many concessions from a contractor as possible. This is again a risky move for all parties, including the subcontractors.

Bid projects are often accompanied with bid protests. A bid protest is a complaint from an unsuccessful GC (or subcontractor) indicating that the low-bidder did not follow the rules exactly and should be disqualified. Infractions could be due to a late bid, not completing the bid form properly, or an unqualified bidding contractor. These are very difficult situations for the project owner. If the owner agrees and disqualifies the low-bidder, that firm in turn will now file its own bid protest. This can go on for some time and eventually the entire bid may need to be thrown out. When this happens, unfortunately, all contractors know the price to beat and again this results in a bid auction and the 'successful' re-bidder will have often underpriced the project and will need to find other ways to make up for their losses, such as change orders or claims.

Summary

A contractor's reputation is greatly influenced by the ethical standards of their leaders and employees. In this chapter, we have discussed ethical issues related to cost estimating. For additional discussion of construction ethics, the reader may want to review *Professional Ethics for the Construction Industry*, by Mirsky and Schaufelberger.

During the pre-bid phase, contractors and owners need to practice good faith and fair dealing in the solicitation and preparation of bids and proposals. Project owners need to ensure that all prospective bidders have the same information relative to project scope and conditions. General contractors need to treat subcontractors' price proposals as confidential and not disclose the information to other subcontractors. Anyone engaged in contract procurement needs to avoid the perception of favoritism.

During bid day, GCs must avoid bid shopping by not disclosing the subcontractor ranking for specific scopes of work. Subcontractors' willingness to work with a GC is greatly influenced by their perspectives regarding how they will be treated by the contractor. During the post-bid process, GCs need to be careful about responding to subcontractor and supplier inquiries until final subcontractor and supplier selections have been made. General contractors often select best-value subcontractors rather than the lowest-price subcontractors. Private owners may wish to engage in post-bid negotiations with the selected GC. The contractor needs to decide which issues they are willing to concede and which they are not.

Review questions

1   Describe a method that the owner might use to get a contractor to lower their bid after the time the bid was tendered.
2   Why would a subcontractor give a lower price to one GC and/or, conversely, a higher price to another GC?
3   Why would an unsuccessful bidding GC share subcontractor quotes?
4   How should a young project engineer who has been asked to assist a bid team and receive subcontractor telephone quotations be instructed to respond to subcontractors attempting to find out where their price stands?
5   In a private owner project, is it acceptable to use a subcontractor whose quote was received after the GC has forwarded its bid to the bid runner?

Exercises

1   The important topic of ethics has been threaded throughout this book. Look back now on our previous chapters and create a table of ethical issues and/or questions.

Propose 'answers'; however, realize that there are many potentials, and some are better than others. Present and debate these in class, maybe through the use of small groups. Can you develop a consensus for each one?

2 State your thoughts about the following ethics issues. Include the reasoning for your conclusions.

    a   Bid shopping,
    b   The post-bid auction process, and
    c   Issuing subcontracts for amounts below the low bidder's price.

3 What is a 'reverse bid auction'? Have you been involved in one of those? Was it successful for the project owner? Was it successful for the GC?

4 What are the differences and similarities between these terms: collaborative, contractual, ethical, fair, friendly, legal, moral, and responsible?

5 Assume that only one subcontractor bid was received for roofing on a project. After award to the general contractor, that subcontractor indicates they have a bid error. What should the general contractor do? If they re-bid that section, is that ethical? Should they just try to negotiate with one favorite subcontractor, or maybe another bidder? Should the general contractor notify the owner? If they rebid that portion of the work and the new bid price is less than the pulled quote, should the GC or owner realize the benefit? What if the new bid was higher?

6 Sixteen boxed-in ethical questions and scenarios were raised in this chapter. Prepare answers for each either as topics to debate in the classroom or as homework assignments to be turned in.

7 When a subcontractor asks how their price stacked up on bid day, how should this question be answered?

8 If the subcontractor is low and inquires about their price after the bid has been tendered, how should the question be answered?

9 Which subcontractors and suppliers would a project manager buy out quickly, not because their price was too low but because of schedule?

10 List at least three steps involved in an ethical buyout process. You may need to look ahead to Chapter 24 to answer this question, or to another project management book.

11 What would happen to a GC who violated some of our ethical questions, such as contacting an owner during the bid on a public works project?

22 Technology tools

Introduction

Technology has completely changed how estimators do their jobs. Though the basic concept of estimate creation has not changed, computers, digital databases, web-based applications, design and construction integrated software, cloud access and capability, electronic contract documents, and mobile devices are now being used for cost estimating. Consequently, estimating efficiency and speed have improved dramatically from this adoption of technology. The purpose of this chapter is to introduce some of the technology tools that can be used to develop a cost estimate. The principles used in developing an estimate are similar to those discussed in previous chapters, but use of technology has accelerated the process.

Design drawings and specifications typically are issued as electronic PDF files. The drawings can be scaled and quantities of work measured, as discussed in Chapter 6. On-Screen Takeoff (OST) and Bluebeam can be used to determine quantities of work from the PDF drawings, eliminating the need for manual take-off. Both tools are discussed in this chapter. A multidimensional building information model (BIM) can also be used to determine quantities of work and is also discussed. Specialized estimating software or Microsoft Excel with an estimating database of unit prices can then be used to prepare the estimate.

To assess the use of estimating technology tools, the authors conducted a survey of 38 estimators in the Northwest to determine which tools they are using. Most were using OST and Bluebeam for quantity take-off (QTO), and both are discussed in this chapter. Excel was the most popular software for developing estimates using techniques discussed in previous chapters. Sage Estimating was the most popular specialized estimating software, followed by WinEst and Beck Technology's Destini Estimator. Each of these tools, except Excel, is also discussed. Use of drones for estimating, estimating databases, and subcontractor solicitation are the final topics addressed. New technology tools and upgrades to existing tools are being launched frequently, so estimators need to be aware of the changes on the market as they select appropriate tools for their work.

Estimating software

Estimating software is designed to streamline and improve the process of creating estimates. It is available in many varieties, ranging from simple spreadsheet templates to cloud-based software promising transparency and collaborative experiences. Estimating software companies can give customers access to their technology either by installing their products on computers or servers the customer owns or by providing web-based applications that can be assessed by multiple mobile devices.

The power of an estimating software lies in the features that improve the take-off and pricing processes for users. Different software companies package and design these features differently. Digital take-off capabilities often are sold in separate software packages from estimating tools. The core technology of take-off tools is to use digital copies of the project plans to measure material, quantity, and part requirements and organize the data in a list or spreadsheet. Estimating capabilities typically provide pre-built estimating templates and a pricing database that can be combined with the take-off quantities to create estimates of direct costs. Estimates for jobsite general conditions and markups discussed in Chapters 17 and 18 would still need to be added.

Sage Estimating

Sage Estimating is an on-premises estimating software that enables users to build estimates using data from their QTO or by connecting with eTakeoff Dimension for two-dimensional (2D) take-off. The estimator can send the take-off information directly to the estimate through eTakeoff Bridge. Using the Bridge, the estimator can send the take-off information directly to the estimate without manual intervention like copying or pasting. Default costs for labor, material, subcontractor, equipment, and other categories are automatically calculated and included in the estimate, as shown in Figure 22.1.

Once the items are on the spreadsheet, the estimator can customize the estimate to conform to conditions of the project. Labor and equipment production rates can be manipulated to reflect jobsite conditions, and the Sage Estimating database allows the estimator to enter their own pricing and provides tools for updating the pricing. Users may also use updated cost data from commercially available cost databases, such as *RS Means Data Online* or *The Guide*, discussed later in this chapter.

Advanced versions of the estimating software link to *RS Means Data Online* and integrate with three-dimensional (3D) modeling software by Autodesk to use a BIM for QTO. Sage Estimating, eTakeoff, and Autodesk have combined efforts to enable estimate development by concurrent use of both 2D and 3D content. The look and feel of the software are very similar to that of Excel, because it is a spreadsheet-based product. The use of eTakeoff Bridge to extract model data is shown in Figure 22.2.

In addition, Sage Estimating has a buyout module that can be used to solicit quotations from suppliers and subcontractors. Additional information on Sage Estimating can be found at www.sage.com/en-us/products/sage-estimating/. The QTO tool eTakeoff Dimension is similar to On-Screen Takeoff and Bluebeam, discussed later in this chapter. Users perform the take-off with their mouse after scaling the drawings. Information on eTakeoff Dimension can be found at https://etakeoff.com/products/dimension-overview/.

WinEst

WinEst is an on-premises database-driven estimating system that helps estimators manage and integrate estimates. Key features are QTO, cost estimating, audit trails, and reporting. Users can set up estimates from scratch or use a project-type template that follows

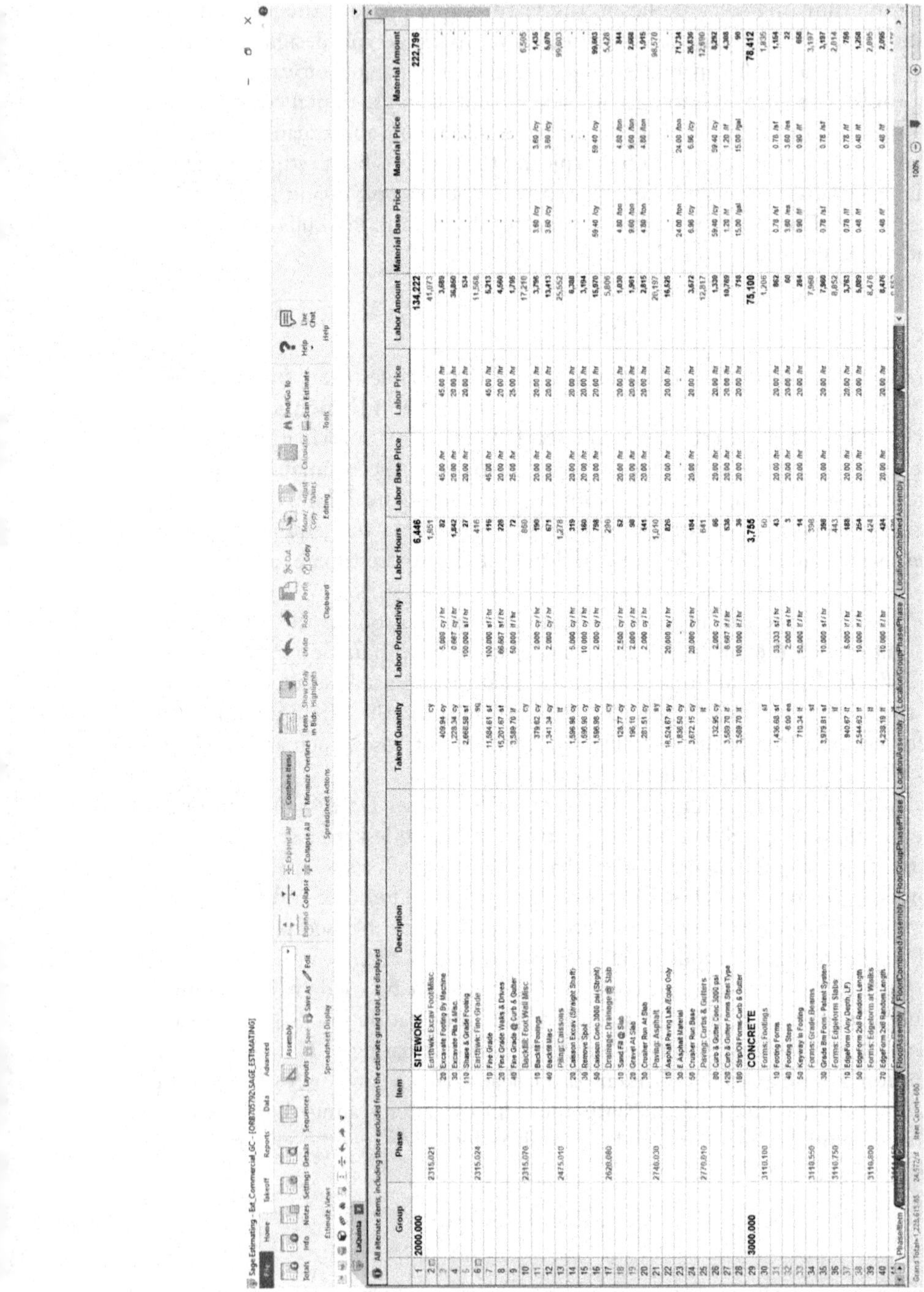

Figure 22.1 Sage Estimating spreadsheet

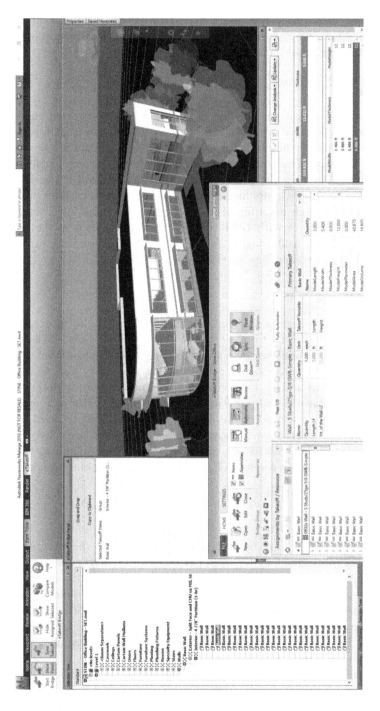

Figure 22.2 QTO with eTakeoff bridge

company standards for general conditions and indirect costs. Estimates can be organized using either MasterFormat or Uniformat. The software has systems for QTO from paper drawings and electronic drawings and an interface with On-Screen Takeoff. Each line item and assembly in WinEst can be assigned to a task/activity identification and exported to Primavera or Microsoft Project for scheduling. Additional information on WinEst can be found at https://gc.trimble.com/product/winest.

Destini Estimator

Destini Estimator is an on-premises or cloud-hosted estimating system that enables multiple users to access and work on estimates at the same time. It supports disconnected workflows so that the estimator has access to the estimate in the office, on the jobsite, or in between. The software integrates QTO with a cost database to create estimates formatted to meet the user's needs. This includes both 2D and 3D take-off. The software also keeps track of which estimate updates were created by which users of the software. The software has internal tools for performing 2D QTO from electronic drawings, and it integrates with BIM 360 or Navisworks enabling the development of a QTO from a BIM. Additional information on Destini Estimator can be found at https://beck-technology.com/destini-estimator/.

On-Screen Takeoff

OST is a software that enables the user to streamline the take-off process by measuring quantities of work directly from electronic drawings with a few clicks and drags of a computer mouse. The user sets the scale of the drawing in the software and traces the area or length to be measured with the mouse. The mouse can also be used to count objects such as windows and doors and measure volumes. The software then determines the length or area measured and populates a table with the appropriate quantity. Once measured, the quantified section of the drawing will be colored to avoid the potential for double counting of any scope. Those sections not colored are clearly identified as needing measurement. OST seamlessly exports data into Excel and other estimating software. Additional information about OST can be found at www.oncenter.com/products/on-screen-takeoff.

Bluebeam

Bluebeam Revu is a PDF markup and editing software that can be used to develop QTOs, including measuring lengths, areas, perimeters, diameters, and volume, as well as count objects. An example is shown in Figure 22.3.

Before any accurate measurements can be taken, the user must calibrate the measurement tool. This can be done by entering the scale of the drawing (if known) or by selecting the distance between two known points. After the drawing scale has been selected, the user determines the measurement desired, such as length or area, and uses the mouse to determine the measurement. The measurements and markups taken are stored in a Bluebeam Revu Markup List that can be linked to an Excel spreadsheet. Once the QTO has been completed, pricing for the general contractor's direct work can be performed as discussed in Chapter 10. Additional information on Bluebeam Revu can be found at www.bluebeam.com/.

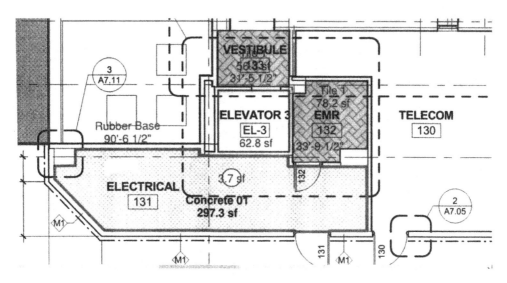

Figure 22.3 QTO with Bluebeam Revue

Building information models

BIM is an industry term for an integrated database containing parametric information and design documents for the entire project. The data stored in a BIM are used to create 3D projections of the project, as illustrated in Figure 22.4. These models contain dimensional properties such as area, count, perimeter, and volume that can be used to develop QTOs of project components needed to create cost estimates. A BIM developed in the early stages of the design will not be highly detailed but can be used for preliminary conceptual cost estimates. As design decisions are made and incorporated into the model, it can be used for estimating quantities of work for final cost estimates.

Assemble, which is one of the products included in Autodesk's Construction Cloud, contains multiple cost tracking functions, including a quantification tool that allows users to extract real-time project updates from the model and organize it for use in QTO and estimating. This instant access to quantities can be organized based on user defined parameters, but the most commonly used organization is the MasterFormat developed by the Construction Specification Institute and discussed in Chapter 2. Once the QTO has been completed, pricing can be added to complete the estimate. Information on Assemble can be found at https://construction.autodesk.com/products/assemble.

Use of drones

A drone is an unmanned aircraft. Essentially it is a flying robot that can be remotely controlled or fly autonomously through software-controlled flight plans in its embedded systems, working in conjunction with onboard sensors and the Global Positioning System (GPS). A drone's capability to accurately survey areas of land is an essential benefit of utilizing drones in construction. On a construction site, drones can assist with pre-construction site review, aerial surveying and mapping, and measurement of excavation

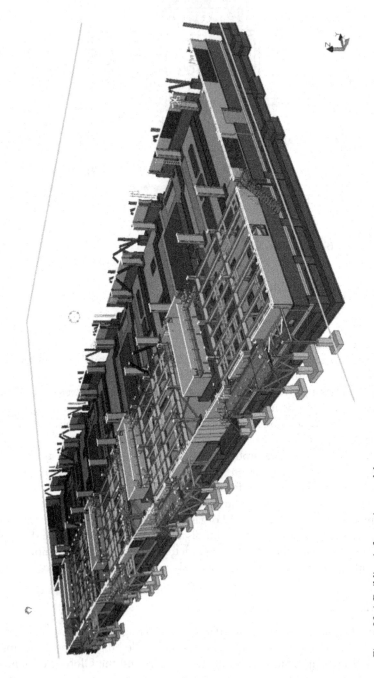

Figure 22.4 Building information model

depths and material stockpiles. They provide a bird's eye view of expansive project sites that is not ascertainable from the ground.

Drone software systems can provide convenient and time-saving desktop or mobile device access to jobsite conditions and help facilitate rapid assessment and response to developments in the field. In addition to GPS and high-resolution cameras, drones can be equipped with advanced technologies, such as LiDAR (light detection and ranging), which is a remote sensing pulsed laser to detect objects and measure distances. With the right software, users can integrate imagery and data into BIM or other 3D models or use drones in conjunction with other tools, such as land-based 3D laser surveying and equipment and inventory tracking devices.

Drones are particularly useful in measuring quantities of work on highway projects, especially those in remote areas. They can also provide aerial photographs of the project site to enable the estimator to fully understand the existing site conditions and any access restrictions to be encountered. The photograph of the tower crane shown in Figure 16.1 was taken from a drone.

Estimating databases

Many estimating databases are commercially available. They include pricing averages for a variety of systems and materials. The collection of equipment, material, and labor costs is especially helpful for developing cost estimates for projects in new markets, for which the estimator may not have historical cost data. Materials, tasks, and equipment are easily searchable inside the database software. An estimating software system can take the quantities of work that have been determined separately and apply the database unit prices to develop direct cost estimates. The challenge for the estimator is to ensure that the unit price data reflects contemporary market conditions at the project location.

RS Means Data Online

Many estimators are very familiar with the RS Means series of construction cost guides. The guides have been combined into a cloud-based database that can be accessed to create unit price, assemblies, or square foot cost models for developing construction project cost estimates. Instead of using city cost index numbers to modify the database values for local conditions, the user can enter the project zip code to convert the cost data to reflect local pricing. The cost data is updated quarterly, and predictive analytics are included to predict project construction costs up to three years in the future. Information on the *RS Means Data Online* can be found at www.rsmeans.com/products/online.aspx.

The Guide

The Guide includes current building material prices organized by the 2014 MasterFormat and based on Northwest prices, current in-place common construction assemblies, and square foot costs for common building types. The data is updated semiannually based on market surveys. The cost and other information contained in *The Guide* are derived from local material suppliers and subcontractors. Also included is a regional directory of material suppliers and subcontractors.

The Guide was introduced in Chapter 11. It was developed for commercial and multi-family construction but is also applicable to residential construction. It is used by many construction education programs because the format is practical and easily translated into use in class projects. Information on *The Guide* can be found at www.bestconstructionsite. com/, and a more detailed description is included on the book's eResource.

Subcontractor solicitation

One of the first things that an estimator should do when preparing an estimate is to get subcontractors and suppliers involved. This involves notifying them that a project is being reviewed and that the general contractor would like their quotations on bid day. In the past, this task was done either by telephone calls or individual e-mails. Today, this notification often comes from the general contractor's website and project management software for managing the solicitation process. One technology tool used by some general contractors in the Northwest is Building Connected. It is a service for prequalification of subcontractors as well as for soliciting and evaluating quotations. Information on the tool can be found at www.buildingconnected.com/subcontractor-construction-network/. Some general contractors post contract drawings and specifications on their websites for subcontractors and suppliers to download and review, while others place the contract drawings and specifications in electronic plan rooms for subcontractors and suppliers to access.

Summary

The use of technology tools to support the preparation of cost estimates has greatly increased during the past decade. As a result, estimating efficiency and speed have improved. Estimating software has been developed that enables improved take-off and pricing processes for users. The three estimating software vendors discussed in this chapter are Sage Estimating, WinEst, and Destini Estimator. Each has different features, but all basically combine a spreadsheet with a cost database to enable estimation of direct costs for individual scopes of work.

On-Screen Takeoff and Bluebeam are software products that are used by many estimators for QTO. Building information models provide 3D projections of construction projects that also can be used for QTO with software such as Assemble. Drones can be used for site investigation and QTO for projects in remote areas or for horizontal projects such as highway construction.

Several electronic cost databases are available for pricing the quantities of work. Both *RS Means Data Online* and *The Guide* are discussed. Subcontractor quotations are also used in developing cost estimates. Procurement systems such as Building Connected can be used by general contractors in the solicitation and evaluation of subcontractor quotations.

The use of technology is always changing. New systems are being developed continually, and the estimator needs to be apprised of the latest technology tools.

Review questions

1   What is the difference between an on-premises estimating system and a cloud-hosted estimating system?
2   What are the two digital QTO tools described in the chapter? What is the primary benefit that they provide to an estimator?
3   What is the basic process for determining a QTO using Bluebeam Revu?

4 Why are drones sometimes used to support preparation of a cost estimate?
5 What challenge does an estimator face when using a commercial estimating database for pricing work items in a cost estimate?
6 What type of digital tool might an estimator use in soliciting price quotations from prospective subcontractors?

Exercises

1 You are preparing a cost estimate for the construction of an office building using a digital estimating software. What factors would you consider in deciding whether to use a MasterFormat or a Uniformat organizational structure for the estimate?
2 Destini Estimator estimating system allows multiple users to access and work on estimates at the same time. Describe a situation where this capability would be useful.
3 You are an estimator providing preconstruction estimating support for a new office building. The designer has developed a BIM to illustrate design concepts as design decisions are made. How could you use the BIM to develop preliminary cost estimates? How would you use the BIM to develop the final cost estimate?
4 You are developing a cost estimate for the construction of a highway in a remote area of Alaska and have obtained a drone to help with preparing the estimate. What type of information would you expect to obtain from use of the drone? How would the information help in the preparation of the cost estimate?
5 Are you currently using any of the technology tools discussed in this chapter or tools that are not addressed? What are the advantages and disadvantages of technology tools that you are using?

23 Other types of estimates

Introduction

Can a house builder develop an estimate for an industrial manufacturing facility? Can an apartment developer create an estimate for a warehouse? Does a contractor who specializes in tenant improvement projects have the tools necessary to estimate a downtown high-rise? The answers: Both yes and no. Many of the procedures utilized to measure the concrete in a spot footing, call a vendor for a current quote, and apply necessary markups are consistent in whichever type of construction project is being estimated.

There are some differences, however, that require consideration. In most instances, these differences are used by a contractor to gain a competitive advantage. If general contractors (GCs) do not have experience constructing oil refineries, they generally are not qualified to bid on this type of project. If contractors had an opportunity to estimate a few similar projects, they would learn that the tools required are similar, except that instead of measuring square yards of carpet, they are measuring lineal feet of pipe. This is, of course, an oversimplification. The message is that estimators should not be afraid of trying. They will often be surprised that many of the tools necessary to estimate new types of projects or materials are already in their toolbox. The business decision to take on a different market, however, typically is made by a company executive, not the estimator.

It would be nearly impossible in one book to cover ALL types of estimates in detail. We have chosen a lump sum commercial process for this book for our mixed-use case study, as it has many interesting elements compared to other types of estimates. In this chapter, we will briefly introduce a variety of other common estimates developed either by the staff estimator or the project manager during the construction process. Estimating is a process, whether the estimate is a budget, a lump sum bid, a heavy civil unit price, or a guaranteed maximum price (GMP) proposal.

Budget estimates are developed for projects early in the design process or for those projects that need to start construction before the design can be finalized. Cost-plus contracts are used when the scope of work cannot be defined. All the contractor's project-related costs are reimbursed by the owner, and a fee is paid to cover profit and contractor overhead. A GMP contract is a cost-plus contract in which the contractor agrees not to exceed a specified cost, but cost accounting is open-book and savings are subject to a pre-agreed split. Unit price contracts are used on heavy civil or industrial projects when the exact quantities of work cannot be defined. The designer estimates the quantities of work, and the contractor submits unit prices for each work item. The actual installed quantities required are measured and multiplied by the bid unit prices to determine the final contract price. Lump sum contracts are awarded on the basis of a single lump sum estimate for a specified scope of work. The lump sum estimating process is the one that was featured throughout this book.

In addition to these major estimate types, this chapter will introduce a few different types of estimates, including pro formas, residential, allowances, bid alternates, pay requests, change order proposals, monthly forecasts, and as-built estimates. Many of these estimates tie into and rely on a current company database maintained by the staff estimator. Much of the material from this and the next project management (PM) chapter has relied upon *Management of Construction Projects, A Constructor's Perspective*. The reader is suggested to review that resource for additional discussions on other types of estimates and more-advanced PM processes.

Budgets and guaranteed maximum price estimates

Budget estimates

In Chapter 2, we introduced and compared the three major types of commercial construction estimates, including budgets, lump sum, and guaranteed maximum prices. Budget estimates are developed throughout the design process of a construction project. Early in the programming phase, the owner usually establishes a tentative project scope and a preliminary budget. The project budget includes both the estimated construction cost and other estimated owner costs. As design decisions are made and drawings created, additional budget estimates are prepared to ensure that the project being designed can be constructed within the owner's established budget. If the budget estimates indicate that the construction costs will exceed the owner's budget, the owner has two choices – either reduce the project scope or increase the budget. These budget estimates may be prepared by the designer, by an estimating cost consultant, by a construction management consultant, or by the GC, as one of the preconstruction services discussed in Chapter 5.

The accuracy of budget estimates is a function of the degree of design completion. Budget estimates are based on $/square foot of floor area (SFF) for the complete project or major Construction Specification Institute (CSI) division or assembly but rarely include any quantity take-off (QTO). Budget estimates are typically prepared as a cost-plus percentage fee or cost-plus fixed fee basis. Because there is considerable uncertainty in preparing budget estimates, the final estimate value should be rounded to the nearest hundred or thousand dollars. Presenting a budget estimate that shows dollars and cents may create a false impression regarding the accuracy of the estimate.

Guaranteed maximum price estimates

Cost-plus construction contracts are often used when the exact project scope of work has not been defined. Many project owners select the construction contractor early in design development to take advantage of the contractor's expertise and to foster early teaming between the design and the construction organizations. Most cost-plus contracts are awarded using the negotiated procurement process that was introduced in Chapter 3.

In a cost-plus contract, all of the contractor's project-related costs are reimbursed by the owner, and a fee is added to cover profit and company overhead. The fee may be a fixed amount or a percentage of project costs, or it may include an incentive component. Many cost-plus contracts contain a GMP that is negotiated between the contractor and the owner. Any costs exceeding the GMP are borne by the contractor. Some of these contracts contain a cost-sharing clause that provides for a sharing of cost savings between the owner and the contractor. If the final project cost is less than the GMP, both parties benefit and share the savings, such as 75% to the owner and 25% to the contractor.

The GMP is often developed using drawings that are about 90% complete for the site work and substructure and about 50 to 60% complete for the superstructure and finishes. This would correspond to design development documents discussed earlier in the book. The GMP includes the estimated cost of constructing the project and the fee requested by the contractor. Because the drawings are more complete with a GMP than with a budget, the estimator will perform as much QTO as possible, especially related to structural work. Subcontractor plug estimates will be developed on an assembly or $/SFF basis, and the GMP will include allowances for systems not yet defined, such as interior finishes. A GMP estimate for a concrete tilt-up project that connects with some of our previous examples is included as Figure 23.1.

City Construction Company
GMP Estimate Summary

Job: Carpenters Training Center   Bldg Size: 45,000   SF   Date: 8/18/2021

CSI Div.	Description	Labor Hours	Direct Labor	Equip. & Material	Subs	Totals
1	Jobsite general conditions ⋆		$292,900	$142,200	$0	$435,100
3.1	CIP concrete	2,495	$99,800	$134,100	$49,800	$283,700
3.2	Tilt-up concrete	1,482	$54,729	$176,629	included	$231,358
4	Masonry				$27,300	$27,300
5	Structural steel	3,750	$150,000	$223,456	$45,000	$418,456
6.1	Rough carpentry	250	$10,000	$47,800	$0	$57,800
6.2	Finish carpentry	1,877	$75,098	$45,453	$0	$120,551
7	Roofing and insulation	63	$2,500	$4,300	$425,400	$432,200
8	Doors and glazing	63	$2,500	$1,500	$339,600	$343,600
9	Finishes	113	$4,500	$5,670	$806,000	$816,170
10	Specialties	88	$3,500	$2,255	$21,000	$26,755
11	Bridge crane				$175,600	$175,600
12	Furnishings	100	$3,750	$2,575	$0	$6,325
13	Equipment					by owner
14	Elevators				$85,000	$85,000
21	Fire protection				$325,000	$325,000
22	Plumbing				$166,200	$166,200
23	HVAC				$455,200	$455,200
26.1	Line voltage electrical				$156,800	$156,800
26.2	Low voltage electrical				$705,600	$705,600
2	Sitework	350	$13,300	$26,600	$817,500	$857,400
	Subtotals:	10,629	$712,577	$812,538	$4,601,000	$6,126,115
	Labor burden on direct labor		$419,677	55%	$230,822	$6,356,937
	Labor burden on indirect labor		$292,900	35%	$102,515	$6,459,452
	Liability insurance			1%	$64,595	$6,524,047
	Builder's risk insurance			0.25%	$16,310	$6,540,357
	State excise tax			1%	$65,404	$6,605,761
	Estimating contingency			3%	$198,173	$6,803,933
	Home office overhead and profit			5%	$340,197	$7,144,130
	Total GMP Estimate:					**$7,144,130**
	Wage Check: $/Hour:	$39.48	OK		$/SF Check:	$158.76
						OK, Checks

⋆ Notes: See separate detail for GCs and recaps for each CSI division

Figure 23.1 Guaranteed maximum price estimate summary

Developer's pro forma

One use of early general contractor budgeting services is to support the owner with early overall budget preparation. A project owner who is in the business of designing and constructing or improving property for the purpose of renting or sale is known as a *developer*. The developer's pro forma is a complicated equation with many variables and is often customized by different developers for different projects. The pro forma can be thought of globally as the financial analysis necessary to determine whether the project 'performs' and is a viable real estate investment. The bottom-line anticipated construction budget prepared by the GC's estimator or PM is just one element in the developer's pro forma. Input to the pro forma is often part of the GC's preconstruction services. The bank requires the pro forma to be included with the developer's construction loan application package. Internal development equity partners and potential external investors also rely on the pro forma as a financial decision-making tool. A very rough 'back-of-the-envelope' pro forma is developed early in the process. The pro forma continues to be refined and modified all the way through the development process and even through construction completion and into occupancy. The pro forma combines many cost estimates and compares the estimated cost of the project to the expected value. If the projected value is 10–15% greater than the estimated cost, then the project is likely a 'go', whereas if the expected costs approach or exceed value, then the project is a 'no-go'. This concept is also known to developers as 'Does it pencil?' – or does the real estate development investment make sense? Either way, the developer may go back into the pro forma and change one or more of the variable values or modify the design to achieve a different result.

Heavy civil unit price bids

Unit price contracts are used when the exact project scope cannot be determined, but a quantity can be estimated for each element of work. The project designer or an estimating consultant provides an estimated quantity, and the GC determines the unit price. The actual contract value for each unit price item is not determined until the project is completed. The actual quantities of work are measured during construction, and the cost is determined by multiplying the quantity used by the unit price submitted by the GC. Unit price contracts are extensively used for heavy civil projects, such as highway, airport, transit, utility, and environmental cleanup projects. Most unit price bid forms contain both lump sum and unit price bid items. It is not unusual to see 100–200 separate line items for which the GC must develop stand-alone pricing. An example unit price bid form for a small utility project is shown in Figure 23.2. The values shown in italics are entered by the GC after completing an estimate for the work.

To determine a unit price for each bid item, the estimator needs to estimate both the direct and indirect costs associated with each bid item. The direct cost includes the labor, material, equipment, and subcontractor costs associated with each item. This involves developing a work package cost estimate, similar to the one shown in Figure 16.2, for each bid item. The estimated direct cost for every line item is then entered on a summary estimate. Figure 23.3 shows a summary estimate for the bid form shown in Figure 23.2. Then the indirect costs are added to determine the final unit price for each bid item. The indirect costs include both general conditions (project indirect) and margin (company overhead and profit). A separate general conditions estimate was developed for the project using the process described in Chapter 17. The estimated value of $37,668 was proportionately distributed across all bid items in Figure 23.3. A margin of 16% was then applied

Bid Proposal
Tri-County Conveyance Pipelines

This certifies that the undersigned has examined the location of the proposed work and the plans, specifications and contract governing the work, and that the method by which payment will be made for said work is understood. The undersigned hereby proposes to undertake and complete the work embraced in this project in accordance with said plans, specifications and contract and the following schedule of rates and prices. Exact quantities will be determined upon completion of the work.

(**Note:** Unit prices for all items, all extensions, and total amount of bid shall be shown. All entries must be typed or entered in ink.)

Item No.	Description	Plan quantity	Units	Unit price	Total amount
1	Mobilization	1	LS	20,544.00	$20,544
2	Install 12″ reclaimed water pipe	1,600	LF	42.29	$67,664
3	Install 12″ gate valve	1	EA	3,729.00	$3,729
4	Install 18″ PVC sanitary sewer pipe	1,600	LF	108.09	$172,944
5	Install air release valve assembly	1	EA	16,875.00	$16,875
6	Install sanitary sewer manholes	5	EA	7,846.59	$39,233
7	Asphalt restoration trench patch	1	LS	14,007.00	$14,007

Total Bid $334,996

Signature of Authorized Official

Date: June 15, 2021   *Jackson Smith,* President

Firm Name: High Country Construction
Address: 1475 First Avenue
             Cascade, WA 98202

Figure 23.2 Heavy civil bid proposal

Tri-County Conveyance Pipelines Summary Estimate

Item No.	Description	Units	Quantity	Direct Cost	General Conditions	Margin	Final Bid	Unit Price
1	Mobilization	LS	1	$15,400	$2,310	$2,834	$20,544	$20,544.00
2	Install 12″ reclaimed water pipe	LF	1,600	$50,724	$7,608	$9,332	$67,664	$42.29
3	Install 12″ gate valve	EA	1	$2,796	$419	$514	$3,729	$3,729.00
4	Install 18″ PVC sanitary sewer pipe	LF	1,600	$129,644	$19,446	$23,854	$172,944	$108.09
5	Install air release valve assembly	EA	1	$12,649	$1,898	$2,328	$16,875	$16,875.00
6	Install sanitary sewer manholes	EA	5	$29,410	$4,412	$5,411	$39,233	$7,846.59
7	Asphalt restoration trench patch	LS	1	$10,500	$1,575	$1,932	$14,007	$14,007.00
				Total	**$37,668**	**$46,205**	**$334,996**	

Figure 23.3 Heavy civil summary estimate

to each bid item by summing the direct cost and the general conditions for the bid item and multiplying the sum by 16%. Company overhead was estimated to be 8%, and the desired profit on this small project was set at 8%, which resulted in a margin rate of 16%.

To complete the bid form in Figure 23.2, the general contractor entered the unit price values from the last column in Figure 23.3 in the unit price column. The planned quantity was then multiplied by the unit price values to determine the total amount values. The total bid was determined by summing the total amount values. The contractor submitting the lowest total bid would be selected for award. If the actual quantities installed during construction vary from the bid quantities, the contract value is adjusted utilizing the unit prices.

Residential

There are many differences between estimating techniques used by the residential construction industry and those of the commercial construction industry. This book has emphasized the tools and techniques most common for the commercial industry. Many of the previous chapters describe this process. Most of these techniques could also be used on residential projects. The differences can best be explained by dividing the residential contracting industry into two significant segments; the small custom builder and the large tract home developer.

The small builder may be a one-person firm who specializes in *custom home* construction or remodeling projects. The owner may also be the lead carpenter and the accountant. He or she may not have the expertise or the staffing to prepare detailed cost estimates. Estimating is limited to the knowledge that the GC estimated the last 2,000 square foot home at $200 per square foot, and therefore this one should be about the same. The builder may self-perform much of the work and does not cost-code time or material invoices. He or she does not estimate or invoice for general conditions. The small residential contractor seldom competitively bids work to owners. A larger fee is customary for smaller GCs, which is expected to cover equipment replacement and home office costs. The contractor will usually have one preferred subcontractor or supplier in each category of work. An independent builder generally does not get competitive subcontractor quotations. The small residential GC may not have an in-house unit price database. At the end of the year, if the contractor made money, he or she will usually proceed in the same fashion the next year. But if they lost money, they would not know exactly where and how. This is one of the reasons that there is such a high failure rate among residential contractors.

A small residential contractor would benefit from the use of a simple estimate summary form, as shown in Figure 23.4. This form is broken down using CSI divisions. Even estimating according to the assemblies method, as was introduced in Chapter 2, for budgets and the GMP proposal would be better than just using a square foot number based on the previous year's projects.

Larger residential contractors or developers may construct several tracts of 200 homes each per year and are known as *speculative (spec) home builders*. They build houses on speculation of sale and are therefore also considered developers. Spec builders generally subcontract most of the work. Their direct crews often are limited to one superintendent per tract and maybe a couple of cleanup or warranty crews. They typically obtain competitive bids on all of the work. They may not have the allegiance to particular subcontractors that is common with smaller custom home builders. These residential contractors usually perform very limited detailed quantity take-off. They generally rely on lumber suppliers to perform their own take-offs from the drawings. The large spec builder has a smaller percentage margin than does the smaller custom builder because it can distribute the overhead costs across more projects. The larger firm's business success tends to be greater than the smaller one-person firm. The larger builder's greatest risk is to become overextended with too many unsold homes and be exposed to adverse economic swings. The

Project: **Residential plan no. 208A**
Date: 5/25/21
Estimator: Terry Jones
Square footage: 1,920

Direct Hours	Old CSI	Description	Labor	Material	Subs	Total	$/SFF
0	1.00	General conditions (CSI 00/01) see below	0	0	0	0	0.00
24	2.00	Sitework	729	3,050	0	3,779	1.97
207	3.00	Concrete	6,200	5,100	0	11,300	5.89
0	4.00	Masonry	0	0	15,000	15,000	7.81
0	5.00	Structural steel	0	0	0	0	0.00
397	6.00	Rough carpentry	11,900	18,436	0	30,336	15.80
145	6.00	Finish carpentry	4,340	16,746	0	21,086	10.98
387	7.00	Siding/Roof/Insulation	11,600	4,500	22,000	38,100	19.84
109	8.00	Doors and windows	3,260	13,840	2,200	19,300	10.05
33	9.00	Finishes	1,000	1,100	28,250	30,350	15.81
8	10.00	Gas fireplace	250	250	6,600	7,100	3.70
24	11.00	Appliances	720	6,200	0	6,920	3.60
12	15.00	Mechanical	360	1,155	11,050	12,565	6.54
17	15.00	Plumbing	500	525	12,250	13,275	6.91
23	16.00	Electrical	700	350	13,350	14,400	7.50
1,385			41,559	71,252	110,700	223,511	116.41
MH	51.00	Labor burden	30%	12,468		12,468	6.49
		Subtotal direct costs:	41,559	83,720	110,700	$235,979	122.91
		General conditions (CSI 00/01)	10.00%			23,598	12.29
		Fee	10.00%			25,958	13.52
		Subtotal				$285,534	148.72
		Contingency	5.00%			14,277	7.44
		Insurance	1.00%			2,855	1.49
		Bond	1.00%			2,855	1.49
		Excise tax	0.80%			2,284	1.19
		Subtotal				$307,806	160.32
		Adjustments:				194	0.10
		Adjusted Total				**$308,000**	**$160**

Figure 23.4 Residential estimate summary

smaller residential GC often can fall back on remodel projects during these situations. The tract developer typically utilizes many more project management tools than the smaller custom home builder, including maintaining a historical estimating database.

The benefit that the larger residential contractor could realize from the detailed estimating techniques discussed earlier in the book would be the ability to prepare better in-house estimates prior to requesting subcontractor bids on a tract of homes. Early estimates would allow them to forecast a development's success better and to guide the designer toward designing homes that fit within a project's pro forma. The contractor would also be better suited to analyze subcontractor change order pricing and to prepare accurate estimates for buyers who desire custom changes to a standard tract home.

The general contractor for the Lee Street Lofts case study project is a custom builder but benefits from the services of an outside estimating consultant. The detailed estimate for that project is included on the book's eResource.

Allowances and alternates

Allowances included in any type of estimate (budget, lump sum bid, or GMP) can be to everyone's advantage. If items of work are not fully designed, such as floor covering or landscaping or window treatment, use of a reasonable allowance within an estimate as a placeholder is better than taking a guess and risk at what the scope might cost once it is fully designed. Then, after the work has been designed and competitively priced by a subcontractor and a firm cost is known and agreed, the allowance is incorporated into the contract value via change order.

Bid alternates were briefly introduced in Chapter 20. Project owners may request bid alternates to be included on the bid form. These items may be added or deducted from the base bid proposal. The purpose of requesting bid alternates may be either to know the true value of a potential added item or area of work or to allow the owner to adjust the project scope if the bids come in under or over the allotted budget. For example, a project owner may not be sure whether or not it has sufficient funding to complete the second-floor interior finishes of a building. As a contingency, the owner could list finishing the second floor as an additive alternate on the bid form. If the bids are higher than anticipated, the owner may choose not to award the alternate and leave the second floor unfinished or 'shelled'. On the other hand, if the bids are lower than anticipated, the owner may choose to award the alternate along with the base bid.

Because the contractors do not know which, if any, of the bid alternates will be accepted, they need to estimate them correctly or 'straight'. Each bid alternate needs to be a stand-alone price including all direct and subcontract costs and markups. On bid day, the contractor needs to set up the bid room to allow for analysis of the bid alternates by a separate bid day estimator. When posted to the bid spreadsheet, subcontractor quotations need to also be reviewed for the value of any alternate pricing. Failure for a GC to submit the requested alternate prices may result in a public project owner rejecting the contractor's bid as nonresponsive. Selection of acceptable bid alternates by the owner may change the order of the general contractors' bids. The initial low bidding contractor may end up as second low after the owner has chosen the acceptable alternates. Private owners may use a choice of additive and/or deductive alternates as a means of selecting the GC they prefer to work with.

Project management estimates

Estimators are not the only ones who prepare cost estimates. There are many team members – including the architect, owner, bank, superintendent, and others – who 'estimate'. Some students will tell us, "I don't want to be an estimator; I just want to be the boss; a project manager." Well project managers estimate every day in a variety of fashions. If you are involved in the construction industry at any level, you will be required to estimate. Here we present just a few estimating activities of the PM, often assisted by the jobsite project engineer and sometimes by the home office staff estimator. A few of these are also expanded on in the next chapter.

Pay requests

Construction project managers are responsible for many jobsite operations, but getting paid for the work performed is one of the most important. A PM may have all of the tools necessary to earn a profit on a job, but if the project owner does not pay for the work, the contractor will not be able to realize a profit. Some PMs do not acknowledge the importance of preparing prompt payment requests. This is especially true with many subcontractors. If a payment request is not submitted on time, the contractor will likely not be paid on time. Cash management is essential, or the GC may find that they are unable to pay suppliers, craftsmen, or subcontractors. The importance of a positive cash flow is also discussed in the next chapter. Project managers must be able to manage jobsite cash flow to be effective contributors to the operation of the construction company.

Processing pay requests is one of the most important aspects of construction financial management for the PM. There are many aspects of construction management that relate to and are involved with the pay request process, including contracts, schedule of values, retention, and lien management. Every GC project manager and subcontractor PM prepares a monthly pay request. This very important PM responsibility is an estimate of work performed or expected to be performed by the end of the month. Some members of the built environment industry may use different terms for this important topic, such as pay estimates, invoices, bills, draws, and progress payments.

The pay request often starts with the schedule of values submitted with the bid on bid day or proposal and expands on that as necessary to support the project owner's requirements. Being paid on time should be one of the top priorities for any contractor, and whatever they need to do to help that process to stay cash-positive should be done.

Change orders

The process of estimating change orders is very similar to that explained in previous chapters for bid and negotiated estimating processes. Detailed quantity take-offs, pricing for the general contractor work, and collection of usually one subcontract price in each area of work is similar. The general contractor gathers all input and prepares the change order proposal (COP). This includes subcontractor estimates as well as estimates for anticipated direct labor and material costs. This takes a form similar to a bid summary, as shown in Figure 23.5. The contract may dictate the form, and if so, the general contractor should have evaluated it prior to submitting the original bid. There are numerous forms the COP coversheet can take. Some of the important points that should be covered on any pricing or proposal form include:

- COPs are numbered sequentially;
- The COP description should be clear;
- Direct labor, hours, labor burden, and supervision costs are itemized;
- Direct material and equipment costs should be summarized;
- Subcontract costs are listed separate from direct costs;
- Markups including overhead (home office vs. field), fees, taxes, insurance, and bond are all clearly shown; and
- A line for the owner or architect to sign approval is provided.

The project manager usually only gets one opportunity for additional compensation for a differing site condition, so the estimate needs to be right the first time. Owners have an

City Construction Company
Change Order Proposal

Project: Dunn Lumber
Change Order Proposal Number: 9821.3   Date: 10/27/21

Description of work: Upgrade finishes in main lobby, stair, and elevator cab
Referenced documents: Construction Change Directive #2, Sketches 3, 4, 5, & 6

	Estimate Summary:		Subtotal	Total
1	Direct Labor (see att'd)	28 hours	1,260	
2	Supervision	15%	189	
3	Labor Burden	56%	667	
4	Safety	2%	42	
5	Total Labor:		2,158	2,158
6	Direct Materials and Equipment (see attached)		4,500	
7	Small Tools	3%	65	
8	Consumables	3%	65	
9	Total Materials and Equipment:			4,629
10	Subtotal Direct Work Items # 1: # 9			6,787
11	Subcontractors (See attached subcontractor quotes)			7,500
12	5% Overhead on Direct Work Items		339	
13	5% Fee on Direct Work Items		356	
14	5% Fee on Subcontractors		375	
15	Subtotal Overhead and Fee:			1,071
16	Subtotal All Costs:			15,358
17	Liability Insurance & Excise Taxes: (2.0%)			307
18	Total this COP			**$15,665**

This added work has an impact on the overall project schedule, full extent to be determined.
Please indicate acceptance by signing and returning one copy to our office within five days of receipt.

Approved by:

Robert Dunn   11/1/2021

Dunn Enterprises   Date

cc: Flad Architects, CCC Superintendent
Job No. 9821/File Code: COP #03

Figure 23.5 Change order proposal

understandably difficult time accepting resubmittals of the same COP because the GC left out one subcontractor's price from the tabulation or a math error occurred. The estimator or PM should do everything possible to 'sell' the COP, similar to pay requests, as mentioned earlier. Some of the attached backup to the one-page COP summary might include:

• The originating request for information or submittal;
• Portions of relevant drawings and specification sections;

- Subcontractor and supplier COPs and quotes;
- All detailed QTO and pricing recap sheets; and
- Related letters, memos, meeting notes, daily diaries, or phone records; also,
- The GC should require the same detail from its subcontractors.

The ultimate goal for the contractor with respect to COPs is to have them approved. The easiest way to achieve this goal is to play it straight with respect to pricing direct and subcontract work. Overly inflated prices will only delay the process. Quantity measurements are to a large degree verifiable, and they should not be inflated. Wage rates paid to craft employees are verifiable and should not be inflated. Subcontractor quotes should be passed through as-is without adjustment (unless incomplete) from the GC to the project owner. The subcontractors and suppliers should practice the same procedures with their second and third-tier firms.

How much is a fair markup? Markups can include a lot of items. Generally, they are percentage add-ons to the direct cost of labor and material associated with change orders. Markups on change orders could include line items such as:

- Fee on direct work,
- Fee on subcontracted work,
- Jobsite general conditions or administration costs,
- Home office overhead,
- Supervision and foreman costs,
- Labor burden, including labor taxes,
- Detailing,
- Builders' risk insurance and liability insurance,
- State sales tax and business or excise taxes,
- Small tools and consumables,
- Bonds,
- Cartage and material handling,
- Cleanup,
- Dumpsters or rubbish removal,
- Safety,
- Cost of preparing the change order proposal estimate, and
- Hoisting.

It is possible to see a series of markups that could double the cost of the direct work. Project owners and architects get frustrated with these add-ons. They do not understand why they have to pay more than the direct costs. Many of these items are required, and sometimes it is just the presentation that makes it difficult to sell. Quite often GCs (and subcontractors) are asked to propose markups with their bid proposal. This helps to minimize a lot of arguing after the first big change order is received. Some markups are listed in the bid documents. This is another reason to carefully read the special conditions of the specifications and contract prior to preparing the estimate. Some designers will try to bundle the subcontractor and GC markups together to prevent markups on top of markups.

Generally, the fee markup for commercial GCs will fall in the range of 4 to 8%. This rate is usually the same percentage fee that was used on the original estimate. Home office overhead costs are usually included in this fee. Jobsite general conditions are not allowed unless they can be substantiated on individual change issues or can be proven to extend the project schedule.

Subcontractors receive higher fees because their volume is generally less and because their risk, which is estimated by the ratio of direct labor to contract value, is higher. Subcontractors may receive 10% fee and an additional 10% overhead. Again, both of these markup calculations will depend upon how many of the items indicated in the suggested list are anticipated to be included in the base estimate, or are in the fee or overhead, or are allowed in addition to the fee.

Rarely is there a project that does not require change orders. Fair play will improve the process for all principal contracting parties including the GC, subcontractors, suppliers, design team, and owner. This can be achieved using open-book estimates with verifiable quantities and pricing systems.

Claim preparation

Claims are the result of either unresolved change order proposals or construction losses discovered after project completion. A subcontractor or GC realizes after the project is complete that they lost money and may submit a claim requesting the difference between the estimate and the actual costs. Claims can be very difficult to prepare and substantiate, and even more difficult to resolve.

Some of the types of costs contractors may attempt to recover in a claim are the difference between estimated and actual costs and losses in productivity due to multiple change orders, schedule compression, or schedule delay. One of the easiest types of costs to recover in a claim situation, if allowed by contract, is extended jobsite general conditions. If the contractor can show that it had estimated and/or was spending $270,000 per month for general conditions and the job was either delayed or put on hold for weeks or months, then a math calculation would likely result in a fair reimbursement for these costs. For example:

Estimated general conditions: $270,000 per month @ 20 months or $5,400,000
Actual project duration due to delay: 22 months
Extended general conditions: 2 months (22 less 20) @ $270,000 or $540,000

Many contractors will also attempt to recover home office overhead costs for schedule delays, asserting that they lost the opportunity for additional work during these types of delays. This is one of the most difficult items for the contractor to successfully negotiate. Many contracts specifically do not allow recovery of these costs through the use of a 'no claim for delay' clause, which is not enforceable in some states.

Monthly cost forecasts

Most GC project managers and subcontractor PMs prepare a monthly cost and fee forecast that is submitted to upper management. This very important PM document reports actual costs incurred and forecasts or estimates the remaining work to be performed.

There is rarely, if ever, a 'perfect estimate'. Many construction projects will not proceed 'exactly' according to the original plan and schedule, despite management's good intentions. Continually comparing actual costs recorded against the estimate will discover variances that warrant attention and potential adjustment by the project team. The contractor cannot wait until the project is finished to find out if it made money. At that time, there is nothing that can be done to fix the problem. Costs may exceed the estimate for a variety of reasons, including estimating errors, subcontractor performance, equipment breakdowns, labor productivity, and others. Once the reason for the cost overrun is understood, the plan or process may need to be adjusted or modified by the jobsite team, and hopefully the cause was discovered soon enough to implement an adjustment. Some of the corrective actions or 'fixes' might include change of means and methods, change of personnel, change orders to the project owner, back charges to subcontractors, and others.

The PM is responsible for developing a monthly forecast for the project that will be shared with the superintendent, officer-in-charge, and the contractor's bonding and banking stakeholders. The owner may be copied with the monthly forecast in the case of a negotiated open-book cost-plus contract. This forecast includes line items for all areas of the estimate, cost to date, and estimated cost to complete. Each of the major areas of work receives a separate forecast page, and each of those are broken down for all categories of work. The monthly forecast estimate should be as detailed as the original detailed estimate was. The forecast is essentially a complete new project estimate developed by the PM each month.

The process in developing the cost forecast is a series of mathematical calculations, and each forecast provides a more accurate projection of what the costs will be at construction completion compared to the original estimate before the project started. The forecast spreadsheet utilizes a simple series of Excel rows and columns. Development of the monthly project management cost forecast is as simple as:

Cost incurred to-date + Cost to-go = Forecast total cost
Current forecast − Original bid or current contract value = Variance +/−
Original fee +/− Variance = New projected fee

It is a good practice for the PM to include a narrative with the monthly forecast explaining the significant differences from the previous month's forecast, as well as a work plan for continuing or improving performance throughout the remainder of the project. One important rule of construction cost reporting is that the financial reports must be consistent to allow the home office a clear picture of project progress, as well as for them to develop confidence in the jobsite team. The management team cannot afford to wait until the project is complete to measure and report the overall project cost. It is not only too late to take corrective action, but it is also too late to accurately determine why the team deviated from the plan.

As-built estimates

Project managers should prepare an as-built estimate before completely demobilizing from the project site. Dividing the actual costs spent or hours incurred by the actual quantities installed creates as-built unit prices. If estimators and managers do not assemble as-built data throughout the course of construction or near the completion of the project, they never will. A lot of work went into tracking actual costs accurately. This is valuable input to the company's ongoing ability to improve its estimating accuracy. Input of as-built unit prices to the contractor's estimating database is necessary on every job if the database is to remain current.

Construction professionals are usually too busy and excited about starting the next project to develop as-built estimates. As-built estimates, like as-built drawings and as-built schedules, are important historical reporting tools. Collecting actual cost data helps to develop better future estimates, but coupling those costs with the actual quantities to develop actual unit prices is better. The units calculated are those that are likely to be used to estimate with in the future, be they $/SF of contact area, $/EA, $/Ton, $/CY, or $/SFF. Assemblies analysis would also be beneficial for future budget estimates. These figures can help with developing the contractor's database as well as providing the estimator with better data to use in the future.

Database updates

An estimating database is a collection of unit prices and productivity rates for a variety of construction materials and systems. The price or rate given is usually an average of many prices input for that specific area. As discussed in the last chapter, there are many different types of databases, which can be used for a variety of different purposes. There are several computer-generated software systems available that allow a contractor to prepare their own customized database. There are also ready-made databases, all filled out, that are commercially available.

The advantages of commercially available databases are that they provide unit prices for firms or individuals that do not have their own unit prices. These figures are of course averages and therefore not necessarily applicable for a contractor to use on a specific project. It is recommended that each estimator develop a database of historical costs. The disadvantage of a company database is that it will have included prices from different types of projects averaged together. An estimator will need to be able to throw out the highs and the lows and sort through the unit rates to select the values that best apply to the specific project being estimated.

The best database is an in-house one that is continually maintained by the construction company's chief estimator. Historical as-built estimating data are input to the database on an ongoing basis. The ability to sort out specific types of projects makes them more usable. Many contractors will have two sets of databases: One used to competitively bid work and one used for budgets, negotiated proposals, or change order estimates.

Summary

Although the case study chosen for this text was a mid-sized commercial construction project, the processes presented for budgeting, lump sum bidding, or proposing a GMP on negotiated projects can be applied to many other different types of projects. Contractors bidding on heavy civil projects customarily utilize quantities prepared by the project owner's estimator, and the GC inputs its unit prices to calculate the total bid. Residential estimators typically do not perform detailed take-offs and estimates but could benefit from many of the detailed steps followed by successful commercial estimators. Estimating bid alternates requires careful organization from the bid team on bid day. The owner may choose an alternate that may alter the order of the successful bidders, so each alternate should be estimated accurately and stand-alone.

The monthly pay request is one of the most important estimates any PM prepares. Being paid on time facilitates a positive cash flow, on which all contractors rely. Change orders should be estimated and detailed to allow the PM or estimator to successfully sell the change to the owner. The PM also prepares a revised project estimate each month that forecasts the total cost and updates the expected fee. This monthly fee forecast is an important cost management tool prepared for the contractor's front office and its financial stakeholders. Project managers should take the time to develop as-built estimates, which can be input back into the company's estimating database. Databases that are too generic or mix together too many different types of construction peculiarities may not be accurate enough to use as a tool to prepare competitive lump sum bids.

Review questions

1 Why should bid alternates be estimated straight?
2 How does a small residential general contractor estimate?
3 Why do larger residential developers not have select subcontractors they use on every project?
4 What similarities might exist between commercial and residential estimates?
5 What is the difference between a lump sum bid and a unit price bid?
6 How does an estimator determine the unit price for an individual bid item on a unit price contract?
7 List four types of markups that may be found on a change order.
8 Why should change orders have extensive backup attached to them?
9 What is the difference between a change order proposal and a claim?
10 What is an as-built estimate? Why is one prepared?
11 Who should prepare an as-built estimate?
12 How are as-built estimates related to databases?

Exercises

1 Using a cost data guide, such as *Means* or *The Guide*, prepare an assemblies estimate for a 2,000 square foot single story residence with at least 20 line items. Apply all of the appropriate markups.
2 For the residential assemblies estimate prepared in Exercise 1, prepare a list of at least five assumptions and allowances that were used to develop the estimate.
3 Without the benefit of drawings or specifications, would the estimate from Exercise 1 be considered a budget, or a guaranteed maximum price, or a lump sum bid? If your answer was budget, what changes to the estimate and documents would be necessary to transform the estimate into a GMP? What additional changes would be necessary to transform the GMP into a lump sum bid?
4 Prepare a change order for the case study project for an additive scope change valued at greater than $100,000. Include all backup necessary to sell the change.
5 Assume the case study was put on hold for three months during the middle of construction (after the roof is on but before the interior finishes are started). Prepare a claim requesting three months' worth of additional jobsite overhead and impact costs due to the work stoppage.
6 Prepare a stand-alone estimate for a bid alternate such as adding a freight elevator for our case study, from the lowest garage floor up to and including the roof.
7 Is it ethical for a private project owner to select a combination of additive and deductive alternates to move their preferred bidder to first place?
8 Why might a project owner want to see the negotiated GC's monthly cost forecast?
9 Would a project owner be allowed to see a lump sum GC's monthly cost forecast?
10 If you were presenting an open-book negotiated estimate to a repeat client, what further breakdown in costs would you make to the GMP presented in Figure 23.1?

24 Construction project management applications of estimates

Introduction

The preparation of a successful estimate for some estimators may be the end of their task. But for the construction company, the estimating step is just one of the very early phases of the construction management process, as is reflected in Figure 24.1. The goal of most project managers (PMs) is to return a clear profit to their firms for their respective projects. Estimating is one of the most important steps to obtain work and therefore is an important element for generating profits for the construction firm. A jobsite team may not be able to provide a profit if the estimator is not successful in the preparation of a competitive and complete bid or proposal. This chapter discusses the use of estimates as PM tools. The estimate document itself is an important tool for the PM and the superintendent. It is important that they participate in developing the estimate. Their individual inputs regarding constructability and their personal commitments to the estimate are essential to ensure not only the success of the estimate but also the ultimate success of the project. This has been a book focusing on the estimating process and not the PM process, but the two are connected, as this chapter shows. Project management and financial management concepts are explained more fully in several books focused just on those topics, such as *Management of Construction Projects, A Constructor's Perspective* and *Cost Accounting and Financial Management for Construction Project Managers*, and the interested reader may look to resources such as those for expanded coverage.

In this, our final chapter, we will introduce many of the PM tools and processes that are a direct byproduct of the construction cost estimate, including:

- Subcontractor buyout,
- Scheduling,
- Pay request,
- Cost loaded schedule and cash flow curves,
- Cost control, and many others.

The buyout process

Project buyout is the process of awarding the subcontracts and procuring materials being furnished by the general contractor and comparing the actual costs to the estimated costs. The purpose is to determine the status of the project with respect to the contractor's original estimate. A form such as the one illustrated in Figure 24.2 can be used to buy out the project and compare buyout costs with estimated costs.

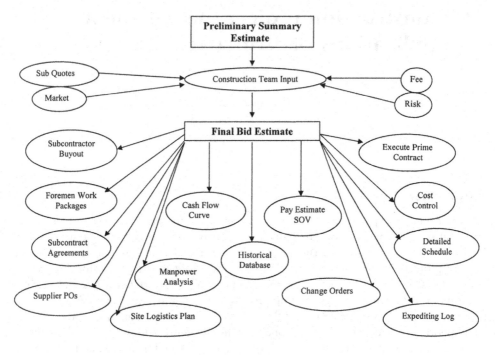

Figure 24.1 Estimate transition into project management

City Construction Company
Buyout Log

Project: Carpenter's Training Center　　　　　Project Manager: Rhonda Martin

Spec Section	Description	Estimate	Buyout	Variance	Comments
050000	Supply Steel	$ 175,000	$165,000	−$10,000	
074000	Roof and Flashing	$ 310,000	$317,500	$7,500	Low bidder dropped
096800	Carpet	$ 91,045	$40,000	−$51,045	Carpet separate from others
096520	VCT, Vinyl, & RB	w/carpet	$43,095	$43,095	
215000	Fire Protection	$ 325,000	$338,900	$13,900	Weekend & evening premium
312200	Mass Excavation	$ 584,600	$552,200	−$32,400	

Figure 24.2 Buyout log

　　Receiving numerous subcontractor quotations on bid day can be a hectic and error-prone event. Even with good estimating practices, it is still possible to make errors receiving, reviewing, posting, and selecting the proper subcontractor bids. Several decades ago, the introduction of the fax machine and, later, electronically transmitted quotes and submission of early bid qualifications has helped the general contractor (GC) minimize these errors.

Because of the volume of subcontractor quotations received at the last minute – and some are received after the bid has been forwarded to the bid runner – it is possible for the GC to receive subcontractor quotations that are lower than the ones used in the bid. If the GC is successful with its bid and receives some late quotations, the company may improve its position during the buyout process. In addition, some subcontractors may not get their quotations out to all of the bidding contractors before the bids are due to the project owner. After the successful GC has been declared, it may also benefit from bids that were not received at all on bid day, especially on private projects.

Is it ethical for the contractor to use these late quotations? Most in the private industry believe that it is acceptable. What is not ethical is bid shopping or asking subcontractors to drop their price to meet or beat another quotation. This was discussed in Chapter 21. Some public project owners require the GC to name its subcontractors with the bid, and therefore these late bids provide the contractor with minimal benefit.

Subcontractors should be selected based on the value that they contribute to the project team. Selection solely on price often leads to problems with quality and timely execution. A subcontractor proposal analysis form, or bid tab, like the one illustrated in Chapter 20 can be used to assist in subcontractor selection. The PM wants to ensure that the intended scope of work is included in the price quotation and uses an analysis form to compare all proposals for each scope of work. Prior to selecting the successful specialty contractor for each scope, the PM should conduct pre-award meetings for the major or risky subcontracted areas. Some of the firms he or she should buy out first include:

- Those with the highest values, such as mechanical and electrical;
- The contractors that will mobilize first, such as earthwork and utilities;
- The longest material delivery lead times, such as elevators, hollow metal door frames, and storefront glass;
- The direct materials that will be needed first, such as structural steel embeds and concrete reinforcement steel; and
- The suppliers with the longest shop drawing process, such as structural steel.

The project manager's goal is to employ the specialty contractor or supplier whose bid matches the estimate and does not have obvious errors in its estimate. The PM and the estimator, and potentially the project superintendent, should meet with one specialty firm at a time and discuss bid-related issues:

- Review the drawings and specifications to ensure the correct scope of work was considered in preparing the proposal.
- Review any exclusions listed on the proposal.
- Discuss any questions submitted with the proposal or raised by a same-discipline firm during the bid cycle.
- Review all the addenda.
- Discuss the size of the specialty contractor's proposed work force and anticipated schedule.

Based on the results of the pre-award meetings and annotated subcontractor proposal analysis forms, best-value subcontractors are selected for all scopes of work that will not be performed by the general contractor's workforce. Each subcontract value is then entered in the project buyout log shown in Figure 24.2.

Correcting the estimate

Is any estimate perfect? Generally, the answer is no. Subcontractor posting errors were discussed previously. Even with good estimating practices, there may still be errors within the direct work portion of the estimate as well. Before the estimate is input into the accounting system and the construction team begins to monitor and control costs against it, the estimate should be corrected. It will be much easier and more fruitful for the job-site project management team to monitor, control, and report against an estimate that has been adjusted and input as correctly as possible.

If a surplus results from the buyout and the estimate correction process, the PM can set up a cost code designated as 'yet-to-buy' to reflect the dollars that have not yet been assigned to a cost control activity code. Even long after the buyout process is complete, additional errors or missed activities may show up, to which some of this surplus can be applied.

Unfortunately, the subcontractor buyout and estimate input process may result in the identification of an estimate shortfall. This is not good news to the construction team, but it is valuable information to have early. It is not wise to reduce the value of work activities just to cover this shortfall. Trying to meet an estimate that cannot be met will discourage the construction team. The shortfall should be reported in the same 'yet-to-buy' category, but as a negative entry. It is important that the home office management team knows that there may be a problem and that it is not being hidden. Ultimately the original estimated fee may therefore be at risk.

Construction schedule

In addition to estimating, scheduling is one of the primary functions of the construction project management team, as well as quality and safety controls. The detailed construction schedule is reliant on a detailed estimate. The project superintendent should be involved with development of a construction schedule, because it is his or her 'plan' or 'means and methods' that will ultimately play an essential role in the success of a construction project. Summary schedules for proposal purposes can be produced without completion of the detailed schedule, but utilizing the detail for backup makes the summary that much more accurate. Superintendents and schedulers rely on the estimated direct work man-hours, which were discussed in Chapter 10. These hours, in combination with anticipated crew size, allows the superintendent to dial in on accurate schedule durations. Input from subcontractors is essential to achieve buy-in for the schedule. This is a collaborative versus a top-down approach to construction scheduling that most of today's successful construction superintendents will practice. A summary schedule for the commercial case study was included as a figure earlier in the book. The detailed schedule for that project, along with a summary schedule for the industrial project, are included on the book's eResource.

Schedule of values and pay request

One of the most important project management functions is to get paid for the work performed. A project manager may have all the tools necessary to earn a profit on a job, but if the owner does not pay for the work, the contractor will not be able to realize a profit. Some project managers do not acknowledge the importance of preparing prompt payment requests. This is especially true with many subcontractors. If a payment request is not submitted on time, the GC will not likely get paid on time.

The first step in developing the pay request is establishing an agreed-upon breakdown of the contract cost, or schedule of values (SOV). An example schedule of values is shown in Figure 24.3. Often the contract will require that a schedule of values be submitted for approval within a certain time after executing the contract, for example, two weeks. This schedule should be established and agreed upon early in the job, well before the first request for payment is submitted. But forcing the issue too early, before accurate subcontract figures have been established because of the buyout process, may cause excessive revisions and explanations in the SOV. If this is done, the PM may be spending the entire job reconciling the payment requests. Unfortunately, many lump sum bid projects require the SOV to be submitted along with the bid, and any subsequent adjustment or correction may not be allowed.

Development of the SOV begins with the summary estimate. That estimate is first corrected for the actual buyout values, which were shown in our Figure 24.2 example. These calculations are shown in the left three columns of Figure 24.3. This third column would be the schedule of values used on a guaranteed maximum price (GMP) negotiated contract because the general conditions and fee are listed separately. On a lump sum contract, the general conditions and fee would be distributed proportionately across all payment items. The SOV the PM would submit in the lump sum case is the far-right column of the Figure 24.3.

Some contractors try to combine scopes into single line items. In this way, they believe that they can possibly overbill, or they can hide the true subcontract values from the project owner. The SOV should be as detailed as is reasonable. The project manager should do all that is possible to assist the owner in paying completely and promptly; nothing should be hidden. At a minimum, the major Construction Specification Institute (CSI) divisions should be used as line items. Major subcontractors should be listed where possible. Separate building components, building wings, distinct site areas, and separate buildings, phases, or systems should be individually shown in a detailed SOV.

Some project managers advocate hiding the fee and general conditions or front loading them. This is more prevalent with bid contracting than with negotiated work. In the case of an open-book project, these items should be listed just as they would be in the project cost accounting system. The schedule of values should look like the final estimate. Trying to explain during an audit or a claim situation why the cost of the concrete tilt-up panels was stated as $200,000 in the pay estimate but was only $100,000 in the original estimate will be difficult. Spreading, but still hiding, the fee and general conditions as a weighted average over the schedule of values may be fair, but it will be more difficult for the owner to track lien releases.

Receipt of timely payment is one of the most important responsibilities of the project manager. The exact format for submitting payment requests will vary depending on the type of contract. A schedule of values is used to support payment applications on lump sum contracts and fee payments on cost plus contracts. The PM is responsible to develop the payment request, make sure payment is received, and subsequently see that the subcontractors and suppliers are paid. The most common billing practice is to request payment at the end of each month. The PM must gather all of the costs and prepare the monthly request that is submitted to the architect or owner for approval. This process should start on the 20th of the month. Subcontractors and major suppliers should be required to have their monthly invoices to the PM at that time. If payment has not been

City Construction Company
Schedule of Values

Project: Carpenters Training Center

Project Manager: Rhonda M.

CSI Division	Description	1 Original Estimate	2 Buyout Adjust	3 GMP SOV Adjusted Estimate	4 % of Subtotal	5 Add-ons Prorated	6 Lump Sum SOV Adjusted Totals
1	Jobsite General Conditions	see below		0			
2	Demolition	0	0	0	0	0	0
3	Concrete	515,058	0	515,058	9.05%	131,512	646,570
4	Masonry	27,300	0	27,300	0.48%	6,971	34,271
5	Structural & Misc. Metals	418,456	−10,000	408,456	7.35%	106,846	515,302
6	Wood & Plastic	178,351	0	178,351	3.13%	45,539	223,890
7	Thermal & Moisture Protection	432,200	7,500	439,700	7.59%	110,356	550,056
8	Doors, Windows, Glass	343,600	0	343,600	6.04%	87,733	431,333
9	Finishes	816,170	−7,950	808,220	14.34%	208,397	1,016,617
10	Specialties	26,755	0	26,755	0.47%	6,831	33,586
11	Equipment: Bridge Crane	175,600	0	175,600	3.09%	44,837	220,437
12	Furnishings	6,325	0	6,325	0.11%	1,615	7,940
13	Equipment	0	0	0	0.00%	0	0
14	Conveying Systems (Elevator)	85,000	0	85,000	1.49%	21,703	106,703
15	Mechanical Systems:	0	0	0	0.00%	0	0
21	Fire Protection	325,000	13,900	338,900	5.71%	82,984	421,884
22	Plumbing	166,200	0	166,200	2.92%	42,437	208,637
23	HVAC	455,200	0	455,200	8.00%	116,228	571,428
26	Electrical Systems	862,400	0	862,400	15.15%	220,201	1,082,601
31	Sitework	857,400	−32,400	825,000	15.07%	218,924	1,043,924
	Subtotal w/o GC's and Fee:	5,691,015	−28,950	5,662,065	100.00%	$ 1,453,115	$7,144,130
	General Conditions	435,100	0	435,100			
	Labor burden	333,337	0	333,337			
	Insurance and taxes	146,308	−724	145,584			
	Contingency	198,173	0	198,173			
	Fee	340,197	29,674	369,871			Buyout improves fee position!
	Subtotal Add-ons	1,453,115	28,950	1,482,065			
	TOTAL GMP:	$7,144,130	0	$7,144,130			

Figure 24.3 Schedule of values

received on time, the project manager should contact the owner to determine the cause. The financial relationship with the owner and the project is the PM's responsibility. The same scenario holds true with respect to subcontractors and suppliers. The GC's PM ensures that they are paid promptly. Owners may withhold a portion of each payment to ensure timely completion of the project. This is known as retention. The retention rate is specified in the contract. Liens can be placed on a project if subcontractors or suppliers are not paid for their labor or materials. To preclude liens, owners require lien releases with payment applications.

Cost-loaded schedules and cash flow curves

Cash management is essential, or the general contractor may find that they are unable to pay suppliers, craftsmen, or subcontractors. Good cash management skills, just like good communications skills, are essential if one is to be an effective project manager. The cash flow curve and the cost-loaded schedule are direct by-products of a detailed estimate combined with a detailed construction schedule. There are several reasons for cost loading the schedule and producing a cash flow curve. Often, it is one of the first things the project owner will ask of the PM and may be specifically required by the construction contract. The cash flow curve is a graphical estimate of when the contractor expects to have work in place and the estimated cost of that work. The curve provides a forecast to the owner and the bank of anticipated monthly pay requests to be made to the contractor for the completed work.

The first step in development of the cash flow curve is the cost-loaded schedule, which is very easy to prepare. The estimated costs (as corrected, bought out, and input to the accounting system) are applied across the schedule activities. If an activity will take five months and its value is $100,000, $20,000 is spread over each of the five months. Costs such as insurance, taxes, and fees should be distributed proportionately over the entire project. Jobsite general conditions can be factored in a variety of ways, but we will leave that to its own chapter in another PM or financial management book. The cost data are summed at the bottom of the schedule for each month to develop anticipated monthly expenses. The likelihood of the GC being billed by each material supplier and subcontractor according to any anticipated schedule is somewhat remote. The result of spreading the estimate across the construction schedule is a document known as a cost-loaded schedule. This is often an Excel spreadsheet or may be produced by scheduling software programs such as Microsoft Project or Primavera P6 by Oracle. Plotting the monthly or cumulative cash flow curve is then a simple next step from the cost-loaded schedule.

Cost control

Cost control begins with assigning cost codes to the elements of work identified in the work breakdown phase of developing the original cost estimate. Work breakdown was discussed in Chapter 1. These cost codes allow the PM to monitor project costs and compare them to the estimated costs. Cost codes usually are a numerical assignment to work activities that allow accurate tracking of estimated and scheduled costs and times to which activities can be recorded. These numbers generally are consistent throughout a construction firm. The objective is not to keep the cost of each element of work under its estimated value but to ensure that the total cost of the completed project is under the estimated cost. The cost control process involves the following series of steps:

- Cost codes are assigned to each element of work in the cost estimate.
- The cost estimated is corrected based upon buyout costs.
- Actual costs are tracked for each work item using the assigned cost codes.
- The construction process is adjusted, if necessary, to mitigate cost overruns.
- Actual quantities, costs, and productivity rates are recorded, and an as-built estimate is prepared and input back into the company's estimating database for use on the next estimate.

While all costs should be monitored, the items that generally involve the greatest risk are:

- Direct labor,
- Equipment cost or rental, and
- Jobsite general conditions or administration.

It is possible to lose money on material purchases, but with good estimating skills, it is not probable, and the risk is not as great as with direct labor. The same holds true with subcontractors. They have quoted prices for specific scopes of work, and the subcontractors therefore bear the risk associated with their labor and equipment.

It is difficult to control costs if the project manager does not start with a detailed estimate. For example, let us assume there was a $10,594 labor cost overrun on the case study's 56,160 square foot slab-on-grade. The assemblies cost analysis shown in Figure 24.4A does not provide enough detail to identify the cause. Project managers should use a more detailed cost breakdown, as shown in Figure 24.4B, to determine the cause of the cost overrun.

Now it is easy to see that the problem is not with the laborers placing the slab; the majority of the overrun is with rebar placement and concrete finishing. Why did this happen? Maybe the rebar had fabrication errors. Maybe the slab got rained on. Maybe personnel changes were necessary. Maybe the estimate was too low. There could be a variety of reasons. The point is that the PM and the superintendent can now focus on evaluating these specific issues.

To be able to control costs, they must be tracked and compared against the corrected estimate. The first step is to record the actual costs incurred and input the information into a cost control database. Cost codes are used to allow comparison of actual cost data with the estimated values. There are several types of cost codes used in the industry. The best system to use on most projects is the coding system selected for the detailed estimate. This will also be the same coding system utilized for the as-built estimate and input into the company historical estimating database. Many construction firms have adopted a version of the CSI MasterFormat system for their internal cost codes.

Depending upon the size of the construction firm, the type of work, and the type of owner and contract agreement, the construction company may perform job cost accounting in the home office or in the field. Generally, the smaller the firm and the smaller the contract value, the more likely all accounting functions will be performed in the home office. On larger projects, the project may have a jobsite accountant or cost engineer. The type of contract and how it addresses reimbursable costs may affect where the construction firm performs the accounting function. For example, assume a project is a $15 million mid-rise speculative office building that has a negotiated GMP contract that allows

A: Assemblies Cost Control Analysis: Labor Only

Project: Dunn Lumber

Cost Code	Description	Quantity	Units	Estimated Labor U.P.	Estimated Cost	Actual Unit Price	Actual Cost	Variance
03.04.00	Slab-on-grade (SOG)	56,160	SF	$1.19	$66,907	$1.38	$77,501	$10,594 Over

B: Detailed Cost Control Analysis: Labor Only

Cost Code	Description	Quantity	Units	Estimated Labor U.P.	Estimated Cost	Actual Unit Price	Actual Cost	Variance
03.04.01	Edge forms	960	LF	$2.46	$2,362	$0.31	$3,540	$1,178
03.04.02	Blockouts	78	EA	$41	$3,198	$7.20	$2,500	−$698
03.04.03	Const & Expansion Joints	3,000	LF	$2.46	$7,380	$2.84	$8,525	$1,145
03.04.04	**Rebar**	25.5	Ton	$516	$13,158	$8.74	$18,200	**$5,042**
03.04.05	Place concrete	721	CY	$14	$10,094	$11.82	$8,525	−$1,569
03.04.06	**Finish SOG**	56,160	SF	$0.34	$19,094	$0.43	$24,149	**$5,054**
03.04.07	Miscellaneous SOG	721	CY	$16.12	$11,621	$16.73	$12,062	$441
	Total System:	56,160	sf	$1.19	$66,907	$1.38	$77,501	$10,594 Over

Figures 24.4A and B Cost control analysis

for all on-site accounting to be reimbursed. It may be more cost effective to perform accounting out of the home office with the assistance of the accounting department, but according to the terms of the contract, the owner will not pay for it. The interested reader may look to a construction cost accounting and financial management book for an in-depth discussion of the financial operations that happen at the jobsite level.

Regardless of where the cost data are collected and where payment is made, most accounting functions on a project are the same. The process begins with a corrected estimate. Actual costs are then incurred, either in the form of direct labor, material purchase, or subcontractor invoice. Cost codes (those matching the estimate) are recorded on the time sheets and invoices and routed for approval. After the invoices are approved, the cost data are input into the cost control system.

One important aspect of this phase of cost recording is the accurate coding of actual costs. If costs are intentionally or accidentally coded incorrectly, the project team will not know how they are doing on that specific item of work. All costs should be recorded correctly to provide the PM an accurate accounting of all expenditures. Accurate recording will not only help with cost control, but it will also provide a more realistic as-built estimate, which will be used to update the company's estimating database and help prepare better estimates in the future.

Control of direct labor and equipment rental costs is the responsibility of the superintendent. The key to getting the field supervisors involved in cost control is to get their personal commitment to the process. One successful way for the PM to do this is to have the superintendent actively involved in developing the original estimate. If the superintendent said it would take four people working two weeks to form a concrete foundation wall, he or she will often see to it that the task is completed within that time.

Work packages are a method of breaking down the estimate into distinct packages or systems that match measurable work activities. For example, spread footings, including forming, reinforcing steel, and concrete placement, could be a work package. The work is planned according to the number of hours in the estimate and monitored for feedback. When a system is complete, such as footings or slab-on-grade, the project manager and the superintendent immediately know how they are doing with respect to the overall estimate. To be able to do this, the field supervision needs to have been provided the estimated costs, both in materials and man-hours, for each work package. The advantages of estimating using unit man-hours over unit prices for labor were previously discussed in Chapter 10. Field supervisors think in terms of crew size and duration. They do not think in terms of $1.19 per square foot. Estimating in this manner also makes for an easy transition into project management.

Which items should the project team track? The 80–20 rule applies here, as it does throughout our discussion on construction cost estimating. Twenty percent of the direct work activities account for 80% of the man-hours and therefore present 80% of the risk for the estimator and the contractor. The PM should prioritize those activities with the greatest risk. The estimate should be reviewed to identify those items that have the most labor hours or, in the case of equipment rental, the most cost. Work packages should be prepared for those items that the project manager and the superintendent believe are worthy of tracking and monitoring. The foreman who is responsible for accomplishing the work should develop and take ownership of each work package. Foremen are the *last planners* and in the best position to achieve cost objectives.

Additional project management applications

There are many different aspects of PM that rely on detailed and accurate construction cost estimates. This is not a book on PM, as stated earlier, and we have limited space to guide the reader through this natural transition. The previous sections cover the primary PM tools of the project estimate, and here we list a few other and more current PM estimate and cost related topics.

Activity-based costing is the process of applying home office and jobsite indirect costs to departments and projects and direct construction activities if possible. Activity-based costing focuses on identifying what cost drivers have impact on cost objects and the degree to which they add value to the project.

Lean construction techniques focus on processes to improve construction costs and eliminate waste by incorporating efficient methods during design and construction. Whereas activity-based costing identified what the cost drivers are, lean focuses on how those costs can be reduced or managed efficiently. There are many aspects of lean, including just-in-time deliveries, target value design, value engineering, pull planning, last planner, supply chain material management, and others.

Time value of money is an economic study comparing past, present, and future values of money using a series of formulas and spreadsheets and relying on several factors such as interest rates, inflation rates, investment years, and other variables. This is relevant to estimating and project management, as monetary factors such as inflation, escalation, and interest factor into the preparation of a cost estimate – knowing when the work will be performed and management of pay requests, change orders, and financial close-out – or when the contractor expects to receive payment for work performed.

Summary

Developing a good estimate is the most important activity for the estimating team. Without securing a contract, the project manager will not have a project to manage. The ultimate success of the estimating efforts, though, will be measured during the construction of the project. If the estimate was prepared properly, the project management activities such as schedule and cost control are much easier. There is not any such thing as an error-free estimate or schedule. One of the first responsibilities of the project construction team is to buy out the project and to correct the estimate so that it is a workable tool during construction. Getting paid on time is an important PM function, and development of an open-book detailed SOV with the corrected estimate is a good means of facilitating timely payments. This same concept holds true with change order development and negotiation.

The cash flow curve is developed from the estimate and the schedule, which is a tool used by the project owner, who may in turn need to provide a pay request forecast to the bank. The corrected estimate and schedule will result in a more accurate cash flow curve. The cost control management aspect of the construction project takes the corrected and bought-out estimate and monitors it against actual costs incurred and adjusting as required. The foreman's work package is an important tool in managing field labor costs.

Review questions

1   What is the difference between bid shopping and buyout?
2   Why might a project manager want to buy fast?
3   Why is it to everyone's benefit to postpone establishing the pay request schedule of values until after buyout is complete?
4   Why might the bid estimate not be correct?
5   Why should both the project manager and superintendent be involved with creating the original bid estimate and schedule?
6   The cost codes on a project should be the same as what other types of codes? There are several.
7   Who prepares a cost control work package?
8   Do craftspeople care if a project or an activity is completed within budget and on time?
9   How does the 80–20 rule relate to cost control?
10 What is the riskiest and most difficult part of the estimate to control?

Exercises

1   Assume the following buyout values for the case study project. Begin with the final bid estimate developed in Chapter 20 and develop a buyout log similar to Figure 24.2. You will have to look at examples presented earlier in the book and make a few derivations and assumptions to complete this exercise.

• Buy reinforcement steel:	$1,150 per ton
• Install structural steel and metal deck:	$2.7 million
• Landscaping and irrigation:	$790,000
• Unit price for supplying concrete:	Average $110 per cubic yard
• Concrete finishing subcontractor:	U.P. of 45 cents per SF of slab

- Package plumbing and HVAC: $3,960,000
- Package roofing, traffic control, and waterproofing: $1,625,000
- Storefront, excluding purchase of punch windows: $3.2 million
- Elevators, including cab upgrades: $1.2 million

2   Using the buyout values in Exercise 1, develop a new schedule of values for our case study project that was shown in Chapter 20.

3   Revise the lump sum schedule of values (as shown in the right column of Figure 24.3) by combining items to create a 10 line-item schedule of values. This is exclusive of Exercise 1.

4   Revise the lump sum schedule of values (as shown in the far-right column of Figure 24.3) by front-end loading all the fee and general conditions and other markups to the areas of self-performed general contractor work. This is exclusive of Exercise 1.

5   Utilizing the buyout log developed in Exercise 1, if the project was to finish with exactly those buyout values equaling as-built costs, and all other estimated costs were to equal actual costs (which is unlikely), prepare a fee forecast for the project. What would the final fee for City Construction Company be? Would the owner realize any project savings? How would those savings be reflected in the final contract amount?

6   How would your answers differ to Exercise 5 if this were an open-book negotiated project with a 75–25% savings split favoring the project owner?

7   If you were presenting an open-book pay request to a repeat client, what additional detail/line items would you make to the GMP SOV Figure 24.3, Column 3?

8   Assume you are the project manager for the Dunn case study project. Utilize the direct work example and subcontract options presented earlier in this book and make an argument why you would perform the (A) elevated CIP deck formwork with your direct work forces and/or why you would choose to subcontract that work, or (B) install the doors, frames, and hardware, either self-performed or subcontracted.

Appendix
Estimating resources

A logical step from the initial drawing review is to work breakdown schedule (WBS) and quantity take-off (QTO) and cost estimating. An understanding of the language and abbreviations associated with contract documents and estimating is critical. All construction estimators must be able to successfully speak this language.

- Table A.1: Estimating symbols and math conversions
- Table A.2: Inch to foot conversions
- Table A.3: Waste and lap estimating allowances

Table A.1 Standard conversions

- Area = Length × Width
- Volume = L × W × Height (or depth)
- 12 inches (12 in or 12″) = 1 Foot (1FT or 1′) or 1.0 Feet
- 1′ × 1′ = 1 Square Foot (SF)
- 12″ × 12″ = 144 Square Inches = 1 SF
- 1′ × 1′ × 1′ = 1 Cubic Foot (1 CF)
- 10′ × 10′ = 100 SF = 1 CSF = 1 Square (SQ), common with roofing
- 3′ = 1 Yard (YD) = 36″
- 3′ × 3′ = 9 SF = 1 Square Yard (1 SY), common with floor covering or asphalt
- 1 YD × 1 YD = 1 SY
- 3′ × 3′ × 3′ = 27 CF = 1 Cubic Yard (1 CY), common with concrete and earthwork
- 1 YD × 1 YD × 1 YD = 1 CY
- 1″ × 12″ × 12″ = 1 Board Foot (1 BF), common with framing lumber
- 1000 BF = 1 MBF (thousand board feet)
- 4′ × 8′ = 32 SF, standard size of one sheet of plywood
- W30 × 80 is a structural steel wide flange beam or column with a depth of 30″ and weighs 80#/Lineal Feet (LF), formerly WF or 'I' or 'H' beams
- 6″ × 6″ × ½″ Angle Iron or <, or "L" weighs 19.6 #/LF
- HSS: Hollow Steel Structure, formerly Tube Steel (TS), [], such as 4 × 4 × 1/4
- 43,560 SF = 1 Acre (AC)
- 2,000 Pounds (#) = 1 Ton (TN)
- 5,280′ = 1 Mile

Table A.2 Inch to foot conversions

Inches	Feet	Inches	Feet
1	0.08	7	0.58
2	0.16	8	0.67
3	0.25	9	0.75
4	0.33	10	0.83
5	0.41	11	0.91
6	0.50	12	1.0

For example if we measured 55'-2" this would converted to 55.16'
Or 2'-10" would be converted to 2.83'

Table A.3 Miscellaneous waste and lap allowances

Material	Add*
Concrete	3 to 5 to 7% for waste
Wire mesh	10% for lap
Rebar	10% for lap for rough estimates, see specifications for exact lap
Structural steel	15% for gussets and plates
Framing lumber	10% for waste
Plywood	5 to 10% for waste
Building papers and vapor barriers	10% for lap
Finishes such as carpet and tile	10% for waste
Gravel	20% to convert from bank yards (BY) to truck yards (TY)
Earth backfill	30 to 50% to convert from BY to TY

* Note: Waste and lap are dependent upon selected materials and contractor methods

References

The construction estimator or construction management enthusiast may want to refer to some of these other resources. We referenced many of them in the book and utilized others for our research.

Allen, E. and Lano, J. (2019). *Fundamentals of Building Construction: Materials and Methods* (7th ed.). Hoboken, NJ: John Wiley and Sons, Inc.

Assemble (https://construction.autodesk.com/products/assemble).

Bluebeam Revu (www.bluebeam.com/).

Building Connected (www.buildingconnected.com/subcontractor-construction-network/).

Destini Estimator (https://beck-technology.com/destini-estimator/).

eTakeoff Dimension (https://etakeoff.com/products/dimension-overview/).

Guide Building Construction Material Prices (www.bestconstructionsite.com).

Holm, L. (2019). *101 Case Studies in Construction Management*. New York: Routledge.

Holm, L. (2020). *Construction Contract Documents, Including Plan Reading Essentials and Extensive Lists of Abbreviations and Construction Glossary Terms*. Washington, DC: Amazon.

Holm, L., Schaufelberger, J., Griffin, D. and Cole, T. (2018). *Construction Cost Estimating: Process and Practices* (2nd ed.). Buford, GA: LAD Custom Publishing, Inc. Predecessor for this book.

Holm, L. and Schaufelberger, J. (2020). *Construction Superintendents, Essential Skills for the Next Generation*. New York: Routledge.

Holm, L. (2019). *Cost Accounting and Financial Management for Construction Project Managers*. New York: Routledge.

Migliaccio, G. and Holm, L. (2018). *Introduction to Construction Project Engineering*. New York: Routledge.

Mirsky, R. and Schaufelberger, J. (2014). *Professional Ethics for the Construction Industry*. New York: Routledge.

On Screen Takeoff (www.oncenter.com/products/on-screen-takeoff).

Peterson, S. and Dagostino, F. (2019). *Estimating in Building Construction* (9th ed.). New York: Pearson.

RS Means Construction Cost Data. Rockland, MA: Gordian, Inc. Published annually.

RS Means Data Online (www.rsmeans.com/products/online.aspx).

Sage Estimating (www.sage.com/en-us/products/sage-estimating/).

Schaufelberger, J. and Migliaccio, G. (2019). *Construction Equipment Management* (2nd ed.). New York: Routledge.

Schaufelberger, J. and Holm, L. (2017). *Management of Construction Projects, A Constructor's Perspective* (2nd ed.). New York: Routledge.

Schaufelberger, J. and Lin, K. (2014). *Construction Project Safety*. Hoboken, NJ: John Wiley and Sons, Inc.

Steel Construction Manual (www.aisc.org/Steel-Construction-Manual).

Ticola, V., editor-in-chief. (2017). *Walker's Building Estimator's Reference Book* (31st ed.). Lombard, IL: Frank R. Walker Company.

WinEst (https://gc.trimble.com/product/winest).

Glossary

Many of the terms used in the book are expanded on here. A few additional industry-standard construction management terms have been included as a valuable tool for the construction estimator.

Activity-based costing process of applying home office and also jobsite indirect costs to departments and projects and direct construction activities if possible

Addenda additions to or changes in bid documents issued prior to bid and contract award

Additive alternates alternates that add to the base bid if selected by the owner

Agreement a document that sets forth the provisions, responsibilities, and obligations of parties to a contract. Standard forms of agreement are available from professional organizations; also known as contract agreement

Allowance an amount stated in the contract for inclusion in the contract sum to cover the cost of prescribed items, the full description of which is not known at the time of bidding. The actual cost of such items is determined by the contractor at the time of product selection by the architect or owner, and the total contract amount is adjusted accordingly

Alternates selected items of work for which bidders are required to provide prices outside of the base bid amount

Amendments see addenda

American Institute of Architects a national association that promotes the practice of architecture and publishes many standard contract forms used in the construction industry

Application for payment see progress payment request

As-built drawings contractor-corrected construction drawings depicting actual dimensions, elevations, and conditions of in-place constructed work

As-built estimate assessment in which actual costs incurred are applied to the quantities installed to develop actual unit prices and productivity rates

Assemblies analysis determining the cost estimated per unit of work or assembly, such as dollars per cubic yard of concrete, and comparing the result with similar data from other projects

Assemblies cost estimate a semi-detailed estimate that requires quantity take-off of bulk in-place systems, such as foundations, and applies a unit price for all work associated with that system, including labor, material, and subcontractor

Associated General Contractors of America a national trade association primarily made up of construction firms and construction industry professionals

Bank cubic yard   a measure of the volume of soil in its natural, undisturbed condition

Base hourly wage   that portion of a worker's gross hourly pay that does not represent deductions for fringe benefits or payroll taxes

Best and final proposal   the final proposal submitted by a contractor in a negotiated procurement process after the owner has discussed the previous proposals received with each contractor submitting one and has issued any clarifications or changes to the request for proposal

Bid bond   a surety instrument that guarantees that the contractor, if awarded the contract, will enter into a binding contract for the price bid and provide all required bonds

Bid documents   drawings, specifications, and contract terms issued to general contractors to be used for developing a bid for the defined work. Bid documents are not always the same as the final construction documents

Bid form   the form issued by a project owner on which contractors submit their bids

Bid procurement process   selection of a contractor based upon a lump sum bid

Bid room   location where a general contractor's bid team processes subcontractor and supplier quotations and determines the final bid price

Bid security   money placed in escrow, a cashier's check, or a bid bond offered as assurance to an owner that the bid is valid and that the bidder will enter into a contract for that price

Bid shopping   unethical contractor activity of sharing subcontractor or supplier bid values with competitors in order to drive down prices

Bid summary   the estimating form onto which all previously priced work and subcontractor bid totals are entered and markups and the final price calculated

Bid tab   box on subcontractor posting spreadsheet for a specification section or system and also location of vendor bids for a single category of work

Bin   physical location where posted subcontractor and supplier bid forms are stored on bid day

Boilerplate   standard contract language that owners include in most of their contracts

Bridging delivery method   project delivery method where the owner or architect engages designers to create a conceptual design and design-build subcontractors finish the design conforming to that criteria; also known as design-assist delivery method

Budget control log   spreadsheet used to monitor changes of materials or scope throughout the design process and the budget implications of those changes

Budget cost estimate   preliminary estimate based on early design documents

Builder's risk insurance   protects the contractor and owner in the event that the project is damaged or destroyed while under construction; similar to the owner's property insurance

Building information models or modeling   computer design software involving multi-discipline three-dimensional overlays improving constructability and reducing change orders

Burden   see labor burden

Buyout   the process of awarding subcontracts and issuing purchase orders for materials and equipment

Buyout log   a project management document that is used for planning and tracking the buyout process

Cash flow curve   a plot of the estimated value of work to be completed each month during the construction of a project

Change order   modifications to contract documents made after contract award that incorporate changes in scope and adjustments in contract price and time

Change order proposal   a request for a change order submitted to the owner by the contractor or a proposed change sent to the contractor by the owner requesting pricing data

Chief estimator   the head of the estimating department in a construction firm

Claim   an unresolved request for a change order

Close-out   the process of completing all construction and paperwork required to complete the project and close out the contract

Conceptual budget estimate   cost estimates developed using incomplete project documentation; may also be a programming budget estimate

Concrete reinforcement steel   round steel bars or rods that handle tension loads imposed on concrete elements

Construction documents   detailed drawings and specifications corresponding with lump sum bidding

Constructability analysis   an evaluation of preferred and alternative materials and construction methods

Construction documents   the agreement, general conditions, special conditions, drawings, and specifications

Construction joint   the interface or meeting surface between two successive placements of concrete

Construction management agency delivery method   a delivery method in which the client has three contracts: one with the architect, one with the general contractor, and one with the construction manager. The construction manager acts as the client's agent but has no contractual authority over the architect or the general contractor

Construction management at-risk delivery method   delivery method in which the owner engages a construction management company early during design, and when the design is complete, that contractor guarantees the final cost of the project – subject to approved change orders; also known as construction manager/general contractor delivery method

Construction Specifications Institute   the professional organization that developed the 16-division MasterFormat that is used to organize the technical specifications, now 50 divisions

Contingency   amount applied to an estimate to cover unknown issues and incomplete bid documents

Contract   a legally enforceable agreement between two parties; also, an agreement

Control joint   joint cut into a concrete slab to control where cracking occurs

Corrected estimate   estimate that is adjusted based on buyout costs

Cost codes   codes established in the firm's accounting system that are used for recording specific types of costs

Cost estimating   process of preparing the best educated anticipated cost of a project given the parameters available

Cost indices   numerical values that reflect the variation in price levels at different geographic locations, economies of scale, and/or different time frames

Cost-loaded schedule   a schedule in which the value of each activity is distributed across the activity, and monthly costs are summed to produce a cash flow curve

Cost-plus contract   a contract in which the contractor is reimbursed for stipulated direct and indirect costs associated with the construction of a project and is paid a fee to cover profit and company overhead

Cost-plus contract with guaranteed maximum price a cost-plus contract in which the contractor agrees to bear any construction costs that exceed the guaranteed maximum price unless the project scope of work is increased

Cost-plus-fixed-fee contract a cost-plus contract in which the contractor is guaranteed a fixed fee irrespective of the actual construction costs

Cost-plus-percentage-fee contract a cost-plus contract in which the contractor's fee is a percentage of the actual construction costs

Cost-reimbursable contract a contract in which the contractor is reimbursed stipulated direct and indirect costs associated with the construction of a project. The contractor may or may not receive an additional fee to cover profit and company overhead

Craftsmen or craftspeople nonmanagerial field labor force who construct the work, such as carpenters and electricians, also crafts or craftspeople or tradesmen

Davis-Bacon wage rates prevailing wage rates determined by the U.S. Department of Labor that must be met or exceeded by contractors and subcontractors on federally funded construction projects. Other public agencies may also require prevailing wages

Deductive alternates alternates that subtract from the base bid if selected by the owner

Design-build delivery method or contract delivery method where the project owner has only one contract with a designer-builder that is responsible for the complete design and construction of a project

Design development documents the plans and specifications when they are about 75–80% complete

Detailed cost estimate extensive estimate based on definitive design documents. Includes separate labor, material, equipment, and subcontractor quantities. Unit prices are applied to material quantity take-offs for every item of work

Differing site condition some condition of the project site that is materially more adverse than as depicted in the contract documentation and could not be seen by a visit to the site, for example, encountering a buried water line where none was shown in the contract drawings

Digitizer computer tool used to measure quantities from two-dimensional drawings

Direct construction costs labor, material, equipment, and subcontractor costs for the contractor, exclusive of any markups

Eighty-twenty rule on most projects, about 80% of the costs are included in 20% of the work items; also 80-20 rule

Electronic mail internet tool for sending communications and attached documents; also e-mail

Escalation percentage added to an estimate to account for anticipated cost increases over time

Estimate schedule management document used to plan and forecast the activities and durations associated with preparing the cost estimate; not a construction schedule

Estimating database collection of historical labor and material unit prices, which can be applied to quantities in future estimates

Estimator a person charged with preparing a cost estimate

Fast track schedule a schedule that shows construction of a project starting before the entire design is completed. Construction of the foundation would be started as soon as the foundation design is complete, even though the remainder of the building design is not finished

Fee   contractor's income after direct project and job site general conditions are subtracted; generally includes home office overhead costs and profit

Footprint   the ground upon which a building is situated

Foreman   direct supervisor of craft labor on a project

Free on board   an item whose quoted price includes delivery at the point specified. Any additional shipping costs are to be paid by the purchaser of the item; also known as freight on board

Fringe benefits   that portion of the gross hourly pay that is deducted for payment of benefits such as retirement, health insurance, and life insurance

General conditions   a part of the construction contract that contains a set of operating procedures that the owner typically uses on all projects. They describe the relationship between the owner and the contractor, the authority of the owner's representatives or agents, and the terms of the contract. The term also is used to describe jobsite overhead costs

General contractor   the principal party to a construction contract who agrees to construct the project in accordance with the contract documents

General liability insurance   protects the contractor against claims from a third party for bodily injury or property damage

Geotechnical report   a report prepared by a geotechnical engineering firm that includes the results of soil borings or test pits and recommends systems and procedures for foundations, roads, and excavation work; also known as a soils report

Guaranteed maximum price contract   a type of cost-plus contract in which the contractor agrees to construct the project at or below a specified cost

Home office overhead   contractor's operating costs that are not related to specific projects

Indirect construction costs   expenses indirectly incurred and not directly related to construction activity, such as project or home office overhead

Indirect equipment   construction equipment that is used for multiple purposes on a project, such as a tower crane, and cannot be charged to a single construction activity

Integrated project delivery method   fairly new contracting method where the project owner, architect, and general contractor all sign the same contract agreement and share risks equally in financial, safety, schedule, and quality performance

Invitation for bid   a portion of the bidding documents soliciting bids for a project; also instructions for bidders, also invitation to bid

Jobsite general conditions costs   field indirect costs that cannot be tied to an item of work but that are project specific and, in the case of cost reimbursable contracts, are considered part of the cost of the work

Jobsite overhead   see jobsite general conditions cost

Just-in-time material delivery   a material management approach in which supplies are delivered to the jobsite just in time to support construction activities. This minimizes the amount of space needed for on-site storage of materials; an element of lean construction

Labor and material payment bond   a surety instrument that guarantees that the contractor (or subcontractor) will make payments to his or her craftspeople, subcontractors, and suppliers

Labor burden   markup of labor wage rates to account for worker benefits and labor taxes

Labor taxes   markup of labor wage rates to account for payroll taxes required by the government such as workers' compensation insurance, unemployment insurance or tax, Social Security, and Medicare; see also labor burden

Laydown areas   areas of the construction site that have been designated for storage of construction materials that are being stored until they are used in the construction of the project

Lean construction techniques   process to improve costs and eliminate waste incorporating efficient methods during both design and construction; includes just-in-time deliveries, target value design, value engineering, pull-planning, and other elements

Letter of intent   written authorization from a project owner to a general contractor or general contractor to a subcontractor directing them to proceed with some facet of construction, often preconstruction services or material ordering, after which a construction contract will follow

Liquidated damages   an amount specified in the contract that is owed by the contractor to the owner as compensation for damaged incurred as a result of the contractor's failure to complete the project by the date specified in the contract

Loose cubic yards   a measure of the volume of soil after it is excavated and expands or swells; also known as truck cubic yards

Lump sum contract   a contract that provides a specific price for a defined scope of work; also known as fixed-price or stipulated-sum contract

Major supplier   vendor who supplies fabricated material or a large amount of material for a construction project

Markup   percentage added to the direct cost of the work to cover such items as overhead, fee, taxes, and insurance

MasterFormat   a numerical system of organization developed by the Construction Specifications Institute that is used to organize contract specifications and cost estimates; formerly 16 divisions and now 50 divisions

Material supplier   vendor that provides materials but no on-site craft labor

Mixed use development   real estate development or project that involves three or more uses or functions such as office, retail, hotel, apartments, and/or entertainment, among others

Negotiated procurement process   selection of a contractor based on a set of criteria the owner selects

Nonreimbursable costs   contractor costs that are not reimbursed by the project owner under the terms of a cost-plus contract

Notice to proceed   written communication issued by the owner to the contractor, authorizing the contractor to proceed with the project and establishing the date for project commencement

Officer-in-charge   general contractor's principal individual who supervises the project manager and is responsible for overall contract compliance

Open shop   a nonunion firm, or one that has not signed a contract with a labor union, also known as merit shop

Order-of-magnitude estimates   general contractor's cost estimates for subcontracted scopes of work

Overhead   expenses incurred that do not directly relate to a specific project, for example, rent on the contractor's home office

Payment bond   see labor and material payment bond

Payment request   see progress payment request

Payroll taxes   that portion of the gross hourly pay that is used to pay federal and state unemployment insurance or taxes, Social Security, Medicare, and workers' compensation insurance; also known as labor taxes

Performance bond   a surety instrument that guarantees that the contractor will complete the project in accordance with the contract; it protects the owner from the general contractor's default and the general contractor from the subcontractor's default

Plan center or room   location where bid documents are available for review by both general contractors and subcontractors

Plug estimates   see order-of-magnitude estimates

Post-tension concrete   concrete that utilizes post-tension cables in lieu of, or in addition to, rebar; the PT cables function similar to rebar and handle tension

Pre-bid conference   meeting of bidding contractors with the project owner and architect; the purpose of the meeting is to explain the project and bid process and solicit questions regarding the design or contract requirements

Pre-cast concrete   concrete cast or fabricated, typically in an offsite yard or warehouse, that is then delivered to the jobsite as needed and erected with methods similar to erecting structural steel

Preconstruction agreement   a short contract that describes the contractor's responsibilities and compensation for preconstruction services

Preconstruction conference   meeting conducted by owner or designer to introduce project participants and to discuss project issues and management procedures

Preconstruction cost or fee   the amount of money charged by a general contractor or preconstruction agent to perform services such as budgeting and scheduling

Preconstruction services   services that a construction contractor performs for a project owner during design development and before construction starts

Premium time   that portion of a worker's wage that represents the cost of overtime

Pre-proposal conference   meeting of potential contractors with the project owner and architect; the purpose of the meeting is to explain the project, the negotiating process, and selection criteria and solicit questions regarding the design or contract requirements

Prequalification of contractors   investigating and evaluating prospective contractors based on selected criteria prior to inviting them to submit bids or proposals

Pricing recaps   form used by estimators to price the quantities of work determined during quantity take-off

Profit   the contractor's net income after all expenses, including home office overhead, have been subtracted

Pro forma   financial analysis of a real estate development, such as a tract of homes, to predict the anticipated costs and revenues

Programming budget estimate   budget estimate prepared during the programming phase to assess the financial feasibility of a project and to identify anticipated funding requirements; may also be a conceptual budget estimate

Progress payment request   document or package of documents requesting progress payments for work performed during the period covered by the request, usually monthly

Progress payments   periodic (usually monthly) payments made during the course of a construction project to cover the value of work satisfactorily completed during the previous period, also pay estimates

Project control   the methods a project manager uses to anticipate, monitor, and adjust to risks and trends in controlling cost and schedule

Project engineer   project management team member who assists the project manager on larger projects. More experienced and has more responsibilities than the field engineer, but less than the project manager. Responsible for management of technical issues on the jobsite

Project manager individual on the jobsite team responsible for overall project performance especially those tasks related to administration and documentation; may also be the project estimator

Project manual a specification volume usually containing the instructions to bidders, the bid form, general conditions, and special conditions; it also may include a geotechnical report

Property damage insurance protects the contractor against financial loss due to damage to the contractor's property

Public–private partnership delivery method a construction delivery method in which a public agency partners with a contractor or developer to reduce costs and lawsuits and ultimately save taxpayer money

Purchase orders written contracts for the purchase of materials and equipment from suppliers

Quantity recaps form used by estimators to group like items, such as reinforcing steel or concrete, from several quantity take-off sheets

Quantity take-off one of the first steps in the estimating process to measure and count items of work to which unit prices will later be applied to determine a project cost estimate

Recapitulation the form upon which tabulated quantities or work are listed and priced

Reimbursable costs costs incurred on a project that are reimbursed by the owner. The categories of costs that are reimbursable are specifically stated in the contract agreement

Request for proposal document containing instructions to prospective contractors regarding documentation required and the process to be used in selecting the contractor for a project

Request for qualification a request for prospective contractors or subcontractors to submit a specific set of documents to demonstrate the firm's qualifications for a specific project

Request for quotation a request for a prospective general contractor, subcontractor, and/or supplier to submit a quotation for a defined scope of work

Rough order-of-magnitude cost estimate a conceptual cost estimate usually based on the gross size of the project. It is prepared early in the estimating process to establish a preliminary budget and decide whether or not to pursue the project

Savings split a clause in a guaranteed maximum price contract that provides a formula for sharing the savings between the owner and the contractor if the final cost of construction is less than the guaranteed maximum price

Saw joint slab-on-grade concrete cracking method involving sawing recently placed concrete into grids and then often associated with caulking the saw joints at a later time

Schedule of values an allocation of the entire project cost to each of the various work packages required to complete a project. Used to develop a cash flow curve for an owner and to support requests for progress payments; may also be required to accompany a bid or negotiated cost proposal

Schematic design budget estimate a budget estimate that is prepared at the completion of schematic design

Schematic design documents   the plans and specifications early in the design process. They typically consist of sketches and preliminary drawings

Self-performed work   project work performed by the general contractor's work force rather than by a subcontractor

Semi-detailed cost estimate   an estimate that is prepared during design development that includes estimates for some components based on quantity take-off and estimates for other components based on order-of-magnitude estimates

Short list   the list of best-qualified contractors developed after reviewing the qualification of prospective bidders from a long list. Only the contractors on the short list are invited to bid or submit a proposal

Soils report   see geotechnical report

Special conditions   a part of the construction contract that supplements and may also modify, add to, or delete portions of the general conditions; also known as supplementary conditions

Specialties   building materials specified in CSI division 10 such as toilet accessories and fire extinguishers

Specialty contractors   construction firms that specialize in specific areas of construction work, such as painting, roofing, or mechanical. Such firms typically are involved in construction projects as subcontractors; also known as trade contractors

Statement of qualification   documentation submitted by a contractor in response to a request for qualifications; the statements of qualification are evaluated by the owner to determine the best qualified contractors from whom to solicit proposals

Subcontractors   specialty contractors who contract with and are under the supervision of the general contractor

Subcontractor call sheet   a form used to list all of the bidding firms from which the general contractor is soliciting subcontractor and vendor quotations

Subcontracts   written contracts between the general contractor and specialty contractors who provide craft labor and usually material for specialized areas of work; also known as subcontract agreements

Substantial completion   state of a project when it is sufficiently completed that the owner can use it for its intended purpose

Substructure   the portion of a building that is constructed below ground, usually the foundation, basements, and potentially the slab-on-grade

Summary schedule   abbreviated version of a detailed construction schedule that may include 20 to 30 major activities

Superintendent   individual from the contractor's project team who is the leader on the job site and who is responsible for supervision of daily field operations on the project

Superstructure   the portion of a building that is constructed above ground

Supplier   see material supplier

Target value design   lean construction process where the project budget is established before design begins and each element of the construction project is given a portion of the design, like a piece of pie, and the each design discipline must stay within that budget

Technical specifications   a part of the construction contract that provides the qualitative requirements for a project in terms of materials, equipment, and workmanship

Tilt-up concrete construction   a method of concrete construction in which members, usually walls, are cast horizontally and then tilted into place after the forms have been removed

Time and materials contract   a cost-plus contract in which the owner and the contractor agree to a labor rate that includes the contractor's profit and overhead. Reimbursement to the contractor is made based on the actual costs for materials and the agreed labor rate times the number of hours worked

Traditional project delivery method   a delivery method in which the client has a contract with an architect to prepare a design for a project; when the design is completed, the client hires a contractor to construct the project

Truck cubic yards   see loose cubic yards

Uniformat   a system for organizing a cost estimate that is based on building systems

Unit price contract   a contract that contains an estimated quantity for each element of work and a unit price; the actual cost is determined once the work is completed and the total quantity of work measured

Value engineering   a study of the relative value of various materials and construction techniques to identify the least costly alternative without sacrificing quality or performance

Work breakdown structure   a list of significant work items that will have associated cost or schedule implications

Workers' compensation insurance   protects the contractor from a claim due to injury or death of an employee on the project site

Work package   a defined segment of the work required to complete a project

Index

Printed in the United States
By Bookmasters